Biomedical Imaging in Neurodegeneration

Patricia A. Broderick
Editor

Biomedical Imaging in Neurodegeneration

Springer

Editor
Patricia A. Broderick
Department of Molecular, Cellular and
Biomedical Science
The City University of New York Medical
School
New York, NY, USA

ISBN 978-3-031-95877-9 ISBN 978-3-031-95878-6 (eBook)
https://doi.org/10.1007/978-3-031-95878-6

This Springer imprint is published by the registered company Springer Nature Switzerland AG
The registered company address is: Gewerbestrasse 11, 6330 Cham, Switzerland

Preface

The book, *Biomedical Imaging in Neurodegeneration*, seamlessly sews the intricate and delicate threads between and among three threatening brain dementia diseases shrouded behind and within the screen of Alzheimer's (AD). The purpose of the book is to design overlying and underlying organized patchwork matrices that meld biomarkers for AD, the many disorders of Parkinson as well as those for the epilepsies. Indeed, the authors reveal theragnostics for these dementias that shape layer after layer of newly found biomarkers to manipulate the molecular and neuromolecular bases of these neuronal threats to our very own quality of life. Systematic delineation of what underlies the disorders of Parkinson and the epilepsies, as well as Alzheimer's, is met with point of care innovative theragnostics that meet translational success in the clinic to varying degrees. In-depth and beautifully illustrated data share the proof-of-concept biomarkers discovered by neuroimaging. This multinational scholarly effort aims to bring a cohesive, revolutionary aspect of neurodegeneration to the globe.

We share keen insight into the inner mechanics of the biomedical neuroimaging art, the beginning of the art, its exact mechanics, equipment, complex know-how, and its drastically needed applications to the health, personalized medicine, the electronics, and artificial intelligence industry. Progressing to places in the mind where the proverbial angels "fear to tread," we bring elaborate biosensing, biochemistry, biophysics, and quantum mechanics to the table while we discuss the ever-present vitality that surrounds the topic of neurodegeneration. Thus, the material presented in this book discerns intraoperative imaging; structural and functional imaging in magnetic resonance (MR); functional nuclear, molecular, and neuromolecular imaging; and stimulation techniques. A multitude of probes and tracers are introduced that profoundly change the way that we think about the archetypal Alzheimer's deadly dementias and their offspring.

The book maintains its focus on the application of magnetic resonance imaging (MRI) and dopamine transporter single-photon emission computed tomography (DAT-SPECT) in degenerative parkinsonism and related disorders. Whereas MRI uses strong magnetic fields and radio frequency waves, DAT-SPECT uses nuclear imaging for differential diagnoses of PD from parkinsonian disorders. Innate facility with these Radioactive Tracers brings exciting critiques of multicenter studies on efficacy and nuclear reactions at atomic nuclei to treat PD. Scans without apparent evidence of dopaminergic deficits, despite the clinical assessment of the patient as Parkinson's, are not uncommon as clinical assessment remains the gold standard. Differential diagnosis between PD and essential tremor, (ET), parkinsonian syndrome (PS), and dystonia tremor (DT) is clarified.

Neuroimaging via a DA tracer brings remarkable advances to functional molecular imaging, and differential diagnosis further separates PD and parkinsonian disorders, as well as other atypical parkinsonian syndromes including multiple system atrophy (MSA), progressive supranuclear palsy (PSP), and corticobasal degeneration (CBD) from Atypical PD. Therefore, functional molecular nuclear imaging studies use single-photon emission computed tomography (SPECT) or positron emission tomography (PET); pre-synaptic imaging of dopamine (DA) synthesis, via vesicular monoamine transporter 2 (VMAT2), and dopamine transporter (DAT) and post-synaptic functional imaging involving DA receptors on the post-synaptic membrane are imaged. ^{18}F-DOPA PET/SPECT imaging reflects the process of ^{18}F-DOPA-blood-brain barrier uptake through large neutral amino acid transporters and subsequent conversion by aromatic amino acid decarboxylase (AADC) in monoaminergic neurons into ^{18}F-dopamine. The identification of prodromal PD via REM sleep behavior disorder (RBD) in ^{123}I-FP-CIT PET studies elicits the urgency of early testing since this happens 20 years before PD begins.

Further, there are cutting-edge technologies that emanate in the area of nanotechnology that are valuable in the neuroimaging technologies for PD. Acting at the nanoscale, these sensors and biosensor probes are attracting a great deal of attention because the materials are smaller, lighter, more durable, and are better electrical conductors and semiconductors. Potentiometric polymers, comprising carbon, lipids, and phosphotidyl saturated and unsaturated fatty acids, bond covalently to neuroimage PD in a facile, fast redox potential energy nanotechnology which is molecular and does not exert its electrochemical power by using nuclear energy. Neuromolecular Imaging with the BRODERICK PROBE® family of sensors image the Tau peptide as a metallopolymer in the striatum of the living and mobile PD animal model, while motor and hand skills algorithms enable brain and behavioral outcome as the quantum sensing ultra-small probes quantify phosphorylated Tau. The Broderick nanoprobe acts to capacitate redox matching of the metallopolymer of the Tau peptide, LIVE, inferably at one of the oxidation states of the transition metal polymeric compounds of, but not limited to, cobalt, to further polymerize with the phosphorylated Broderick nanoprobe. Studies of the striking consequences among intensities of phosphorylated Tau are assessed.

PD patients may develop a form of a speech defect, called hypokinetic dysarthria, characterized by reduced voice quality, mono loudness, and monotonicity. Therefore, we introduce in this book, a way to detect speech and language in early PD patients as a novel diagnostic that allows treatment to ensue early on. Convolutional Neural Networks, CNN, a subset of machine learning that is at the heart of deep learning algorithms is used with image classifications and object recognition tasks. Thus, a unique diagnostic approach is presented in which speech recordings of PD patients are modeled with speech embeddings resulting from Convolutional Neural Networks (CNNs) and the corresponding transliterations of those recordings are modeled with Bidirectional Encoder Representations from Transformer (BERT) embeddings, and a CNN architecture is further adapted to language analysis. The PD patient data from a large cohort are contrasted and compared with that of a non-PD cohort, and the outcome is remarkable.

While epilepsy is primarily a disease of brain function and not structure, association with a structural abnormality is quite common, and precise localization of the structural abnormality is often critical to successful epilepsy surgery. Not uncommonly, however, structural abnormalities that can be seen on MRI scan are not grossly apparent to the surgeon in the operating room. The ability to identify abnormal MRI signals in real-time and correlate these signals with visualized brain structures in the OR is a powerful advance in our ability to surgically treat chronic seizure disorders that is offered by intraoperative MRI (iMRI). The Chapter is a frank treatise on the surgical intervention for epilepsy using iMRI and fMRI and the authors await the next discovery made by MR Imaging in the epilepsy patient.

This book highlights the next discovery for MR imaging in the epilepsy patient. It couples sensing nanotechnologies to intraoperative MR imaging. These studies begin with electroencephalographic (EEG) monitoring wherein the site of cortical resection is defined with subdural grid epilepsy electrodes, placed on the surface of the patient's brain with epilepsy strip electrodes that are placed deeper in the temporal lobe (i.e., the hippocampus). Guided by iMRI, the anterior temporal neocortex of the epilepsy patient is imaged intraoperatively in the epilepsy patient by the Broderick nanoprobe. Broderick polymeric biosensors, BRODERICK PROBE® with a diameter five times less than epilepsy depth electrodes for invasive EEG. A nanoprobe is placed by direct visualization in the exposed cortical region with and without epileptic spike activity in regions destined for resection. About 6 to 10 recordings with γ-irradiated (11.6–12.7 kilograys), laminar biocompatible carbon-based laurate biosensors are taken at a cortical depth of microns to less than 2 mm for a 20–30 minute time period. Glial Fibrillary Acidic Protein (GFAP) stains of tissue were post-surgically conducted in order to assess gliosis or other neuronal damage after electrode placement. Based on the NYU pathology report, "the BRODERICK PROBE® biosensors did not cause gliosis, sclerosis, brain/neuronal damage, or histiocyte infiltration." Thus, the data provides evidence that the nanoprobe by Prof. Broderick is a safe biomedical tool. An operationally stable voltaic image of the soluble healthy Tau Metallopolymer at an electrochemically positive waveform is reported.

Status epilepticus (SE) comes in several forms, and functional molecular and nuclear imaging technologies are used in SE and epilepsy theragnostics. Neuroimaging's most important function in clinical practice is to support the diagnosis and determine the cause of SE, and in fact, neuroimaging is recommended for all SE patients. When it comes to detecting SE in blood vessels, Computerized Axial Tomography (CAT or CT), CT scans are more easily available in emergency rooms than are MRI scans, and approximately 15–33% of patients had abnormalities found on CT scans, such as mass lesions, trauma, edema, and bleeding. When CT and

MRI are combined, the diagnostic yield is slightly higher, with abnormalities being identified in 30–36% of cases. Moreover, the superiority of MRI in identifying inflammatory changes and subtle structural abnormalities accounts for a significant percentage of this improvement. The most frequent acute and distant symptomatic causes of SE, according to one prospective investigation that combined CT and MRI, were cerebral dysgenesis and CNS infections, respectively. The most common MRI abnormalities linked to SE are Increased T2 or fluid-attenuated inversion recovery (FLAIR) signal and/or diffusion limitation, focal diffusion restriction and an elevated T2 or FLAIR signal that may last for days to weeks, and importantly, the signals may link to unilateral or focal SE as well as lateralized periodic discharges seen during electroencephalograms (EEG).

Childhood epilepsy is addressed in this book as well, and new data from this group show tuber and focal cortical dysplasia by high AMT uptake on PET imaging, introducing the PET tracer alpha[^{11}C]-methyl-L-tryptophan (AMT) to epileptogenesis in children. Detection of an epileptogenic tuber and focal cortical dysplasia by high AMT uptake on PET imaging is imaged. Axial FLAIR MRI showed multiple lesions in both hemispheres, suggesting epileptogenicity in the lesion. In a young child with suspected frontal lobe seizures but no clear lesion detected by MRI, although mild blurring of the gray/white matter junction was noticed in the left frontal lobe, FDG-PET showed mild hypometabolism in the left frontal cortex; in the same area, AMT-PET showed increased interictal uptake, suggesting epileptogenicity in a small dysplastic lesion that remained undetected by MRI. The PET tracer alpha[^{11}C]-methyl-L-tryptophan (AMT) has emerged as an important tool in the study of certain types of epileptogenic lesions and foci. In particular, AMT PET has emerged as one of the best currently available methods for the noninvasive localization of epileptogenic tuber(s) in a subset of children with tuberous sclerosis and intractable seizures. Unlike most other clinically tested PET tracers, which show predominantly decreased uptake in epileptogenic cortical areas interictally, AMT PET scanning reveals locally increased uptake of AMT in and/or around epileptogenic tubers, while showing normal or decreased uptake in non-epileptogenic tubers.

^{1}H MRS (Proton Magnetic Resolution Spectroscopy) is a diagnostic nuclear imaging technique that provides information on neurochemical alterations linked to molecular-level pathological processes. In this chapter, the AD patient is imaged. ^{1}H MRS has been used to monitor the AD continuum, including the preclinical (i.e., cognitively unimpaired), prodromal, and AD dementia stages. ^{1}H MRS can measure the level of proton-containing molecules, such as neurotransmitters and metabolites, using the chemical shift information of molecules. The ^{1}H MRS metabolites related to the AD continuum, which can be reliably quantified up to 3 Tesla (T) magnetic field strengths, include N-acetyl aspartate (NAA), myo-inositol (mIns), choline-containing metabolites (total choline [tCho]=phosphocholine [pCho] + glycerophosphocholine [GPC]), and glutamine + glutamate (Glx). In addition, alterations in glutamate (Glu), glutamine (Gln), glucose (Glc), gamma-aminobutyric acid (GABA), and glutathione (GSH) have also been reported in preclinical, prodromal, and dementia stages at 3 T and higher magnetic field strengths.

By examining the impact of Cannabis Compounds (CBD) on human brain function through neuroimaging studies, the chapter develops the biochemistry underlying neuroimaging. The chapter aims to provide a mechanistic understanding of CBD's potential role as a therapeutic strategy with a neuroprotective effect, via biochemistry, the key to neurophysiology and pathophysiology. The chapter shows how structure and function of molecular moieties transform the biochemistry of the cannabinoids and it is all-inclusive. Recent studies show that Aβ peptide exposure activates glycogen synthase kinase 3β (GSK-3β) in hippocampal and cortical neurons, impairing the Wnt signaling pathway and contributing to tau protein hyperphosphorylation and neurofibrillary tangles in AD brains. CBD has been shown to mitigate Aβ peptide toxicity in PC12 cells via the Wnt/β-catenin pathway. Specifically, CBD enhances T-cell factor/lymphocyte enhancement factor (TCF/LEF)-mediated transcription by modulating the Wnt/β-catenin pathway, inhibiting GSK-3β phosphorylation.

Subclinical Epileptiform Activity (SAE) of the AD patient during mild cognitive impairment (MCI) in AD is introduced as a likely phenotype for epilepsy in the AD patient that is akin to interictal epileptic spiking in the epilepsy patient and new streams of thought compel us to discuss neurodegenerative diabetes, the veritable tip of the iceberg that emerges from diabetic neuropathy in Type 2 diabetes mellitus (T2D), as a novel mechanism of neurodegeneration. This may lend an explanatory note to AD pathophysiology via *signaling* insulin/IGF (insulin-like growth factor), coupled with impaired glucose metabolism. The data show that electrical or chemical stimulation techniques of the Locus Coeruleus (LC) can reduce diabetic neuropathic pain. Therefore, therapies targeting the LC could be beneficial in attenuating or preventing the development of persistent pain and related comorbid and neuropathic pain connects to the *modus operandi* of AD. Thus, neurodegenerative diabetic neuropathy is becoming a riveting area of focus for the dementia diseases of AD.

The book is a sweepingly intense and exceptionally educating resource book about biomedical imaging within the diseases of dementia in the patient and in proven animal models while streaming novelty at its best.

New York, NY, USA Patricia A. Broderick

Contents

About the Editor

Patricia A. Broderick PhD, is a Full Medical Professor with Tenure at the City University of New York Medical School, Dept. of Molecular, Cellular and Biomedical Science, USA, where she also holds faculty positions at the CUNY Grad Ctr.: Psychology: Cognitive and Behavioral Neuroscience as well as at CUNY Neuroscience Collaborative Graduate Program: Biology. Patricia served as the Director of Pharmacology at the City University of New York Physician Assistant/MS program at the CUNY Medical School and Director of the Neurobiology of Drugs of Abuse Course as well as Adjunct Professor in Neurology and the Comprehensive Epilepsy Center at NYU Langone Medical Center, USA. As graduate, doctoral, medical, physician assistant, and high school mentor, her innovative teaching is deeply appreciated. As principal and founder of Eazysense Nanotechnologies Inc., the Broderick Brain Foundation Inc., and inventor of the BRODERICK PROBE® series of nanobiosensors and circuits, Patricia has garnered respect and admiration worldwide. Achieving her PhD in Pharmacology from St. John's University, College of Arts and Sciences, NY, USA, her expertise in the field of atomic absorption photometry and fluorescent spectrometry, her inventive neuroimaging art in the biosensing and bioimaging field at the Albert Einstein College of Medicine and Cornell University, NY, USA, makes sense. Indeed, as postdoctoral fellow, Dr. Broderick was awarded the Rosenbaum Award from Albert Einstein Medical College/Montefiore Hospital and her patents which began at the Albert Einstein College of Medicine are now held by CUNY. Patricia has published and patented extensively and has garnered respect for her novel research around the world. The BRODERICK PROBE® trademark, named after her father, brings a solid brand as the nanoprobe is poised for marketing her unique cutting-edge nanotechnology in the field of Neuroscience. She is the recipient of multiple grants, awards, and prestigious honors from organizations such as Inner Circle Executives, Acquisitions International Global/Corporate America, International Association of Professionals, National Association of Distinguished Professionals, and Top Industry Experts from Patient Preferred Physicians and Practitioners was celebrated in 2024. A pivotal figure in biosensing technology and the brain, spinal cord, and the skin, she served as Session Chair, at the Virtual International Electronic Conference on

Brain Sciences, Beijing, October 2024, and her pending 2025 patent application for replenishing skin and her multiple patents are recognized by major diagnostic and pharmaceutical markets, worldwide. She serves on the editorial boards of several prestigious scientific and medical journals and her work as editor, author, and lecturer is renowned. No stranger to the airways, she is a sought-after speaker and her weekly talk show, The Eazysense Show, is at the top of the list at Bold Brave Media Productions and advances to The PodUnity Network at Landford Productions. Her unwavering dedication to advancing the understanding of neurological, psychological, and mental health is reflected in her latest launched textbook, Broderick PA (2025) NEUROIMAGING: Sensing Biochemistry in the Brain, Jenny Stanford Publishing, Pte Ltd., Singapore https://doi.org/10.1201/9780429029745. The Editor thanks Springer Nature for the opportunity to lead the most talented and influential physicians and scientists in the world via this book, Biomedical Imaging in Neurodegeneration.

Impact of Different Time-Spans in Multimodal Representations with Speech and Language Embeddings for Parkinson's Disease Detection

D. Escobar-Grisales, C. D. Rios-Urrego, and J. R. Orozco-Arroyave

Abstract Parkinson's disease (PD) affects millions of people worldwide. Among others, its symptoms negatively impact speech and language production. Therefore, the research community has worked on modeling these modalities (or sources of information) to perform the automatic detection and evaluation of the disease. Although language encodes relevant information regarding the cognitive status of the patients, most works have focused attention on speech production, and few studies have considered language abnormalities. This chapter presents a methodology where speech recordings of PD patients are modeled with speech embeddings resulting from Convolutional Neural Networks (CNNs), and the corresponding transliterations of those recordings are modeled with Bidirectional Encoder Representations from Transformer (BERT) embeddings and CNN architecture adapted to language analysis. First, we evaluated the suitability of these representations to classify PD patients and healthy subjects considering each modality separately, and second, speech and language representations are combined. To perform the combination, speech representations are converted into a static representation following two strategies: statistical functionals and Gaussian Mixture Model Supervectors (GMM-Supervectors). We evaluated the fusion of both information sources on different time-span levels (namely, granularities) to validate whether a finer granularity fusion provides better performance. The results show that speech and language representations are clearly complementary to each other, which motivates us to continue working on different fusion strategies and also on different methods to extract speech and language information.

D. Escobar-Grisales and C. D. Rios-Urrego authors contributed equally to this work.

D. Escobar-Grisales · C. D. Rios-Urrego
GITA Lab. at the Faculty of Engineering, University of Antioquia UdeA, Medellín, Colombia

J. R. Orozco-Arroyave (✉)
GITA Lab. at the Faculty of Engineering, University of Antioquia UdeA, Medellín, Colombia

Pattern Recognition Lab., Friedrich-Alexander Universität, Erlangen-Nürnberg, Germany
e-mail: rafael.orozco@udea.edu.co

P. A. Broderick (ed.), *Biomedical Imaging in Neurodegeneration*,
https://doi.org/10.1007/978-3-031-95878-6_1

Keywords Multimodal representations · Speech analysis · Language analysis · Parkinson's disease · Deep learning

1 Introduction

Speech is a biomarker suitable to support the diagnosis and monitoring of different neurodegenerative diseases [1]. Recent studies have shown that digital measures benefit research and clinical practice to a great extent [2, 3]. Besides the study of recordings at the acoustic level, the transliterations corresponding to those recordings can also be analyzed to model the patients' language. The benefit of this approach is twofold: First, the accuracy of existing methods can be improved, and second, these two modalities (speech and language) allow the modeling of different aspects in the brain, therefore opening the possibility of extending the approach to modeling and understanding several neurological disorders. Parkinson's disease (PD) has been among the main topics in the research of pathological speech and language because it is the second most prevalent degenerative disorder in the world after Alzheimer's [4]. PD is characterized by several motor symptoms including resting tremor, bradykinesia, rigidity, and freezing of gait [5]; with respect to speech deficits, most PD patients develop hypokinetic dysarthria which is characterized by reduced voice quality, mono loudness, monotonicity, imprecise pronunciation of consonants and vowels, lack of fluency, voice tremor, and others [6, 7]. These symptoms limit the possibility of patients having fluent communication and social interaction [8]. Particularly, the ability to produce spontaneous speech is considerably reduced [9], and the use and production of motor and non-motor verbs are altered [10, 11] making deficits in verb inflection [12], verbal fluency [13], and verb generation [14].

A Brief State of the Art on Speech Modeling in PD

Pathological speech analysis is a more popular topic than language evaluation in neurodegenerative diseases, especially PD. For example, in [15], a pilot experiment was implemented to detect the presence of dysarthria in speech and to predict the severity level based on a Deep Learning (DL) model. Automated feature extraction and classification using Convolutional Neural Networks (CNNs) yielded 77.48% of accuracy on test samples of the TORGO database using spectrograms as input. In [16], a Deep Neural Network (DNN) model was proposed to predict PD progression according to the Unified Parkinson's Disease Rating Scale (UPDRS). The authors reported a root mean square error of 2.2 and a coefficient of determination of 0.956. In [17], the authors proposed to apply a bidirectional Long Short-Term Memory (LSTM) model to capture dynamic information of the signals to detect PD. The database considered in the study included a total of 30 PD patients and 15 Healthy Control (HC) subjects. The experimental results showed that the method improved the accuracy of PD detection over traditional machine learning models which are based on static features. The highest accuracy reported in the study was 84%. In [18], the authors proposed a CNN model with a transfer learning strategy where information from different languages and pathologies were considered to classify between PD patients and HC subjects. The authors reported accuracies ranging between 83% and 86% and demonstrated the benefit of incorporating additional information like the one from different disorders and languages. In [19], the authors proposed another methodology based on CNN to predict PD using speech parameters extracted from voice recordings. The approach was evaluated with the dataset available in the UCI machine learning repository. The best experimental setup yielded an accuracy of 87.8%, a sensitivity of 88.7%, and a specificity of 90.5%. Also in 2021 the authors in [11] introduced a framework where different characteristics of dysarthric speech were presented including prosody, articulation, and phonemic identifiability, to classify PD patients and HC subjects. Besides this classification experiment, the authors classified cognitively impaired PD patients and HC subjects with an accuracy of 86.9%. The classification between the cognitively preserved PD patients and cognitively impaired ones yielded an accuracy of 72%, which marked a promising path on the way to detect cognitive decline among PD patients. One year after, a novel strategy based on 1D-CNN LSTM was proposed in [20] for classifying PD patients and HC subjects and also for classifying four different severity levels of dysarthria level of the patients according to the modified Frenchay Dysarthria Assessment (m-FDA) scale [21], which is a modification of the FDA scale that was originally introduced in [22]. The authors reported an accuracy of 89% in the bi-class classification scenario and 60% in the multi-class classification experiment. Finally, the authors in [23] proposed a novel end-to-end DL model for PD detection. The architecture extracts dynamic features using a 2D-CNN and then captures temporal dependencies using a 1D-CNN. The performance of the model ranged between 81.6% and 92%.

A Brief State of the Art on Language Modeling in PD

Some studies consider language analysis to evaluate PD patients. In [24], the authors aimed to assess cognitive decline in PD patients using Natural Language Processing (NLP) techniques, particularly using Global Vector Representations (GloVe), Bag of Words (BoW), Part-of-Speech Tagging (POS-Tagging), and others. Spearman's rank correlation coefficients were computed in the PD group, and results indicated a strong positive correlation ($\rho = 0.79$) between features based on GloVe and the Alternate Uses Test (AUT), which is a cognitive test that measures the ability to produce ideas flexibly [25]. The results in classifying HC subjects and PD patients were not satisfactory, and the authors argued that this was due to linguistic similarities between the two groups of participants. In [26], different NLP methods, including Term Frequency-Inverse Document Frequency (TF-IDF), BoW, and Word2Vec, were considered to classify PD patients from HC subjects. The transliterations from the monologue task were used as input, and a Support Vector Machine (SVM) was used as a classifier. The Word2Vec approach yielded an accuracy of 72% outperforming other NLP methods. In [10], the authors proposed a methodology based on morphological tagging to predict PD in three different languages: Spanish, German, and Czech. Stochastic Gradient Descent (SGD), Logistic Regression (LR), and SVMs were considered to classify monologues from PD patients and HC subjects. Accuracies of up to 71%, 71%, and 80% were achieved for each language, respectively. Recently, the capability to process action concepts has also been studied as a biomarker to model cognitive impairment in PD patients. In [27], the authors modeled the discourse of PD patients using the retelling of action and nonaction stories produced by 80 participants (40 with PD and 40 HC). In each retelling transliteration, the authors weighted action and nonaction concepts using a Proximity-to-Reference-Semantic-Field (P-RSF) metric, which was computed using Latent Semantic Analysis (LSA). These features were used to train an SVM, and accuracies of up to 72.5% were obtained using the retelling generated from action stories, outperforming other approaches based on GloVe. Results of that work indicate that PD patients exhibit an impaired ability to process action concepts compared with nonaction concepts.

A Brief State of the Art on the Fusion of Speech and Language Models
The fusion between speech and language has been explored mainly in the field of emotion recognition and evaluation of depression. However, due to the differences in their nature and time basis (i.e., the speed of speech signals is too high compared to the speed of language embeddings), the fusion of these information sources needs to be better explored to improve and better understand abnormal patterns in the communication of patients with neurodegenerative conditions like PD. For example, in [28], a method for speech emotion recognition based on phoneme sequences and spectrograms was proposed. The authors used different CNN topologies to characterize the IEMOCAP database and classified four emotions based on a joint fusion strategy. The authors reported improvements of at least 4% in the accuracy compared to other studies. Also working with the IEMOCAP database, the authors in [29] proposed a methodology that incorporates acoustic and the corresponding transcripts to classify different emotions. The model proposed by the authors was based on LSTMs for the speech signals and CNNs for the texts; in addition, the fusion was performed following a late fusion strategy. The bimodal system outperformed the acoustic model by 3.4%. In [30], the authors proposed a learning method where hidden representations between speech and text data were considered to feed convolutional attention networks. They used spectrograms and word-embeddings such as GloVe and Word2Vec to create those representations. The fusion of information was performed using an attention matrix. Six emotions of the CMU-MOSEI database were classified with an accuracy of up to 83.1%. In 2021 it was introduced an experimental method for depression detection based on audio and text [31]. The authors extracted 384-dimensional low-level audio features and, at the same time, presented a multimodal fusion method based on DeepSpectrum features and word vector features to detect depression. The authors reported that the multimodal fusion achieved an accuracy of 91.2% and an F1-score of 90.6%.

Regarding contributions to the specific topic of PD, a recent study was presented in [32] where depression in people with PD was modeled using a combination of speech features and language embeddings. The authors used Bidirectional Encoder Representations from Transformer (BERT) to model the transliterations of monologues and acoustic features to model the corresponding speech signals. In addition, they considered early and late fusion strategies. The automatic classification of depressed and non-depressed PD patients yielded F1-scores of up to 77%.

From the above-presented literature revision, it can be observed that contributions to model PD based on the fusion of speech and language information are scarce. Motivated by this evidenced necessity, in this chapter we present a methodology for the classification of PD patients vs. HC subjects using speech and language representations. Speech recordings are modeled using CNNs, and the corresponding transliterations of those recordings are modeled with BERT embeddings together with CNNs with different n-grams relations. We evaluated the fusion of both information sources at four different granularity levels: fine, intermediate, coarse, and full. Speech recordings were modeled via static representations using two different approaches: estimating statistical functionals of low-level features and creating Gaussian Mixture Model Supervectors (GMM-Supervectors). Our experiments show that the intermediate granularity level is the one that yields the best results.

2 Data

A total of 165 participants were included in this study (80 PD patients and 85 HC subjects). All subjects are Colombian Spanish native speakers, balanced in age and gender. Each recording was captured under controlled acoustic conditions with a sampling frequency of 44.1 kHz and 16-bit resolution. The recordings were normalized using a GSM full-rate compression technique and down-sampled to 8 kHz [33]. In addition, all patients were recorded in ON state during the session, i.e., under the influence of medication. A neurologist expert evaluated each PD patient to determine the disease severity level according to the third section of the Movement Disorders Society—Unified Parkinson's Disease Rating Scale (MDS-UPDRS-III) [34]. Concerning the recordings included in this work, participants were asked to talk for about 90 seconds about their daily routine. In addition, the transliterations of each recording were generated using the Amazon transcribe service. Table 1 shows additional information about the speakers.

Table 1 Clinical and demographic information of the subjects. [F/M]: female/male. Values in terms of mean ± standard deviation

	PD patients	HC subjects
Gender* [F/M]	38/42	43/42
Age** [F/M]	63.7±7.3/64.5±10.2	60.9±8.2/64.8±10.5
Range of age [F/M]	51–81/45–86	49–83/42–86
MDS-UPDRS-III [F/M]	34.6±19.9/38.5±19.6	
Range of MDS-UPDRS-III [F/M]	9–106/7–92	
Hoehn and Yahr [F/M]	2.4±0.6/2.3±0.7	
Range of Hoehn and Yahr [F/M]	1–4/1–5	

*Gender matching between PD and HC subjects with a p-value $= 0.81$ calculated through a chi-square test

**Age matching between PD and HC subjects with a p-value $= 0.38$ calculated through a t-test

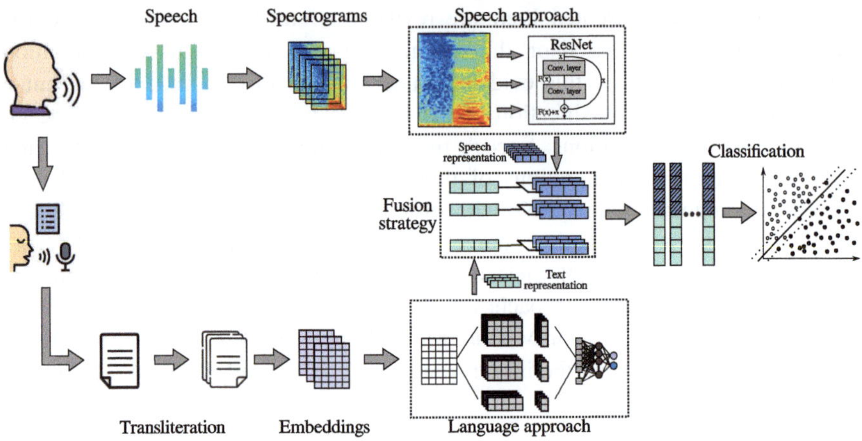

Fig. 1 General methodology proposed in this study

3 Methods

Figure 1 summarizes the methodology proposed in this work. Speech and language data are processed independently. For speech signals, spectrograms derived from the utterances are used to feed the convolutional layers and obtained the speech representation. For the language embeddings, the representations are extracted using BERT and a CNN architecture that includes different kernel sizes. Then, the fusion of both representations is performed considering four granularity levels. Finally, each bimodal representation is moved to the classification stage for the discrimination between PD patients and HC subjects. Details of the methods considered in each stage are presented below.

3.1 The Language Model

Figure 2 summarizes the methodology to model the language modality. One document is generated from the transliteration of the complete recording produced by each patient. Numerical representations for the words in the documents, i.e., embeddings, are generated such that they keep the semantic relation between the words. In this case, we used a BERT pretrained model to compute one word-embedding per document [35]. BERT is based on the decoder of a transformer-based architecture, and it is one of the most popular context-dependent embeddings. The source code to compute BERT embeddings is also available online[1] [36].

[1] https://github.com/PauPerezT/WEBERT

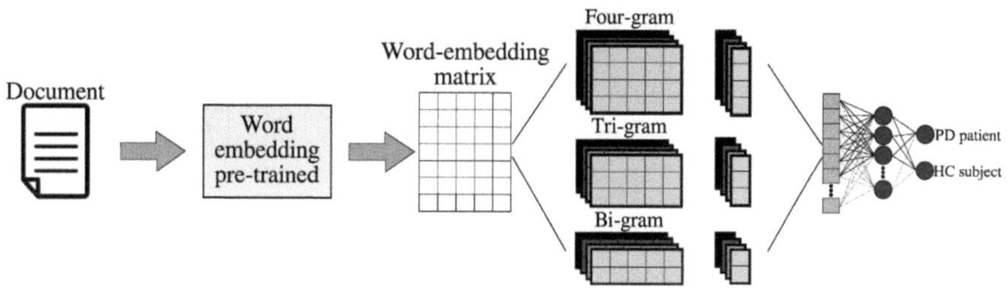

Fig. 2 Language architecture

These word-embeddings were concatenated vertically to form the real-valued embedding matrix $[\mathbf{X}_{\text{text}}]_{n\times d}$ which was used to feed a CNN adapted to analyze embedding matrices for NLP tasks. This architecture was formed with three parallel filters to map different semantic relationships among the words in the document. We used $2 \times d$, $3 \times d$, and $4 \times d$ filters in the convolutional layers for bigram, trigram, and four-gram relations, respectively. The convolution process was developed in one dimension; therefore each filter gradually moved down one word at a time along the sequence of words. The max-pooling layer was applied to the output of each filter, and then all output vectors were concatenated. Finally, the resulting vector was used as input to a fully connected layer, which made the final decision using a softmax activation function. This architecture was trained using an Adam optimizer and a sparse categorical cross-entropy loss function. Regularization strategies such as dropout and early stopping were used to fine-tune the model. The training process was stopped when validation loss did not improve after ten epochs. Typically, the embedding layer is jointly optimized with the CNN as it is shown in [37]; however, we used a pretrained word-embedding model (BERT) to reduce the number of trainable parameters, which were required to be kept small due to the low amount of available training data.

In this work, the transcription generated for each participant was divided into small documents, where the number of generated documents depends on the granularity level. Therefore, following the methodology described previously, we obtain different predictions per subject (one per document). Thus, the label's subject is estimated by computing the mean classification score. Details about the strategy with different granularity levels are presented in Sect. 3.3.

3.2 The Speech Model

Each recording was segmented into chunks of 500 ms with 250 ms time-shift. Next, each chunk was transformed into a time-frequency representation using the Short-Time Fourier Transform (STFT). Each spectrogram was obtained using Hanning windows of 32 ms and a step-size of 8 ms, forming groups of 63 time frames per chunk. Finally, each representation is transformed into a Mel-scale spectrogram

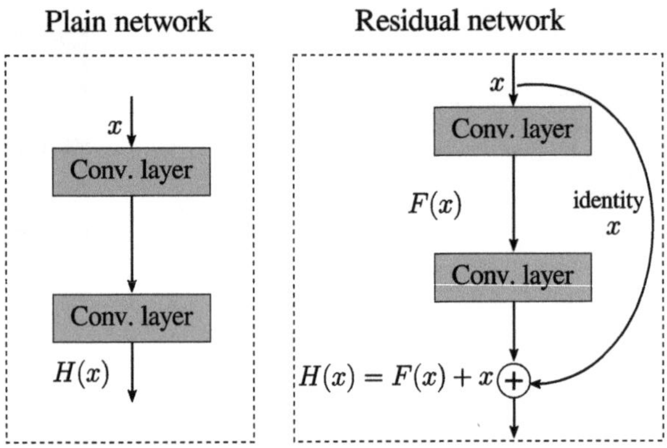

Fig. 3 Identity mapping in residual blocks

with 128 Mel filters; therefore, the final representation to feed the CNN is 128×63. Regarding the DL model, we adopted a CNN architecture based on a ResNet topology, which is the architecture where the "identity shortcut connection" concept was introduced. It allows the model to skip one or more layers, and its main motivation is to solve the vanishing gradient problem [38].

A ResNet consists of a series of layers and an identity mapping that adds the input from the block to the output, i.e., instead of trying to learn from a direct mapping of x with a function $H(x)$, a residual function $H(x) = F(x)+x$ is defined, where $F(x)$ represents the stacked layers of the neural network, and x is the identity function, which is passed directly to the output of the residual block (see Fig. 3). This characteristic makes it the ResNet appropriate to train robust models with a small amount of data and avoiding the vanishing gradient problem.

We implemented a ResNet architecture in this work with six residual blocks and three main blocks with 16, 32, and 64 feature maps, respectively. In addition, we incorporated a convolutional layer with 16 channels at the input to adjust the spectrograms to feed the first main block. With respect to the output of the last main block, we performed an average pooling between channels to get a speech representation with 64 features per spectrogram. Finally, we used a fully connected layer to classify whether each representation came from a PD patient or a healthy subject. The mean value over the spectrograms of each speaker was computed to determine the final prediction per subject. Figure 4 shows the architecture implemented in the speech model.

3.2.1 Static Representation of Speech Embeddings

Static representations are generated with two different strategies. Figure 5 illustrates the two approaches to convert a dynamic into a static representation: (i) A GMM is generated per speech embedding, and the GMM-Supervector is created by stacking the means ($\boldsymbol{\mu}$) and the diagonal of the covariance matrix ($\boldsymbol{\Sigma}$). The resulting static representation depends on the speech dimension (64 due to the output in the ResNet

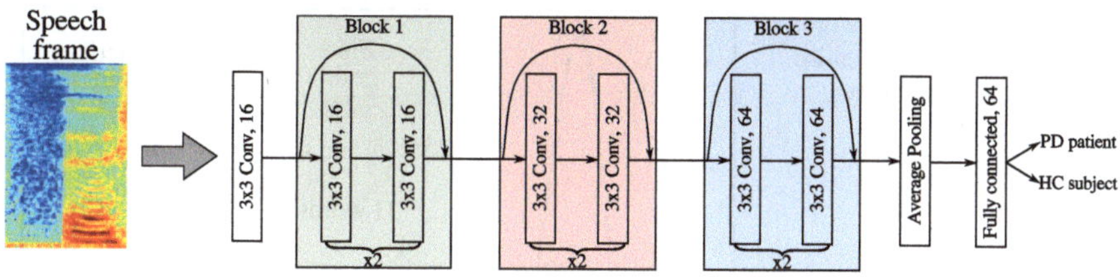

Fig. 4 Speech architecture

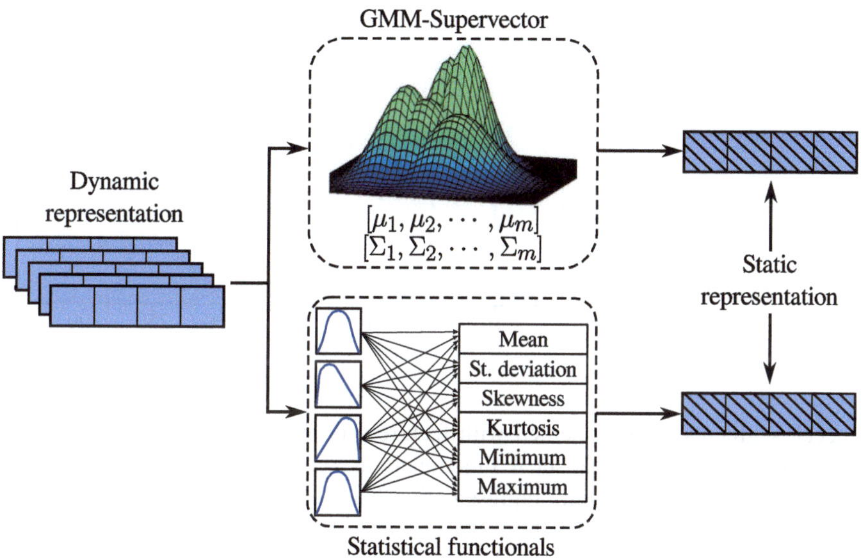

Fig. 5 Process to obtain the static representations from the speech model

architecture) and the number of Gaussians (parameter to optimize), and (ii) different statistical functionals are estimated from the speech representations including mean, standard deviation, skewness, kurtosis, minimum, and maximum. The resulting functionals are concatenated forming 384-dimensional feature vectors.

3.3 The Fusion Model

The typical way to perform the fusion of information from speech and language representations consists of concatenating them in an early fusion strategy.

Text representations are typically created considering the number of words in the 95th percentile of the document set of the corpus. Texts with less than the resulting number are adjusted via a padding procedure. On the other hand,

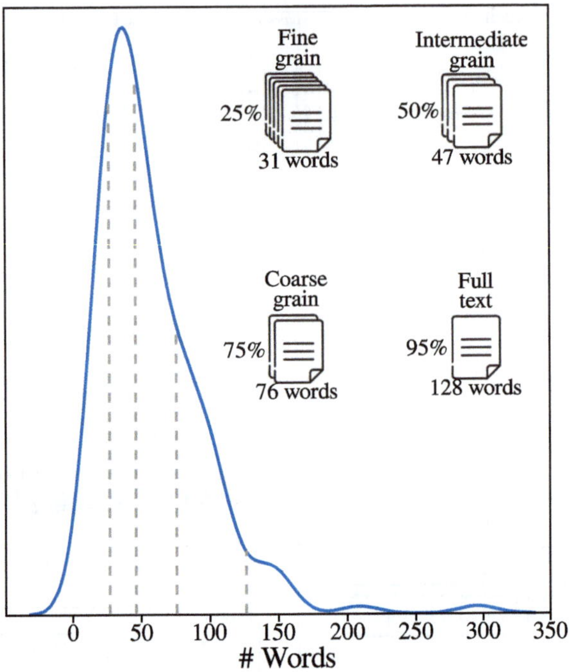

Fig. 6 Distribution of words per document over the entire database and granularity levels selected

speech segments are represented with feature sets, which directly yield frame by frame representations. Statistical functionals are computed upon these dynamic representations to get a static representation from the complete recording. Notice that in speech there is a reduction that produces a loss of information. This work aims to generate dynamic representations of speech and language to combine both information sources on a frame basis. One of the questions we address in this chapter is how can we define the size of those frames?

To answer the above question, we propose to evaluate the early fusion of speech and text representations at different granularity levels defined according to the corresponding percentiles chosen from the distribution of the words in the documents set: 25th (fine), 50th (intermediate), 75th (coarse), and 95th (full). Figure 6 indicates the distribution of words per document over the entire database, selected percentiles, and their corresponding number of words.

To synchronize the resulting language representations with the corresponding speech representations, the embeddings obtained from the audio recordings were split uniformly according to the number of documents in a given granularity. The transliteration resulting from a given subject is divided into a different number of documents depending on the granularity. Similarly, for each granularity level, the dynamic audio representation is divided into a different number of groups. Static representations (based on functionals or GMM-Supervectors) are generated per group. The example in Fig. 7 shows the case in which the full text of a transliteration is considered to create the language representation and a total of 16 speech

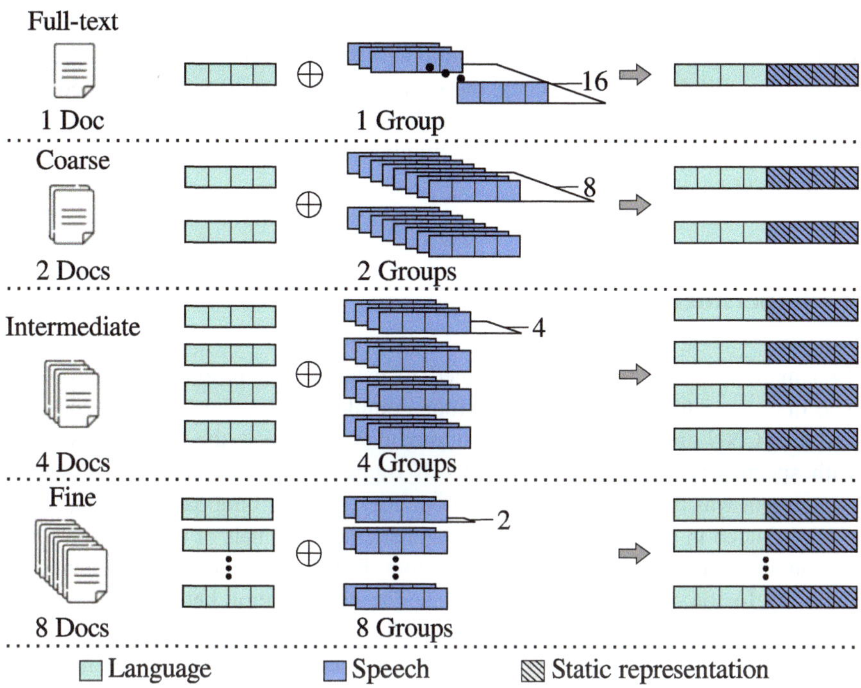

Fig. 7 Fusion with different granularity levels

representations are generated (one per spectrogram) to create one group. Then the two versions of the static representation of speech are generated (for comparison purposes) to proceed with the merging step. For the case of coarse granularity, transliterations are split into two documents, and the speech representations are divided into two groups, each of them with eight speech representations. Similarly, the intermediate and fine granularities are generated with four and eight documents, respectively, and the corresponding speech representation is divided into four and eight groups of speech representations, respectively. The final prediction for each subject is performed by computing the mean prediction score between the resulting static representations (speech and language).

3.4 Parameter Optimization and Classification

An SVM is used to classify PD patients vs. HC subjects. This method allows the discrimination of N samples by finding a separating hyperplane that maximizes the margin between classes. We used a radial basis function as the kernel for the SVM with the aim to improve its generalization capability. Its parameters were optimized upon a grid search where $C \in \{0.001, 0.01, \cdots, 100\}$ and

$\gamma \in \{0.0001, 0.001, \cdots, 100\}$. Finally, the optimal hyperparameters were obtained in the training process as the mode computed over the repetitions of the process. Concerning the fusion approaches, the statistical functionals were computed from the dynamic speech representations as indicated in Sect. 3.2.1. For the case of the GMM-Supervectors, we optimized the number of Gaussians as 2, 4, and 8, based on the accuracy in test.

4 Experiments and Results

Two main experiments were performed to evaluate the proposed methodology. The first consisted of evaluating the four granularities for the language approach and the CNN model for the speech modality in which the network was fed with spectrograms resulting from speech frames of 500 ms length. The second experiment was the fusion of the two modalities, i.e., the bimodal approach, which was evaluated in two variants: one with statistical functionals of the feature vectors and another one with the GMM-Supervectors. Both fusion variants were generated via early fusion as explained in Sect. 3.3. The classification between PD and HC subjects was performed with an SVM which was optimized following a stratified k-fold cross-validation strategy with ten folds. Different performance measures were used to evaluate the proposed methodology, including accuracy, sensitivity, specificity, and F1-score.

4.1 Unimodal Approach

Table 2 shows the results obtained for each granularity level in the language approach and also the results obtained with the speech model. It is important to mention that each approach was evaluated individually but considering the same folds in the cross-validation procedure. Notice that the best performance was obtained in the speech approach with an accuracy of 84.4%. Conversely, for the language models, the best result was obtained considering the granularity that

Table 2 Classification between PD patients and HC subjects for the two unimodal approaches. Values reported in terms of mean ± standard deviation

	Language				
	Fine	Intermediate	Coarse	Full text	Speech
Accuracy	67.7 ± 8.9	67.6 ± 10.5	70.6 ± 12.9	72.8 ± 9.8	84.4 ± 8.8
Sensitivity	82.5 ± 10.0	62.5 ± 27.4	73.8 ± 20.5	72.5 ± 16.6	81.3 ± 15.0
Specificity	52.9 ± 21.8	72.6 ± 20.1	67.4 ± 18.6	73.1 ± 18.0	84.6 ± 15.8
F1-Score	71.2 ± 6.0	61.0 ± 23.8	69.9 ± 14.0	71.6 ± 11.5	84.3 ± 9.2

corresponds to the full text, i.e., 95th percentile in the distribution of the words, which corresponds to a total of 128 words (see Fig. 6). Notice that the other granularity levels did not yield better results in this case, where speech and language modalities are considered separately.

4.2 Bimodal Approach

The bimodal approach considers the early fusion between language and speech models/modalities. The four different granularities defined above are evaluated: fine (25%), intermediate (50%), coarse (75%), and full text (95%). Two strategies to create the speech models were considered: statistical functionals and GMM-Supervectors. The results are presented in Table 3. Notice that the intermediate granularity yielded the best results in all cases, with an accuracy of 86.1% for the statistical functionals and 81.9% for the GMM-Supervectors. These two results are above those obtained with the typical approaches where speech and language are considered separately. The absolute improvement observed in the experiments with the functionals is 1.7% and 13.3% with respect to the results obtained with speech and language, respectively (Table 2). Regarding the results with GMM-Supervectors, an absolute improvement of 9.1% is observed with respect to the best language model reported in Table 2. When comparing the speech model with respect to the bimodal approach with GMM-Supervectors of speech, the results are 2.4% below the ones obtained individually.

Perhaps the main aspect to highlight from these results is the fact that language models with intermediate granularity (50%) seem to preserve more information than the classical approach which is based on the 95th percentile in the distribution of the words. This result is relevant especially for applications where accessing large amounts of data is not possible or very difficult and also for cases in which the length of the documents or transliterations is highly unbalanced. Another aspect to highlight is that the fusion of speech and language information clearly provides better results with higher accuracies.

When comparing the results of GMM-Supervectors and the statistical functionals, it is observed that the second ones are always better for fine and intermediate granularities but do not so for the case of the coarse and full ones. The reason for this is that the GMM-Supervector approach requires more samples (longer recordings) to generate better representations. This observation might help other researchers to make better decisions about which approach to use depending on the length of the signals.

Finally, Fig. 8 shows the histograms and the probability density distributions of the scores obtained from the SVM when classifying the samples, i.e., the distance to the separating hyperplane (only the best results of Table 3 are considered). The left-hand side corresponds to the results obtained with the statistical functionals (Acc.: 86.1%), and the right-hand side corresponds to the best result using GMM-Supervectors (Acc.: 81.9%). Both cases correspond to intermediate granularity.

Table 3 Classification between PD patients and HC subjects for the different granularity levels for the fusion of speech and language. Values reported in terms of mean ± standard deviation

	Statistical functionals				GMM-Supervectors			
	Fine	Intermediate	Coarse	Full text	Fine	Intermediate	Coarse	Full text
Accuracy	80.0±8.7	**86.1±9.0**	78.7±8.8	79.4±10.7	79.5±10.1	**81.9±11.6**	79.4±14.1	79.5±12.1
Sensitivity	75.0±23.1	**80.0±15.0**	75.0±18.5	78.8±15.9	68.8±18.8	**73.8±19.7**	73.8±21.3	75.0±17.7
Specificity	85.0±12.4	**92.1±8.7**	82.2±16.2	80.1±13.6	89.9±11.6	**89.9±10.5**	85.0±11.3	83.8±12.4
F1-Score	76.4±15.0	**84.3±10.4**	76.6±11.8	78.4±12.4	75.2±14.1	**78.7±15.0**	76.5±17.8	77.4±14.3

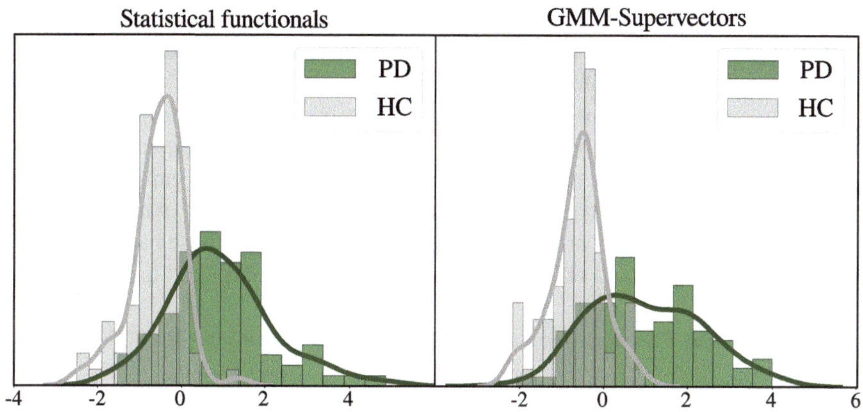

Fig. 8 Distribution of the SVM scores obtained with the bimodal approaches with intermediate granularity

Notice that in both figures, the error when classifying healthy subjects is low, which corresponds to a specificity of 92.1% with the statistical functionals and of 89.9% for the GMM-Supervectors. Conversely, in both cases, the sensitivity was lower than the specificity, which suggests that further research is required to create models better adapted to capture patterns possibly associated with PD.

5 Discussion

This chapter introduced an early fusion strategy where speech and language representations are merged considering different levels of granularity according to the distribution of the number of words obtained from transliterations of audio recordings collected from 80 Parkinson's patients and 85 healthy subjects. Four granularity levels were considered according to the percentile in the distribution: 25th (grain), 50th (intermediate), 75th (coarse), and 95th (full). Speech recordings were modeled using a CNN architecture based on a ResNet topology. The corresponding transliterations were modeled with BERT embeddings together with a CNN with different kernel sizes. To synchronize speech and language representations, speech representations were segmented in the same proportions as the ones given by the granularity levels (see Sect. 3.3). The resulting dynamic speech representations were converted into static representations following two approaches: (i) statistical functionals and (ii) GMM-Supervectors. Then, each static representation of speech is concatenated with its corresponding language representation, generating a new bimodal representation. The results suggest that the proposed methodology is more accurate than the typical approach with full text. Particularly, we observed that the intermediate granularity yielded the best performance with an absolute improvement of 6.7% in the statistical functionals and

2.4% using GMM-Supervectors. We believe that this work opens a new gap where different granularity levels might be explored in cases where sources of information with different sampling rates are considered.

We are aware that partitioning the speech representations uniformly implies ignoring differences in the speech rate and rhythm among speakers, therefore generating a loss of information. However, this work opens the gap to combining speech and language representations considering granularities, which is a step forward in the state of the art. Synchronization strategies like those based on forced alignment techniques and other fusion strategies might be explored in the future.

Acknowledgement This work was funded by CODI at UdeA grant number IAPFI23-1-01.

References

1. Boschi V, et al. Connected speech in neurodegenerative language disorders: a review. Front Psychol. 2017;8:269.
2. Babrak LM, et al. Traditional and digital biomarkers: two worlds apart? Digit Biomarkers. 2019;3(2):92–102.
3. Coravos A, Khozin S, Mandl KD. Developing and adopting safe and effective digital biomarkers to improve patient outcomes. NPJ Digit Med. 2019;2(1):1–5.
4. De Rijk MC, et al. Prevalence of Parkinson's disease in Europe: a collaborative study of population-based cohorts. neurologic diseases in the elderly research group. Neurology. 2000;54:S21–3.
5. Jankovic J. Parkinson's disease: clinical features and diagnosis. J Neurol Neurosurg Psychiatry. 2008;79(4):368–376.
6. Pinto S, et al. Treatments for dysarthria in Parkinson's disease. Lancet Neurol. 2004;3(9):547–556.
7. Rusz J, Cmejla R, Tykalova T, Ruzickova H, Klempir J, Majerova V, Picmausova J, Roth J, Ruzicka E. Imprecise vowel articulation as a potential early marker of Parkinson's disease: effect of speaking task. J Acoustical Soc Am. 2013;134(3):2171–2181.
8. Ho A, et al. Speech impairment in a large sample of patients with Parkinson's disease. Behav Neurol. 1998;11(3):131–7.
9. Liu L, et al. Characteristics of language impairment in Parkinson's disease and its influencing factors. Transl Neurodegener. 2015;4(1):1–8.
10. Eyigoz E, et al. From discourse to pathology: automatic identification of Parkinson's disease patients via morphological measures across three languages. Cortex 2020;132:191–205.
11. García AM, Arias-Vergara T, Vásquez-Correa JC, Nöth E, Schuster M, Welch AE, Bocanegra Y, Baena A, Orozco-Arroyave JR. Cognitive determinants of dysarthria in Parkinson's disease: an automated machine learning approach. Movement Disorders. 2021;36(12):2862–2873.
12. Grossman M, et al. Sentence comprehension in Parkinson's disease: the role of attention and memory. Brain Language. 1992;42(4):347–384.
13. Obeso I, et al. Semantic and phonemic verbal fluency in Parkinson's disease: nfluence of clinical and demographic variables. Behav Neurol. 2012;25(2):111–118.
14. Crescentini C, et al. Supervisory and routine processes in noun and verb generation in nondemented patients with Parkinson's disease. Neuropsychologia. 2008;46(2):434–447.
15. Sonawane B, Sharma P. Speech-based solution to Parkinson's disease management. Multimedia Tools Appl. 2021;80(19):29437–29451.
16. Shahid A, Singh, M. A deep learning approach for prediction of Parkinson's disease progression. Biomed Eng Lett. 2020;10(2):227–239.

17. Quan C, Ren K, Luo Z. A deep learning based method for Parkinson's disease detection using dynamic features of speech. IEEE Access 2021;9:10239–10252.
18. Vásquez-Correa J, et al. Transfer learning helps to improve the accuracy to classify patients with different speech disorders in different languages. Pattern Recognit Lett. 2021;150:272–279.
19. Krishna A, et al. Speech parameter and deep learning based approach for the detection of Parkinson's disease. In: Computer networks, big data and IoT. Berlin: Springer; 2021. p. 507–517.
20. Rios-Urrego C, et al. End-to-end Parkinson's disease detection using a deep convolutional recurrent network. In: International conference on text, speech, and dialogue. Berlin: Springer; 2022. p. 326–338.
21. Vásquez-Correa JC, Orozco-Arroyave JR, Bocklet T, Nöth E. Towards an automatic evaluation of the dysarthria level of patients with Parkinson's disease. J Commun Disorders. 2018;76:21–36.
22. Enderby P. Frenchay dysarthria assessment. Br J Disorders Commun. 1980;15(3):165–173.
23. Quan C, et al. End-to-end deep learning approach for Parkinson's disease detection from speech signals. Biocybernet Biomed Eng. 2022;42(2):556–574.
24. Jessiman L, Murray G, Braley M. Language-based automatic assessment of cognitive and communicative functions related to Parkinson's disease. In: Proceedings of the first international workshop on language cognition and computational models. 2018. p. 63–74
25. Beaty RE, Johnson DR, Zeitlen DC, Forthmann B. Semantic distance and the alternate uses task: Recommendations for reliable automated assessment of originality. Creativity Res J 2022;34(3):245–60.
26. Pérez-Toro PA, et al. Natural language analysis to detect Parkinson's disease. In: International conference on text, speech, and dialogue. Betlin: Springer; 2019. p. 82–90.
27. García AM, Escobar-Grisales D, Vásquez-Correa JC, Bocanegra Y, Moreno L, Carmona J, Orozco-Arroyave JR. Detecting Parkinson's disease and its cognitive phenotypes via automated semantic analyses of action stories. NPJ Parkinson's Disease. 2022;8(1):163–10.
28. Yenigalla P, et al. Speech emotion recognition using spectrogram & phoneme embedding. In: Interspeech. 2018. vol. 2018, p. 3688–3692.
29. Cho J. et al. Deep neural networks for emotion recognition combining audio and transcripts. 2019. arXiv preprint arXiv:1911.00432.
30. Choi W, Song K, Lee C. Convolutional attention networks for multimodal emotion recognition from speech and text data. In: Proceedings of grand challenge and workshop on human multimodal language (Challenge-HML). 2018. p. 28–34.
31. Ye J, et al. Multi-modal depression detection based on emotional audio and evaluation text. J Affective Disorders. 2021;295:904–913.
32. Pérez-Toro P, et al. Depression assessment in people with Parkinson's disease: the combination of acoustic features and natural language processing. Speech Commun. 2022;145:10–20.
33. Huerta JM, Stern RM. Speech recognition from GSM codec parameters. In: ICSLP. 1998. p. 1463–1466.
34. Goetz CG, et al. Movement Disorder Society-sponsored revision of the Unified Parkinson's Disease Rating Scale (MDS-UPDRS): scale presentation and clinimetric testing results. Movement Disorders. 2008;23(15):2129–2170.
35. Devlin J, Chang M-W, Lee K, Toutanova K. Bert: pre-training of deep bidirectional transformers for language understanding. 2018. arXiv:1810.04805.
36. Perez-Toro PA. PauPerezT/WEBERT: Word Embeddings using BERT. 2020. https://doi.org/10.5281/zenodo.3964244.
37. Escobar-Grisales D, et al. Colombian dialect recognition based on information extracted from speech and text signals. In: 2021 IEEE automatic speech recognition and understanding workshop (ASRU). Piscataway: IEEE; 2021. p. 556–563.
38. He K, et al. Identity mappings in deep residual networks. In: European conference on computer vision. Berlin: Springer; 2016. p. 630–645.

Magnetic Resonance Imaging and Dopamine Transporter Imaging in Parkinson's Disease

Stephen Joza and Fang Ba

Abstract Accurate diagnosis of Parkinson's and other degenerative parkinsonian syndromes is important for management and prognostic purposes. Diagnosis can be challenging in early disease and in atypical cases. This chapter focuses on the application of magnetic resonance imaging (MRI) and dopamine transporter single-photon emission computed tomography (DAT-SPECT) in degenerative parkinsonism and related disorders.

For Parkinson's disease (PD) and related conditions, conventional MRI is not diagnostic, but it can be useful to distinguish PD from other atypical parkinsonism. The new developments in the field utilizing T1-weighted structural MRI, iron-sensitive MRI, diffusion-weighted imaging (DWI), and resting-state fMRI (rs-fMRI) have shed light on the underlying pathophysiology of PD. They may also hold promise as both diagnostic and prognostic biomarkers.

The use of DAT-SPECT shows some clinical advantages in early diagnosis for Parkinson's disease and differential from other non-degenerative parkinsonian disorders (i.e., essential tremor, dystonic tremor, drug-induced and in most cases of psychogenic parkinsonism), since it can accurately detect the presynaptic dopaminergic deficit. The test has been shown to have high sensitivity/specificity by multiple studies. DAT imaging may also have some prognostic value for disease progression. However, it has limited value in differentiating among degenerative causes of parkinsonism. Clinical assessment remains the most important aspect in evaluating these patients.

DAT imaging has some limitations. In most studies, true test accuracy is unknown since the gold standard is clinical diagnosis by a movement disorders neurologist. Therefore, the sensitivity of the test cannot exceed that of the clinical diagnosis.

S. Joza
Department of Neurology, McGill University, Montreal General Hospital, Montreal, QC, Canada
e-mail: steve.joza@mail.mcgill.ca

F. Ba (✉)
Parkinson and Movement Disorder Program, Division of Neurology, University of Alberta, Edmonton, AB, Canada
e-mail: fb@ualberta.ca

P. A. Broderick (ed.), *Biomedical Imaging in Neurodegeneration*,
https://doi.org/10.1007/978-3-031-95878-6_2

In addition, false negative scans occur and highlight the need for clinical follow-up. DAT-SPECT is a sensitive modality to detect nigrostriatal degeneration. In spite of increasing data using this technique, however, more long-term clinical studies are required to determine how DAT-SPECT scan can guide decision-making.

Keywords MRI · Dopamine transporter · SPECT · PET · Parkinson's disease · Diagnostic

1 Introduction

Dysfunction of the dopaminergic system is associated with multiple neurological conditions. Parkinson's disease (PD) is one classic example. PD is common: its estimated prevalence is close to 1% in those aged 65–69 years and increases with age [1]. The clinical progression of PD is heterogeneous, and precise diagnosis based on clinical criteria such as those of the UK Brain Bank [2] and Movement Disorders Society (MDS) [3] can be difficult, especially in early stages. Other degenerative causes of parkinsonism, including multiple system atrophy (MSA), progressive supranuclear palsy (PSP), corticobasal degeneration (CBD), and dementia with Lewy bodies (DLB) are also diagnosed primarily on the basis of their clinical features, and accurate diagnosis can be challenging. The time to reach a correct clinical diagnosis for PD can range from 2 months to 18 years depending on the clinical presentation [4]. Misdiagnoses of conditions such as essential tremor (ET), vascular parkinsonism (VP), MSA, and PSP implies that 5–25% of patients with an initial diagnosis of PD may be inappropriately managed [5]. Accurate diagnosis is essential for correct management and prognostication.

Over the past decade, neuroimaging with specific DA-related ligands has been used to assess function at both presynaptic terminals and postsynaptic binding sites. Currently, single-photon emission computed tomography (SPECT) and positron emission tomography (PET) are widely available imaging tools. In this review, we will discuss the utility of *in vivo* presynaptic DA transporter (DAT) imaging as a diagnostic tool for degenerative parkinsonism and related movement disorders in clinical practice. We will subsequently discuss the value of MRI in the diagnosis and progression of PD, including conventional structural MRI and several advanced MRI modalities that may hold promise as diagnostic and longitudinal biomarkers.

2 Nuclear Imaging in Parkinsonism

DAT is a sodium chloride-dependent protein on the presynaptic membranes of the terminals of dopaminergic projections [6]. DAT plays a critical role in the regulation of the extracellular DA concentration and is considered a marker of DA terminal integrity. Degeneration of dopaminergic projections from the substantia nigra to the striatum results in loss of DAT [7]. DAT concentration closely relates to striatal DA levels [8, 9], supporting its use as an imaging biomarker for PD. However, DAT imaging may underestimate the true terminal density due to down-regulation of DAT in remaining dopaminergic neurons in response to the reduced levels of synaptic DA [10].

To assess presynaptic dopaminergic function, both SPECT and PET have been used as *in vivo* tools. SPECT is the more widely available of the two modalities *due to the lower cost and the availability of the tracers.* Several ^{123}I and ^{99m}Tc-labeled DAT-SPECT imaging agents have been developed based on cocaine and other closely related tropane derivatives [11]. A few examples are (1) [^{123}I]FP-CIT (also known as DaTSCAN™ or DaTSCAN™) (N-ω-fluoropropyl-2β-carbomethoxy-3β-(4-[^{123}I]iodophenyl) nortropane); (2) ^{99m}Tc-TRODAT-1 [2-[[2-[[[3-(4-chlorophenyl)-8-methyl-8-azabicyclo]3.2.1]oct-2-yl]methyl] (2-mercaptoethyl)amino]ethyl]amino]ethanethiolato(3-)-N2,N20, S2,S20]oxo-[1R-(exo-exo)]; and (3) [^{123}I]β-CIT (2β-carbomethoxy-3b-(4-iodophenyl)tropane) [12]. Among these agents, [^{123}I]FP-CIT and [^{99m}Tc]TRODAT-1 have more rapid binding kinetics, which allow much earlier imaging (3–6 hours) after administration [12]. The sensitivity and specificity of ^{123}I -FP-CIT SPECT for PD diagnosis is comparable to those of other DAT-SPECT methods [13]. Although [^{123}I]-FP-CIT was not the first agent developed, it was approved by the European Medicines Agency (EMA) as DaTSCAN™ in 2000 [14] with a revised guideline suggesting DaTSCAN™ is indicated in patients with clinically uncertain parkinsonian syndromes (PS). The modality can help differentiate ET from conditions associated with pre-synaptic dopaminergic dysfunction, such as PD, MSA, and PSP. It is also indicated for the differentiation of DLB from other dementias [15]. [^{123}I]FP-CIT (DaTSCAN™) was approved by the Food and Drug Administration (FDA) for clinical use for suspected PS in 2011 [16]. The FDA suggests that it be considered as an adjunct to other diagnostic evaluations when evaluating a tremor disorder. Other approved compounds include [^{123}I]β-CIT (DOPASCAN™) in Europe and Japan; ^{18}F-FP-CIT PET in Korea, and ^{99m}Tc-TRODAT-1 SPECT in Taiwan [13].

PET can provide higher resolution and better quantitative capacity than SPECT but has seen limited use in clinical practice, due to its complex and expensive infrastructure and the lack of commercial availability of dopaminergic ligands. This review will focus on DAT-SPECT imaging, given its wide potential use and availability of the DAT ligands.

2.1 Diagnostic Application of DAT Imaging

DAT imaging allows *in vivo* assessment of presynaptic striatal dopaminergic function in patients with neurological symptoms. Image analysis is typically semi-quantitative, based on the placement of regions of interest (ROI) in the striatum and on a background region devoid of specific binding, enabling the estimation of DAT binding. There is a known gradient of dopaminergic dysfunction in PD, starting in the posterior putamen progressing anteriorly to caudate contralateral to the affected limbs, with later involvement of the putamen and caudate in the ipsilateral side as demonstrated by DAT imaging [17, 18] and postmortem pathological studies [19, 20].

2.1.1 Early Diagnosis

DAT imaging is capable of distinguishing patients with PD from normal subjects even in early disease [21–23]. A decrease in DAT binding at striatal presynaptic terminals is associated with the loss of functional neuronal activity that characterizes PD. Significant DAT changes may precede the onset of clinical symptoms [24]. Decreased striatal [^{123}I]β–CIT binding has been reported in anosmic relatives of patients with PD who do not themselves have clinical evidence of parkinsonism [25], and in individuals with hyposmia and REM sleep behavioral disorder (RBD) [25–27]. Striatal uptake correlates with duration and severity of motor features assessed by the Unified Parkinson's Disease Rating Scale (UPDRS-III) and Hoehn and Yahr scores [8, 28]. Among the motor symptoms of PD, tremor does not correlate well with striatal DAT binding, whereas bradykinesia, rigidity as well as gait, posture, facial expression, and speech show a significant correlation with DAT activity [29–31].

Overall, DAT SPECT is a highly sensitive test for PD and degenerative PS. However, striatal DAT density decrease is not only seen in response to DA neuronal loss in PD [32] but also occurs in the premotor stage of PD and in rare cases of drug-induced parkinsonism [23, 33], when the density of dopaminergic nerve terminals is preserved. When DAT is functionally down-regulated as a compensatory mechanism for reduced levels of synaptic DA [34], DAT SPECT can underestimate true terminal density in remaining dopaminergic neurons [10]. In contrast, ^{18}F-dopa uptake correlates more closely with the number of nigral dopaminergic neurons in humans [35]. Reduced DAT density but preserved ^{18}F-Dopa uptake was found in asymptomatic individuals with LRRK2 mutations, where compensatory changes are thought to maintain stable extracellular DA levels [34].

There is a subgroup of clinically diagnosed PD patients that have been categorized as having scans without evidence of dopaminergic deficit (SWEDDs). The clinical presentations of these patients are consistent with PD, but functional imaging failed to demonstrate dopamine (DA) deficiency suggesting an alternate diagnosis [17, 36, 37].

2.1.2 Diagnostic Accuracy for Nigrostriatal Neurodegeneration

Although not specific for etiology, in clinically uncertain cases, DAT imaging can help provide evidence of presynaptic nigrostriatal pathology. However, multiple studies have revealed some limitations of the test and different sensitivity and specificity of DAT imaging in patients with parkinsonism of uncertain etiology. Representative studies are summarized in Table 1.

The results of these studies indicate that abnormal DAT imaging supports the presence of a neurodegenerative presynaptic dopaminergic deficit. Patients with a clinical diagnosis of PD but normal DAT SPECT imaging are consistent in most cases with the diagnosis of conditions other than a degenerative PS [17, 42, 43].

False positive results have been reported in drug-induced parkinsonism [33]. [^{123}I]FP-CIT SPECT, but not ^{18}F-Dopa PET, was significantly decreased in a patient receiving chronic lithium treatment who presented with bilateral parkinsonism. Imaging and clinical symptoms improved with the discontinuation of lithium. These observations suggest functional down-regulation of membrane DAT density by lithium in spite of the normal DA nerve terminal density indicated by ^{18}F-Dopa imaging [47].

In practice, only patients with a suspected PS, particularly in early stage or with atypical features, may benefit from DAT-imaging. [^{18}F]-β-CIT and PET imaging have demonstrated that striatal DAT uptake was reduced by approximately 13% per year in eight patients with PD, compared with 2.5% per year in six healthy controls [48]. The characteristics of the reduction in striatal uptake (putamen > caudate) in a few cross-sectional studies are consistent with known pathology of PD, and the severity of the imaging deficit correlates with the clinical severity of disease [29, 49]. DAT-imaging, however, does not predict the number of nigral neurons in PD [50].

The presence of SWEDDs amongst patients with a clinical diagnosis of PD raises the question regarding the sensitivity of DAT imaging [8, 51]. Although these may be false-negative scans, further studies have shown that most, if not all, patients with negative scans do not have involvement of the nigrostriatal pathway and therefore do not have PD. In addition, long-term follow-up of patients with SWEDDs indicate a poor response to levodopa in most cases and lack of progression on sequential dopaminergic imaging [17, 45, 52]. However, a sustained response to dopaminergic medications in those with SWEDDs has been reported [42]. A recent study in 23 SWEDDs patients who underwent genetic testing showed a frameshift mutation in the DYT11 gene in one patient accounting for the PD-like phenotype. This patient had a sustained response to levodopa [53]. The electrophysiologic study revealed that the presumed rest tremor was myoclonus-dystonia activity. This study demonstrated that adult-onset monogenic dystonia may underlie a "PD look-alike" clinical phenotype without clinical dystonia and with normal DAT imaging.

Based on the current literature, the consensus is that a normal scan excludes a presynaptic dopaminergic deficit in most clinical cases [12]. However, clinical progression indicative of a neurodegenerative PS has been reported in a small proportion of cases (3%) with normal baseline DAT imaging, indicating that false negative scans can occur and highlighting the need for clinical follow-up [42]. Among all degenerative PS, it has been shown that only corticobasal syndrome can have normal DAT imaging at baseline due to delayed substantia nigra neuronal loss [54, 55]. False negative scans are most likely due to methodological issues in the analysis and interpretation of imaging results.

Table 1 The diagnostic accuracy with DAT imaging studies in patients with nigrostriatal neurodegeneration

Study	Study design	Study subjects	Imaging modality	Main findings	Limitations/concerns of the studies[a]
CALM-PD Study, 2000 [37]	Randomized, multicenter, double-blinded trial comparing pramipexole with levodopa in early PD	82 of 301 patients received DAT-SPECT	[^{123}I]FP-CIT SPECT	Normal scans in 4% of patients (SWEDDs)	Cases diagnosed as SWEDDs lack pathological confirmation of alternate diagnosis
Benamer et al. 2000 [38]	Multicenter trial assessing the reliability of visually assessed DAT imaging in clinically diagnosed PS	158 PD cases, 27 ET, and 35 normal controls	[^{123}I]FP-CIT SPECT	Sensitivity for PS was ≥95%	
Booij et al. 2001 [39]	Prospective clinical follow-up in inconclusive PS cases followed for 2–4 years	33 cases, no controls	[^{123}I]FP-CIT SPECT	Positive scan correctly predicted a diagnosis of presynaptic parkinsonism (100%) Specificity for PS was 92%	Small number of study objects Possible false-negative rate was 8% (one case of MSA, one case of young onset PD had negative scan)
Catafau et al. 2004 [40]	Multicenter, non-randomized, clinical trial of uncertain PS	118 patients, no controls	[^{123}I]FP-CIT SPECT	In 67 presynaptic PS cases, 66% had abnormal scans; in 26 non-presynaptic PS, 54% had abnormal scans, and in 25 inconclusive cases, 68% had abnormal scans DAT-SPECT increased confidence in diagnosis, leading to changes in clinical management in 72% patients	Specific diagnostic criteria were not provided The primary outcome was diagnostic confidence of the neurologists Image analysis was qualitative

Jennings et al. 2004 [41]	Suspected PS cases referred by community neurologists	35 suspected PS cases, no controls	[^{123}I]β-CIT SPECT	Test sensitivity of 92% and specificity of 100% versus 92% sensitivity and 30% specificity for the referral diagnosis. Gold standard was the clinical diagnosis by a second movement disorders expert who was blind to the imaging data and reevaluated the patient 6 months from initial encounter	Small number of study subjects Short follow-up (6 months clinical follow-up by the movement disorders experts) Difficult to comment on the diagnostic accuracy by community neurologists as compared to nuclear medicine experts
Marshall et al. 2006 [42]	Prospective study with DAT SPECT and clinical follow-up for 2.2–3.1 years in PS cases	150 cases with diagnostic uncertainty but normal DAT scans, no controls	[^{123}I]FP-CIT SPECT	3% false negative baseline scans Normal scan in a possible PD case supports the diagnosis of a non-degenerative movement disorder	
Marhall et al. 2009 [43]	Prospective, multicenter, longitudinal study over 3 years to compare clinical diagnosis with DAT imaging	199 patients with diagnostic uncertainty, three healthy controls	[^{123}I]FP-CIT SPECT	Mean sensitivity was 78% and specificity was 97% in degenerative PS, in comparison with values of 93% and 46% for baseline clinical diagnosis. Gold standard was video recording of clinical features scored by two blinded clinicians at 36 months who were also blinded to DAT imaging results	Dropout rate is very high (99 of 199 finished study). 14% (10/71) diagnosed with degenerative parkinsonism had normal SPECT

(continued)

Table 1 (continued)

Study	Study design	Study subjects	Imaging modality	Main findings	Limitations/concerns of the studies[a]
Utiumi et al. 2012 [17]	Case series to demonstrate the usefulness of DAT imaging in atypical PS	Four patients with clinically unclear PS, categorized as SWEDDs, no controls	[^{99m}Tc]-TRODAT-1 SPECT	Only the PD patient (not VP, ET, or DT) had an abnormal scan; scans were normal in the one subject with VP, ET, and DT	Small number of study subjects
De Verdal et al. 2013 [44]	Prospective and retrospective study to analyze DAT imaging in mixed tremors	33 patients with long-term mixed postural/kinetic and rest tremor	[^{123}I]FP-CIT SPECT	In normal DAT-SPECT group, family history was significantly more often present In abnormal DAT-SPECT group, rest tremor was mostly bilateral, and postural tremor was bilateral and asymmetrical	Small number of study subjects Clinical follow-up was only in 20 patients Follow-up was short (average 14 months)
Marek et al. 2014 PRECEPT study [45]	Multicenter study to compare clinical and imaging characteristics of PRECEPT subjects with SWEDDs to those with abnormal DAT uptake at baseline and during a 22-month follow-up	799 patients; 91 were SWEDDs, 708 cases had DAT deficit. 701 subjects had repeat DAT imaging, while 98 patients only had a baseline scan	[^{123}I]β-CIT SPECT	All SWEDDs subjects showed minimal evidence of progression of clinical or imaging parameters SWEDDs showed significantly lower UPDRS score at baseline and minimal change in both striatal binding ratio and UPDRS score at follow-up	Not all patients had a repeat DAT imaging

Marek et al. 2018 PPMI study [46]	Large international multicenter study, designed to identify clinical, DAT imaging, genetic, and biospecimen PD markers to accelerate disease modifying therapeutic trials	More than 400 untreated early PD patients and 196 controls. DAT imaging performed at 12, 24, and 48-month visits for PD subjects, 24-month visits for SWEDD.	[123I]Ioflupane or vesicular monoamine transporter imaging with ^{18}F AV133	PD subjects demonstrated 45% and 68% reduction in mean striatal and contralateral putamen DAT imaging excluded 16% of potential PD subjects with SWEDD	
Isaacson et al. 2021 [23]	Study of the impact of DaTscan results on changing or confirming pre-DaTscan clinical diagnosis and on post-DaTscan treatment changes	201 patients who have (1) early, subtle motor signs, (2) with PS, but suboptimal response to medication (3) with prominent action tremor, (4) clinical diagnosis of PD but stable for 3–5 years (5). Drug-induced PS. (6) Predominant lower extremity parkinsonism or other neurologic disorders with PS	[123I]FP-CIT SPECT	Patients with clinically questionable nigrostriatal dopaminergic degeneration, DaTscan was abnormal in 58.7% of patients, normal in 37.8%, and inconclusive in 3.5%. DaTscan changed clinical diagnosis in 39.8% of patients and led to medication therapy changes in 70.1% of patients	In the stable PD group and PD was diagnosed by movement disorders specialist, 46% had normal DaTscan indicating an incorrect diagnosis of PD. The diagnosis accuracy is lower than usual

[a] All studies used clinical diagnosis as gold standard. None had a tissue diagnosis

SWEDDs scans without evidence of dopaminergic deficit, *PD* Parkinson's disease, *ET* essential tremor, *PS* Parkinsonian syndrome, *DT* dystonia tremor, *PRECPT* Parkinson Research Examination of CEP-1347 Trial

2.2 Prognostic Value of DAT Imaging in PD

DAT-SPECT has been suggested to have a role in prognosis of PD. However, current literature has not established a strong and consistent correlation between tracer uptake and disease severity (i.e., UPDRS-III). Ravina et al. [56] prospectively evaluated 539 patients with [^{123}I] β-CIT SPECT at baseline and after 22 months. With a total clinical follow-up of 5.5 (±0.8) years, and a mean duration of diagnosis of 6.3 (±1.2) years, lower striatal binding at baseline was independently associated with increased disease severity including motor-related disability, falling and postural instability, cognitive impairment, psychosis, and depressive symptoms. However, low DAT binding may not always indicate poor clinical status or prognosis. It has been shown that [^{18}F]-β-CIT tracer uptake ratios correlated with UPDRS-III at baseline, but only weakly correlated with impairment at 44 months follow-up, whereas UPDRS at baseline was highly correlated with impairment [57]. In the longitudinal study by de la Fuente-Fernández et al. [58], younger PD patients showed lower DAT binding at the onset of motor symptoms but slower disease progression when compared with older patients. They suggest better compensatory mechanisms in younger individuals to explain the latency between DA function decline and the onset of motor symptoms between age groups. Compensatory mechanisms to maintain motor function with the loss of striatal DA innervation have been shown postmortem [59], and also with the observations that most of the damage has been done by the time of clinical disease onset, since in the posterior putamen, relatively little further decline occurs beyond 10 years' disease duration [60].

The incongruence between DAT imaging and clinical severity of PD is multifactorial. Issues include a lack of pathological confirmation in most studies, and the application, magnitude, and response to PD medical and other symptomatic treatment. In addition, PD is a complex degenerative condition, which involves non-DA mechanisms. The compensatory changes in different age groups and different stages of disease should also be taken into consideration. Lastly, methodological issues such as the use of qualitative rather than quantitative scan analyses may impact scan interpretation. Therefore, the use of DAT imaging in evaluating disease progression of PD, and in the longitudinal monitoring of the nigrostriatal DA dysfunction in treated PD patients, needs to be better established. Clinicians need to be cautious in interpreting the imaging results.

2.3 Differential Diagnosis Among Parkinson's Plus Syndromes

DAT imaging provides little additional information relevant to differential diagnosis in diseases with presynaptic dopaminergic deficiency [12, 61–64]. The degree of DAT binding does not discriminate among the patient groups comparing PD, MSA, and PSP [64], although DAT-SPECT in PD typically shows an asymmetric loss of DAT binding (more prominent in the putamen), and tracer uptake is more symmetric in MSA [65, 66] and PSP [67]. In one report of pathology-confirmed PD and MSA cases with similar disease duration [68], visual inspection and semiquantitative analysis with [^{123}I]β-CIT SPECT detected bilaterally reduced striatal DAT binding in all patients, but a trend toward greater asymmetry of binding in MSA than in PD. However, this did not help to differentiate between the two conditions. It is also suggested that relatively reduced midbrain DAT binding may aid discrimination of MSA from PD with computer-assisted analysis of [^{123}I] β-CIT SPECT images [69]. In CBD, there may be a unilateral reduction in dopaminergic function, but these findings are non-specific [62] and highly variable [70, 71]. Normal DAT imaging was evident in neuropathologically confirmed CBD cases [55].

To date, DAT imaging still lacks specificity to discriminate between the degenerative PS based on the best current available literature.

2.3.1 Genetic PD

The utility of DAT imaging in patients with genetic forms of parkinsonism has been evaluated [72]. [^{123}I]FP-CIT SPECT was applied to 37 cases of monogenetic PD (7 glucocerebrosidase mutations, 8 alpha-synuclein, 3 LRRK2, 7 PINK1, and 12 Parkin). The asymmetry of radioligand uptake for PD with glucocerebrosidase or LRRK2 mutations was greater than the other forms. The authors postulated that the asymmetry of radioligand uptake in PD associated with glucocerebrosidase or LRRK2 mutations suggests that interactions with additional genetic or environmental factors may be associated with dopaminergic neuronal loss.

2.3.2 PD vs. ET

DAT imaging has been reported to be normal in ET in multiple studies [73–76]. Asenbaum et al. [76] studied PD, ET, and control individuals and observed a 91% sensitivity and 100% specificity for a diagnosis of PD with [^{123}I]FP-CIT SPECT. Isaias et al. [77] reported reduced DAT binding only in PD, with an annual rate of decline of 7.3% in the contralateral putamen, but not in ET or controls. The largest of these studies was a multicenter trial assessing the reliability of visually assessed DAT imaging using [^{123}I]FP-CIT SPECT in patients with clinically diagnosed parkinsonism ($n = 158$) or ET ($n = 27$) and healthy controls ($n = 35$) [38]. An overall sensitivity and specificity of 96% were achieved in PD and ET cohorts. These observations are consistent with previous studies, particularly with the very high specificity to differentiate ET from PD [38, 73, 74, 77]. These data suggest that abnormal DAT-SPECT can be considered exclusionary for ET.

There is some literature suggesting a higher prevalence of PD in patients with ET [78, 79]. Very rarely, there is evidence in some families with a tendency to develop both ET and PD [80]. ET is increasingly recognized as a heterogeneous disorder that can encompass features including rest tremor and mild cognitive impairment, which can challenge accurate diagnosis. DAT imaging may have a role in this population.

2.3.3 PD vs. VP

VP can have a wide spectrum of clinical manifestations, and VP and PD may coexist since both conditions have increased prevalence with age [81, 82]. There are no established diagnostic criteria for VP [83]. PD patients may have coexisting vascular risk factors and vascular changes on structural brain imaging. Therefore, VP cannot always be confirmed on the basis of clinical features or structural neuroimaging. DAT imaging studies may provide some insight, although previous studies have provided inconsistent results. Tzen et al. and Gerschlager's group using [^{99m}Tc]TRODAT-1 SPECT and [^{123}I]β-CIT SPECT, respectively, found near normal DAT binding in patients with VP, but decreased uptake in PD [84, 85]. However, other studies have reported reduced DAT binding in VP with, in some cases, a more diffuse and symmetrical pattern [86, 87]. These observations suggest that a normal DAT imaging study can exclude PD, but an abnormal result could represent either VP or PD. Some specific features, such as a more diffuse uptake, in VP may help to differentiate the two conditions although more studies are needed to confirm this impression.

2.3.4 PD vs. Drug-Induced PS

Drug-induced parkinsonism is the second-most common PS [88] and occurs secondary to the use of DA antagonists. The incidence of extrapyramidal side effects with low-dose neuroleptics is around 30%, and recovery after drug withdrawal can take up to 18 months [89]. In the majority of studies, tracer uptake is normal in pure drug-induced parkinsonism unless there is coexistence of pre-symptomatic PD/neurodegenerative parkinsonism [86, 90, 91]. An abnormal scan suggests a presynaptic DA deficiency and possible unmasking of preclinical PD by these drugs [92]. Such patients warrant regular clinical follow-up, and, if there is persistence of parkinsonism after discontinuation of the presumed offending agent, they may benefit from a trial of dopaminergic therapy. Caution is warranted, however, since false positive results have been reported in drug-induced parkinsonism with functional down-regulation of membrane DAT density, in spite of the normal DA nerve terminal [33].

2.3.5 PD vs. Dystonic Tremor (DT)

DT occurs in a body part affected by dystonia and is often focal with variable frequency. The tremor is mainly postural/kinetic. It has been recognized that patients with DT often have a rest tremor, sometimes even with the classical pill-rolling appearance. Jaw tremor, facial hypomimia, and impaired arm swing have all been observed and could suggest parkinsonism [93, 94]. Tone may increase with dystonia [93]. DT may be a cause of SWEDDs in suspected PD [94]. In ten cases diagnosed as possible PD, patients presented with asymmetric resting arm tremor but had signs of dystonia and the absence of true akinesia after a mean of 5.8 years of follow-up; all had normal baseline [^{123}I]FP-CIT SPECT scans, making a diagnosis of PD unlikely. DAT imaging appears to have some utility in distinguishing these individuals from those with PD.

2.3.6 Psychogenic Tremor

Psychogenic tremor has a variable clinical presentation [95]. It usually has abrupt onset with immediate maximal severity [96]. With psychogenic tremor, a classic but not universal finding is remission or entrainment with distraction maneuvers and the coactivation sign. PD is a consideration in the differential diagnosis, particularly if the tremor is asymmetrical. Normal DAT imaging alone is not sufficient to diagnose psychogenic tremor but helps to rule out a degenerative PS in appropriately selected patients, whereas an abnormal image warrants re-evaluation of the diagnosis [97, 98]. It has been reported that a patient with suspected psychogenic parkinsonism had an abnormal scan but after further screening, the *parkin* gene was identified [98]. PD and psychogenic tremor can also co-exist: in nine patients with a suspect diagnosis of psychogenic PS, five had abnormal [^{123}I]FP-CIT SPECT scans [97]. Caution is important in the setting of normal imaging in cases of possible psychogenic tremor: a normal scan does not exclude alternative organic diagnoses such as ET, DT, or drug-induced parkinsonism.

2.3.7 AD vs. DLB

AD is the most common form of dementia, while DLB accounts for about 7% of dementia in patients younger than 65 [99]. Both AD and DLB may have parkinsonian signs [100, 101] and neuropsychiatric features. Differentiation of these two disorders is important for both treatment and prognostic reasons [102–104]. Compared with neuropathological findings, clinical diagnosis of DLB yields very high specificity (>95%), but rather low sensitivity (32% for pure DLB and 12% for AD plus DLB) [105]. In addition, depending on the diagnostic criteria used, 30–60% of patients with clinically diagnosed AD may fulfill histological criteria for both DLB and AD [106].

DAT imaging may help differentiate patients with PD dementia or DLB from those with AD. There have been multiple studies showing significantly decreased tracer uptake in DLB [19, 107–109]. Walker et al. showed that with [^{123}I]FP-CIT SPECT, the diagnostic accuracy improved significantly [108] with a specificity of 100% for DLB, confirmed by autopsy findings. A multicenter Phase III trial of [^{123}I]FP-CIT SPECT was carried out in 326 patients with clinical consensus diagnoses of probable DLB, possible DLB, or non-DLB dementia at baseline [110]. Sensitivity of [^{123}I]FP-CIT SPECT imaging for clinical diagnosis of probable DLB in this group was 77.7%, while the mean specificity for excluding non-DLB was 90.4%. The diagnostic accuracy is sufficiently high for this technique to be clinically useful in distinguishing DLB from AD. Given these observations, low DAT tracer uptake in the basal ganglia demonstrated by SPECT imaging has been included as a suggestive feature in the most recent diagnostic criteria for DLB [111].

2.3.8 Dopa-Responsive Dystonia vs. Juvenile PD

Dopa-responsive dystonia (DRD) and juvenile PD (JPD) are both young-onset hereditary disorders with parkinsonism and dystonia. Accurate clinical diagnosis is important as prognosis of these conditions is different and levodopa responsiveness can impact patient care. Patients with DRD can present with isolated parkinsonism [112]. Although imaging with [^{123}I]β-CIT SPECT and [^{123}I]FP-CIT has shown normal DAT density in DRD and abnormal activity in JPD in multiple studies [80, 113–117], recent data demonstrated that relatives of patients with DRD who have GTP cyclohydrolase 1 (GCH1) gene mutation can present with pure "PD" clinically and have decreased DAT density on SPECT consistent with PD [118]. In addition, Mencacci et al. reported that the frequency of putative GCH1 mutations was significantly higher in patients with PD than in controls [118]. Therefore, to distinguish DRD from PD may require systematic genotype–phenotype evaluations. Clinical assessment with DAT imaging may still be inadequate for differentiation.

2.4 *DAT Imaging on Non-motor Symptoms of PD*

Although not used as a routine test for PD non-motor symptoms, it has been observed that lower [^{123}I]β-CIT SPECT at baseline is independently associated with non-motor features such as cognitive impairment, psychosis, and depression [56]. Higher incidence of anxiety was observed to associate with a more severe reduction of striatal DAT uptake [119]. Similar correlations were seen between nigrostriatal dopamine transporter availability and visual hallucinations [120], apathy [121], urinary symptoms [122], and impulse control disorders [123, 124].

2.5 *Factors That May Impact on DAT Imaging Interpretations*

In most publications, the sensitivity/specificity of [^{123}I]FP-CIT, although high, is not 100% accurate in predicting diagnosis. However, it is important to recognize that the true accuracy of DAT imaging is unknown due to a lack of pathological diagnoses in most cases. The gold standard has been the clinical diagnosis of PD or PS by movement disorders experts. If the clinical diagnosis is incorrect, it is not viable to determine the test accuracy. So far, there is generally good agreement between the

[^{123}I]FP-CIT result and the clinical diagnosis [125]. Comparing the DAT imaging data with the few available autopsy-confirmed studies [68, 108], the sensitivities and specificities are also consistent.

False negative DAT-SPECT imaging can occur [43, 126]. It has also been reported that readers of the DAT-SPECT performed in multiple trials did not achieve total agreement on interpretation of images [127]. This can be due to the sole application of qualitative visual interpretation [38, 43, 68]. Quantitative evaluation can improve diagnostic confidence [128]. Patient positioning, motion, color-scale, and loss of DAT receptor density with aging should all be considered when interpreting data [129]. Establishing and choosing the proper database for normal controls is important.

DAT binding reflects uptake in presynaptic nerve endings of the dopaminergic nigrostriatal pathway. DAT expression is a well-regulated process in dopamine signaling. Membrane trafficking is a key mechanism for DA reuptake [130]. Many agents are known to have a significant influence on DAT imaging and may impact scan interpretation. These include dopaminergic central nervous system stimulants (cocaine, amphetamine-related compounds, methylphenidate, mazindol, etc.) [93]. In addition, DAT ligands can bind non-selectively to noradrenaline and serotonin transporters [10]. Therefore, multiple aminergic medications can also interfere with DAT imaging interpretation, but with minor effects [93]. It is believed that dopaminergic therapy can be continued because it has only minimal or no influence on DAT binding [93], while the FDA has stated that the impact of dopamine agonists and antagonists upon DaTscan™ imaging results has not been established.

3 The Role of MRI in the Diagnosis and Progression of PD

While the diagnosis of PD remains primarily clinical, neuroimaging techniques have played a pivotal role in elucidating the underlying pathophysiology and aiding in the differential diagnosis of this complex disorder. Among the various imaging modalities available, MRI has emerged as a powerful tool for investigating the structural, functional, and metabolic changes associated with PD. MRI offers the unique advantage of providing high-resolution anatomical images, allowing for the visualization and assessment of brain structures implicated in PD, such as the substantia nigra (SN), basal ganglia, and cortical regions. Additionally, advanced MRI techniques enable the evaluation of other important aspects of PD pathology, including alterations in white matter integrity, cortical thickness, and connectivity patterns.

In this section, we focus on the application of MRI in understanding disease pathophysiology, early diagnosis, and monitoring disease progression. We will discuss the key MRI techniques employed in PD research, including T1-weighted structural MRI, iron-sensitive MRI, diffusion-weighted imaging (DWI), [131].

3.1 T1-Weighted Structural MRI

Conventional T1-weighted structural MRI does not show a consistent pathology in PD, particularly when compared with other neurodegenerative diseases, such as Alzheimer's disease or atypical parkinsonism [132, 133]. Therefore, routine MRI is most useful as a means of investigating abnormal clinical findings or red flag features that would suggest coexisting pathology or other cause of parkinsonism [133].

On an experimental basis, a number of analytical methods have been applied to assess local structural differences in brain tissue between PD and normal tissue. Voxel-based morphometry (VBM) has demonstrated basal ganglia atrophy in PD as compared to healthy controls, consistent with degeneration of the dopaminergic nigrostriatal system [134]. In early PD, deformation-based morphometry (DBM) has shown PD-specific areas of atrophy involving the midbrain, basal ganglia, basal forebrain, medial temporal lobe, and some discrete cortical regions, which correlated with clinical measurements of disease severity and progression [135, 136]. Moreover, consistent with the more severe and aggressive clinical phenotype observed in those with PD and concomitant RBD more extensive cortical thinning is observed as compared with those with PD without RBD [137]. In fact, microstructural damage, including reduced brain volume and cortical thickness, has also been detected during the prodromal PD period in those with idiopathic RBD [137–142]. A number of smaller studies have observed a relationship between gray matter atrophy in discrete regions and motor and non-motor symptoms of PD, including olfactory loss, respiratory dysfunction, freezing of gait, and overall disease severity [143–146].

3.2 Iron and Neuromelanin-Sensitive MRI

Early structural changes in SN morphology may be one promising diagnostic biomarker for PD. The substantia nigra is functionally and structurally divided into two parts: the SN pars compacta (SNc) and the SN pars reticulata (SNr). While the SNr is formed by loose aggregations of GABAergic medium and large neurons, the SNc contains a high density of neuromelanin-containing dopaminergic neurons that are further anatomically subdivided into six subgroups: a calbindin-rich zone known as the "matrix" and five calbindin-poor regions termed "nigrosomes" [147].

Iron and oxidative stress are thought to be important contributors toward the pathogenesis of PD [148]. The degeneration of dopaminergic neurons in PD is associated with increased concentrations of iron and neuromelanin (a strong iron chelator) within the SNc [223]. Whether this is of primary or secondary pathological importance in PD is unclear; however, it is thought that as neuromelanin's iron-protective properties become overwhelmed, increasingly non-sequestered neuronal iron may catalyze the production of free radicals, leading to increased oxidative stress and consequent neuronal death [224].

The association between iron accumulation and PD has led to investigations to use iron as a biomarker for both diagnosis and progression of PD. Iron quantification within the SNc can be reliably estimated using a number of MRI techniques, including apparent transverse relaxation rate (R2*), susceptibility weighted imaging (SWI), and quantitative susceptibility mapping (QSM) modalities [225], which correlate well with postmortem histological measures in PD [149]. Moreover, there is increasing evidence that detectable accumulation of iron in the SNc occurs at the prodromal and early stages of PD. Several studies have similarly shown increased iron in the SN during the prodromal stages of PD, including in those with RBD and asymptomatic carriers of familial PD [150–152]. For differential diagnosis, quantitative ROI analyses in iron-sensitive MRI have shown increased red nucleus susceptibility as a distinguishing feature of PSP compared with PD, MSA, and control subjects, whereas putaminal susceptibility is increased compared to PD and controls [153–156], although more research is needed.

The use of iron-sensitive MRI as a marker of progression is less clear. Some studies have shown cross-sectional and longitudinal SN iron changes over time [157], although this may not be linear, with rapid early increases within the first 3–5 years and then more slowly in advanced disease [158]. By contrast, others have found that levels do not vary significantly among patients in different disease stages [159] or with longitudinal follow-up [160].

3.2.1 Nigrosome-1

Intriguingly, dopamine-containing neuronal cell loss within the SNc follows a stereotyped temporospatial progression related to disease duration, beginning first in nigrosome-1, which is located in the caudal and mediolateral part of the SNc [161]. This property has made nigrosome-1 a possible early diagnostic biomarker in PD. Advances in high or ultra-high field (3 T or above), iron-sensitive MRI sequences have recently made it possible to identify nigrosome-1 on T2*/SWI-weighted images as an ovoid hyperintense signal area embedded within the midbrain region of the SNc of healthy subjects [162–164], resembling the bifurcated tip of a swallow's tail (i.e., "swallow tail sign"), although the exact concordance with pathology has recently been called into question [165]. A number of studies have demonstrated the disappearance of the hyperintense signal (i.e., loss of the swallow tail) in PD, presumably resultant from abnormal accumulation of iron in this region [166]. It is estimated that nigrostriatal dopaminergic input (as assessed ^{18}F-FP-CIT) must decrease to approximately half of normal for the swallow tail sign to disappear [167], which is coincident with the threshold for the motor symptoms of PD to appear [168]. In fact, loss of the swallowtail sign has been observed in prodromal PD (i.e., in subjects with RBD), which corresponded with significantly lower putaminal dopaminergic activity.

With respect to diagnostic accuracy, recent meta-analysis have demonstrated that loss of the swallowtail sign can accurately distinguish between healthy controls and those with idiopathic PD or atypical parkinsonism with pooled sensitivity and specificity of 0.94 and 0.90, respectively, with comparable accuracy when using 7 T or 3 T [169]; by contrast, 1.5 T MRI suffers from poor specificity [170]. In one small study in early PD (mean disease duration of 9 months), loss of the swallowtail sign still performed with excellent specificity (1.00) and sensitivity (0.85) [171]. From a practical clinical perspective, a number of smaller studies have demonstrated the ability to distinguish PD from essential tremor, vascular parkinsonism, and drug-induced parkinsonism [172]. On the other hand, loss of the swallowtail sign is also seen at comparable rates in atypical parkinsonism, including DLB, PSP, and MSA-C and -P [172], and therefore cannot be used to distinguish between PD and atypical parkinsonism.

3.2.2 Neuromelanin

Neuromelanin is a product of dopamine metabolism and acts as an iron chelator within the SN [148]. Neuromelanin-sensitive signaling makes use of the ferrous properties of neuromelanin when it is bound to metals such as iron and copper [173]. Neuromelanin-MRI correlates well with postmortem midbrain neuromelanin concentration, and may be useful as a proxy for dopamine function in the nigrostriatal pathway [174].

Reduced neuromelanin signal in the SN appears to be not only a robust early stage marker of neurodegeneration in the SN, but also disease progression [150, 175]. Areas of reduced neuromelanin signal in RBD and early PD are observed in the posterolateral SN, with the loss of signal extended into the medial and anterior areas as the disease progresses [150]. Importantly, the severity of motor, cognitive, and limbic symptoms correlate with reduced neuromelanin signal in distinct spatial areas within the SN, reflecting the spatiotemporal pattern of neurodegeneration seen as PD progresses [150, 176].

A recent meta-analysis examining the use of neuromelanin-MRI as a diagnostic tool showed good accuracy (pooled sensitivity 0.89, pooled specificity 0.83) when differentiating between PD and healthy controls [177], with some additional evidence that it may distinguish between PD and atypical parkinsonism to some degree [178]. Limited study also shows possible utility in distinguishing PD from essential tremor [179, 180]. The diagnostic accuracy in distinguishing between PD and atypical parkinsonism, however, appears to be marginal [178].

3.3 Diffusion Tensor Imaging

Diffusion tensor imaging allows for the visualization and quantification of the flow of water molecules through a given tissue [181]. Since diffusion is constrained by the local tissue and cellular environment, DTI can therefore be used to make inferences upon the microstructural organization and integrity of the brain.

Many early studies with single-tensor DTI focused upon measuring altered diffusion metrics in segmented ROIs within the nigrostriatal pathway in PD. For example, it is theorized that dopamine cell loss in the SN (particularly the ventrolateral segment) leads to altered microstructural integrity within the gray matter, as reflected by reduced fractional anisotropy and increased diffusivity [182, 183]. While this has been generally observed, a number of meta-analyses have demonstrated some inconsistency in results when comparing between PD and healthy controls, due to a wide degree of signal variability and potentially difficulties in consistently identifying the correct ROIs [184–187].

Further advances in DTI have led to the introduction of free-water imaging, in which a computational approach is used to fit a two-compartment model to separate the diffusion properties of water in brain tissue from those of water in the extra-cellular space [188]. Increases in free water are associated with neurodegenerative and neuroinflammatory changes within the brain [189], and such increases are observed within the posterior substantia nigra in PD [190, 191] as well as in atypical parkinsonism [190, 192]. Mirroring the spatiotemporal pattern of neurodegeneration observed in neuromelanin, when followed longitudinally in PD, free water in the SN increases in a stereotyped pattern involving the posterolateral SN in the earliest stages before extending medially and anteriorly [193, 194]. Further supporting this, increased free water within the SN has been detected in prodromal PD, including in RBD and asymptomatic carriers of familial PD [195–197]. Finally, in a large multicenter study using a machine learning approach, free water measurements within the basal ganglia, midbrain, and cerebellum could correctly differentiate PD from atypical parkinsonism with moderate to high accuracy [198]. Thus, free water imaging holds promise as a diagnostic biomarker for PD and atypical parkinsonism.

3.4 fMRI

fMRI is an advanced non-invasive neuroimaging modality that measures and maps brain activity by detecting changes in blood oxygenation levels [199]. It operates on the principle of neurovascular coupling, wherein alterations in neuronal activation are coupled with regional changes in cerebral blood flow and oxygenation. Resting state fMRI captures spontaneous brain activity and functional connectivity patterns in the absence of explicit tasks or stimuli, providing insight into intrinsic brain organization. In contrast, task-based fMRI involves measuring brain activity while individuals perform specific cognitive or motor tasks, allowing for the examination of task-related activations and functional networks associated with particular mental processes.

There is increasing evidence that network disruption occurs very early in PD and involves multiple anatomical regions and neurotransmitter systems [200]. As such, fMRI may be a particularly useful modality for studying the involvement of network impairment in PD. The diverse clinical presentation of PD (which includes cognitive, psychiatric, and other non-motor disturbances) does not necessarily follow the spatiotemporal pattern of neuropathology (which is theorized to begin in the caudal brainstem and extend to the cortical areas only in later disease stages). This suggests that the dysfunction of widespread functional brain networks—rather than focal pathology—underlie many of the non-motor disturbances in PD [201, 202].

Early studies in PD demonstrated markedly reduced resting-state functional connectivity between the striatum and the extended brainstem (which includes the thalamus, midbrain, upper pons, and cerebellum), and moderately increased striatal correlations with discrete parts of the cerebral cortex (sensorimotor and visual cortices) [200]. Subsequently, a recent large-scale study demonstrated altered functional connectivity in PD involved cortical sensorimotor, thalamic, and cerebellar networks, both within and, especially, between networks [203]. Such disruptions were related to motor and cognitive performance, indicating a potential link to clinical manifestations in PD. Similarly, altered cerebellar circuitry, involving connectivity of the vermis and peristriate visual association cortex was associated with dysfunction in attention, executive functions, and global cognition in PD with cognitive impairment [204].

A number of motor and non-motor symptoms of PD have been linked to the disruption of a variety of circuits, including tremor [205, 206], dyskinesias [207], freezing of gait [159, 180, 208–210], sleep disturbances [211], RBD [212, 213], cognitive dysfunction [214–216], and hallucinations [217]. These studies shed light on the pathophysiology of PD but are of little use in discriminating between diseases. Moreover, because of interindividual variability, fMRI measures are generally not sufficiently robust to be used at the individual level. Therefore, much further research is necessary to ensure the effectiveness of this approach on an individual level.

4 Inter-Relationship Between MRI and DAT-Imaging

Recent emerging interest has focused on the relationship between the structural SN changes on MRI and the functional changes assessed by DAT-imaging. A good agreement between SN-MRI abnormalities and [^{123}I]FP-CIT-SPECT has been reported, with a concordance rate of 86.2% between the two modalities [218]. However, it has also been observed that PD patients can have preservation of nigrosome-1 and degeneration of the nigrostriatal pathway. Such disagreement was also seen in patients with CBD [219], a LRRK2 mutation carrier [220], and RBD cases [221]. Since DAT-imaging acts as a surrogate marker for degenerative PS, the incongruency could question the accuracies of SWI–MRI in diagnosing PD and PS. With SWI-MRI, if iron is assumed to be responsible for the SN signal abnormality, one should note that iron is not invariably associated with degeneration. SWI-MRI and DAT-SPECT do not study nigral neurons at the same levels: DAT-imaging could overestimate degeneration due to the DAT down-regulation in the early stage of the disease. In addition, DAT-imaging reflects synaptic degeneration and is not a direct marker of dopaminergic neuronal death [50, 222].

5 Summary/Conclusion

In clinical practice, the diagnoses of PD and degenerative PS are based on clinical criteria. The use of DAT-SPECT and advanced MRI techniques shows some utility in early diagnosis for PD and differentiation from other non-degenerative parkinsonian disorders. With increased diagnostic accuracy for early PD, patients with some atypical features in response to treatment and/or disease progression, such imaging tools may guide treatment. For pre-clinical detection, it may be important in the future if effective neuroprotective therapies are available. However, clinicians should be aware of issues related to the interpretation of results and consider their impact on patient management. Both DAT and MRI imaging may have some prognostic value; this has limited relevance to current clinical practice but may serve as an inclusion criterion in clinical trials. They have limited value in differentiating among degenerative causes of PS. Clinical assessment remains the most important tool in evaluating patients.

Not only does DAT imaging not differentiate PD from atypical degenerative PS, but it does not also distinguish among conditions in which there is no loss of dopamine neurons (e.g., healthy individuals, ET, DT, psychogenic conditions, drug-induced parkinsonism, etc.). A normal scan does not rule out the presence of some other pathological process. Relying on the imaging result to withhold treatment in certain cases (e.g., DRD) may have a significant negative impact on patient management.

Many of the studies cited have used clinical diagnoses as the gold standards against which to assess sensitivity and specificity of DAT/MRI imaging. Therefore, the documented diagnostic accuracy cannot exceed that of movement disorder experts. If the test adds little to the clinical diagnosis of PD, its utility for that indication is questionable.

DAT imaging and advanced MRI sequences are not widely available in many countries or many centers. Clinicians should consider utilization of healthcare resources, the yield of clinical tests, and the consequences of test results. DAT-SPECT is expensive and is not of superior diagnostic accuracy when compared to movement disorders specialists. Although it can be a useful adjunct to the clinical diagnosis of atypical patients, in current practice, its role in patient management is limited. In addition, potential diagnostic inaccuracies can lead to inappropriate management, or unnecessary follow-up tests [126]. For accurate assessment of PD and its mimickers, a detailed history, careful neurological examination, and response to treatment are still the cornerstones of diagnosis. In spite of increasing data from imaging, more long-term clinical studies are required to determine whether they can accurately guide clinical decision-making.

References

1. Nussbaum RL, Ellis CE. Alzheimer's disease and Parkinson's disease. N Engl J Med. 2003;348(14):1356–64.
2. Gibb WR, Lees AJ. The relevance of the Lewy body to the pathogenesis of idiopathic Parkinson's disease. J Neurol Neurosurg Psychiatry. 1988;51(6):745–52.
3. Postuma RB, Berg D, Stern M, Poewe W, Olanow CW, Oertel W, et al. MDS clinical diagnostic criteria for Parkinson's disease. Mov Disord. 2015;30(12):1591–601.
4. Rajput AH, Rozdilsky B, Rajput A. Accuracy of clinical diagnosis in parkinsonism – a prospective study. Can J Neurol Sci. 1991;18(3):275–8.
5. Newman EJ, Breen K, Patterson J, Hadley DM, Grosset KA, Grosset DG. Accuracy of Parkinson's disease diagnosis in 610 general practice patients in the West of Scotland. Mov Disord. 2009;24(16):2379–85.
6. Huang W-S, Chiang Y-H, Lin J-C, Chou Y-H, Cheng C-Y, Liu R-S. Crossover study of (99m)Tc-TRODAT-1 SPECT and (18)F-FDOPA PET in Parkinson's disease patients. J Nucl Med. 2003;44(7):999–1005.
7. Kish SJ, Shannak K, Hornykiewicz O. Uneven pattern of dopamine loss in the striatum of patients with idiopathic Parkinson's disease. Pathophysiologic and clinical implications. N Engl J Med. 1988;318(14):876–80.
8. Seibyl J, Jennings D, Tabamo R, Marek K. The role of neuroimaging in the early diagnosis and evaluation of Parkinson's disease. Minerva Med. 2005;96(5):353–64.
9. Mozley PD, Schneider JS, Acton PD, Plossl K, Stern MB, Siderowf A, et al. Binding of [99mTc]TRODAT-1 to dopamine transporters in patients with Parkinson's disease and in healthy volunteers. J Nucl Med. 2000;41(4):584–9.
10. Brooks DJ, Pavese N. Imaging biomarkers in Parkinson's disease. Prog Neurobiol. 2011;95(4):614–28.
11. Huang W-S, Lee M-S, Lin J-C, Chen C-Y, Yang Y-W, Lin S-Z, et al. Usefulness of brain 99mTc-TRODAT-1 SPET for the evaluation of Parkinson's disease. Eur J Nucl Med Mol Imaging. 2004;31(2):155–61.
12. Cummings JL, Henchcliffe C, Schaier S, Simuni T, Waxman A, Kemp P. The role of dopaminergic imaging in patients with symptoms of dopaminergic system neurodegeneration. Brain. 2011;134(Pt 11):3146–66.
13. Park E. A new era of clinical dopamine transporter imaging using 123I-FP-CIT. J Nucl Med Technol. 2012;40(4):222–8.
14. Website EMApifD. 2000.
15. Darcourt J, Booij J, Tatsch K, Varrone A, Vander Borght T, Kapucu OL, et al. EANM procedure guidelines for brain neurotransmission SPECT using (123)I-labelled dopamine transporter ligands, version 2. Eur J Nucl Med Mol Imaging. 2010;37(2):443–50.
16. DaTscan Fpif. 2011.
17. Utiumi MA, Felicio AC, Borges CR, Braatz VL, Rezende SA, Munhoz RP, et al. Dopamine transporter imaging in clinically unclear cases of parkinsonism and the importance of scans without evidence of dopaminergic deficit (SWEDDs). Arq Neuropsiquiatr. 2012;70(9):667–73.
18. Cummings JL, Fine MJ, Grachev ID, Jarecke CR, Johnson MK, Kuo PH, et al. Effective and efficient diagnosis of parkinsonism: the role of dopamine transporter SPECT imaging with ioflupane I-123 injection (DaTscan). Am J Manag Care. 2014;20(5 Suppl):S97–109.
19. Piggott MA, Marshall EF, Thomas N, Lloyd S, Court JA, Jaros E, et al. Striatal dopaminergic markers in dementia with Lewy bodies, Alzheimer's and Parkinson's diseases: rostrocaudal distribution. Brain. 1999;122(Pt 8):1449–68.
20. Cheng HC, Ulane CM, Burke RE. Clinical progression in Parkinson disease and the neurobiology of axons. Ann Neurol. 2010;67(6):715–25.

21. Eshuis SA, Jager PL, Maguire RP, Jonkman S, Dierckx RA, Leenders KL. Direct comparison of FP-CIT SPECT and F-DOPA PET in patients with Parkinson's disease and healthy controls. Eur J Nucl Med Mol Imaging. 2009;36(3):454–62.
22. Nikolaus S, Antke C, Kley K, Poeppel TD, Hautzel H, Schmidt D, et al. Investigating the dopaminergic synapse in vivo. I. Molecular imaging studies in humans. Rev Neurosci. 2007;18(6):439–72.
23. Isaacson JR, Brillman S, Chhabria N, Isaacson SH. Impact of DaTscan imaging on clinical decision making in clinically uncertain Parkinson's disease. J Parkinsons Dis. 2021;11(2):885–9.
24. Marek KL, Seibyl JP, Zoghbi SS, Zea-Ponce Y, Baldwin RM, Fussell B, et al. [123I] beta-CIT/SPECT imaging demonstrates bilateral loss of dopamine transporters in hemi-Parkinson's disease. Neurology. 1996;46(1):231–7.
25. Ponsen MM, Stoffers D, Booij J, van Eck-Smit BL, Wolters E, Berendse HW. Idiopathic hyposmia as a preclinical sign of Parkinson's disease. Ann Neurol. 2004;56(2):173–81.
26. Eisensehr I, Linke R, Noachtar S, Schwarz J, Gildehaus FJ, Tatsch K. Reduced striatal dopamine transporters in idiopathic rapid eye movement sleep behaviour disorder: comparison with Parkinson's disease and controls. Brain. 2000;123(6):1155–60.
27. Iranzo A, Santamaría J, Valldeoriola F, Serradell M, Salamero M, Gaig C, et al. Dopamine transporte rimaging deficit predicts early transition to synucleinopathy in idiopathic rapid eye movement sleep behavior disorder. Ann Neurol. 2017;82(3):419–28.
28. van Dyck CH, Malison RT, Jacobsen LK, Seibyl JP, Staley JK, Laruelle M, et al. Increased dopamine transporter availability associated with the 9-repeat allele of the SLC6A3 gene. J Nucl Med. 2005;46(5):745–51.
29. Seibyl JP, Marek KL, Quinlan D, Sheff K, Zoghbi S, Zea-Ponce Y, et al. Decreased single-photon emission computed tomographic [123I]beta-CIT striatal uptake correlates with symptom severity in Parkinson's disease. Ann Neurol. 1995;38(4):589–98.
30. Benamer HT, Patterson J, Wyper DJ, Hadley DM, Macphee GJ, Grosset DG. Correlation of Parkinson's disease severity and duration with 123I-FP-CIT SPECT striatal uptake. Mov Disord. 2000;15(4):692–8.
31. Pirker W. Correlation of dopamine transporter imaging with parkinsonian motor handicap: how close is it? Mov Disord. 2003;18(Suppl 7):S43–51.
32. Lee CS, Samii A, Sossi V, Ruth TJ, Schulzer M, Holden JE, et al. In vivo positron emission tomographic evidence for compensatory changes in presynaptic dopaminergic nerve terminals in Parkinson's disease. Ann Neurol. 2000;47(4):493–503.
33. Cilia R, Marotta G, Belletti A, Siri C, Pezzoli G. Reversible dopamine transporter reduction in drug-induced parkinsonism. Mov Disord. 2014;29(4):575–7.
34. Adams JR, van Netten H, Schulzer M, Mak E, McKenzie J, Strongosky A, et al. PET in LRRK2 mutations: comparison to sporadic Parkinson's disease and evidence for presymptomatic compensation. Brain. 2005;128(Pt 12):2777–85.
35. Snow BJ, Tooyama I, McGeer EG, Yamada T, Calne DB, Takahashi H, et al. Human positron emission tomographic [18F]fluorodopa studies correlate with dopamine cell counts and levels. Ann Neurol. 1993;34(3):324–30.
36. Bajaj NP, Gontu V, Birchall J, Patterson J, Grosset DG, Lees AJ. Accuracy of clinical diagnosis in tremulous parkinsonian patients: a blinded video study. J Neurol Neurosurg Psychiatry. 2010;81(11):1223–8.
37. A randomized controlled trial comparing pramipexole with levodopa in early Parkinson's disease: design and methods of the CALM-PD study. Parkinson Study Group. Clin Neuropharmacol. 2000;23(1):34–44.
38. Benamer TS, Patterson J, Grosset DG, Booij J, de Bruin K, van Royen E, et al. Accurate differentiation of parkinsonism and essential tremor using visual assessment of [123I]-FP-CIT SPECT imaging: the [123I]-FP-CIT study group. Mov Disord. 2000;15(3):503–10.

39. Booij J, Speelman JD, Horstink MW, Wolters EC. The clinical benefit of imaging striatal dopamine transporters with [123I]FP-CIT SPET in differentiating patients with presynaptic parkinsonism from those with other forms of parkinsonism. Eur J Nucl Med. 2001;28(3):266–72.
40. Catafau AM, Tolosa E, Da TCUPSSG. Impact of dopamine transporter SPECT using 123I-Ioflupane on diagnosis and management of patients with clinically uncertain Parkinsonian syndromes. Mov Disord. 2004;19(10):1175–82.
41. Jennings DL, Seibyl JP, Oakes D, Eberly S, Murphy J, Marek K. (123I) beta-CIT and single-photon emission computed tomographic imaging vs clinical evaluation in Parkinsonian syndrome: unmasking an early diagnosis. Arch Neurol. 2004;61(8):1224–9.
42. Marshall VL, Patterson J, Hadley DM, Grosset KA, Grosset DG. Two-year follow-up in 150 consecutive cases with normal dopamine transporter imaging. Nucl Med Commun. 2006;27(12):933–7.
43. Marshall VL, Reininger CB, Marquardt M, Patterson J, Hadley DM, Oertel WH, et al. Parkinson's disease is overdiagnosed clinically at baseline in diagnostically uncertain cases: a 3-year European multicenter study with repeat [123I]FP-CIT SPECT. Mov Disord. 2009;24(4):500–8.
44. de Verdal M, Renard D, Collombier L, Boudousq V, Kotzki PO, Labauge P, et al. I123-FP-CIT single-photon emission tomography in patients with long-standing mixed tremor. Eur J Neurol. 2013;20(2):382–8.
45. Marek K, Seibyl J, Eberly S, Oakes D, Shoulson I, Lang AE, et al. Longitudinal follow-up of SWEDD subjects in the PRECEPT study. Neurology. 2014;82(20):1791–7.
46. Marek K, Chowdhury S, Siderowf A, Lasch S, Coffey CS, Caspell-Garcia C, et al. The Parkinson's progression markers initiative (PPMI) – establishing a PD biomarker cohort. Ann Clin Transl Neurol. 2018;5(12):1460–77.
47. Ferrie LJ, Gartside SE, Martin KM, Young AH, McQuade R. Effect of chronic lithium treatment on D2/3 autoreceptor regulation of dopaminergic function in the rat. Pharmacol Biochem Behav. 2008;90(2):218–25.
48. Nurmi E, Ruottinen HM, Kaasinen V, Bergman J, Haaparanta M, Solin O, et al. Progression in Parkinson's disease: a positron emission tomography study with a dopamine transporter ligand [18F]CFT. Ann Neurol. 2000;47(6):804–8.
49. Asenbaum S, Brucke T, Pirker W, Podreka I, Angelberger P, Wenger S, et al. Imaging of dopamine transporters with iodine-123-beta-CIT and SPECT in Parkinson's disease. J Nuclear Med. 1997;38(1):1–6.
50. Saari L, Kivinen K, Gardberg M, Joutsa J, Noponen T, Kaasinen V. Dopamine transporter imaging does not predict the number of nigral neurons in Parkinson disease. Neurology. 2017;88(15):1461–7.
51. Fahn S, Oakes D, Shoulson I, Kieburtz K, Rudolph A, Lang A, et al. Levodopa and the progression of Parkinson's disease. N Engl J Med. 2004;351(24):2498–508.
52. Ravina B, Eidelberg D, Ahlskog JE, Albin RL, Brooks DJ, Carbon M, et al. The role of radiotracer imaging in Parkinson disease. Neurology. 2005;64(2):208–15.
53. Cilia R, Reale C, Castagna A, Nasca A, Muzi-Falconi M, Barzaghi C, et al. Novel DYT11 gene mutation in patients without dopaminergic deficit (SWEDD) screened for dystonia. Neurology. 2014;83(13):1155–62.
54. Ceravolo R, Rossi C, Cilia R, Tognoni G, Antonini A, Volterrani D, et al. Evidence of delayed nigrostriatal dysfunction in corticobasal syndrome: a SPECT follow-up study. Parkinsonism Relat Disord. 2013;19(5):557–9.
55. Kaasinen V, Gardberg M, Roytta M, Seppanen M, Paivarinta M. Normal dopamine transporter SPECT in neuropathologically confirmed corticobasal degeneration. J Neurol. 2013;260(5):1410–1.
56. Ravina B, Marek K, Eberly S, Oakes D, Kurlan R, Ascherio A, et al. Dopamine transporter imaging is associated with long-term outcomes in Parkinson's disease. Mov Disord. 2012;27(11):1392–7.

57. Vogt T, Kramer K, Gartenschlaeger M, Schreckenberger M. Estimation of further disease progression of Parkinson's disease by dopamin transporter scan vs clinical rating. Parkinsonism Relat Disord. 2011;17(6):459–63.
58. de la Fuente-Fernandez R, Schulzer M, Kuramoto L, Cragg J, Ramachandiran N, Au WL, et al. Age-specific progression of nigrostriatal dysfunction in Parkinson's disease. Ann Neurol. 2011;69(5):803–10.
59. Kordower JH, Olanow CW, Dodiya HB, Chu Y, Beach TG, Adler CH, et al. Disease duration and the integrity of the nigrostriatal system in Parkinson's disease. Brain. 2013;136(Pt 8):2419–31.
60. Stoessl AJ. Developments in neuroimaging: positron emission tomography. Parkinsonism Relat Disord. 2014;20(Suppl 1):S180–3.
61. Cilia R, Marotta G, Benti R, Pezzoli G, Antonini A. Brain SPECT imaging in multiple system atrophy. J Neural Transm. 2005;112(12):1635–45.
62. Sawle GV, Brooks DJ, Marsden CD, Frackowiak RS, Corticobasal degeneration. A unique pattern of regional cortical oxygen hypometabolism and striatal fluorodopa uptake demonstrated by positron emission tomography. Brain. 1991;114(Pt 1B):541–56.
63. Marshall V, Grosset DG. Role of dopamine transporter imaging in the diagnosis of atypical tremor disorders. Mov Disord. 2003;18(Suppl 7):S22–7.
64. Kim YJ, Ichise M, Ballinger JR, Vines D, Erami SS, Tatschida T, et al. Combination of dopamine transporter and D2 receptor SPECT in the diagnostic evaluation of PD, MSA, and PSP. Mov Disord. 2002;17(2):303–12.
65. Varrone A, Marek KL, Jennings D, Innis RB, Seibyl JP. [(123)I]beta-CIT SPECT imaging demonstrates reduced density of striatal dopamine transporters in Parkinson's disease and multiple system atrophy. Mov Disord. 2001;16(6):1023–32.
66. Knudsen GM, Karlsborg M, Thomsen G, Krabbe K, Regeur L, Nygaard T, et al. Imaging of dopamine transporters and D2 receptors in patients with Parkinson's disease and multiple system atrophy. Eur J Nucl Med Mol Imaging. 2004;31(12):1631–8.
67. Filippi L, Manni C, Pierantozzi M, Brusa L, Danieli R, Stanzione P, et al. 123I-FP-CIT in progressive supranuclear palsy and in Parkinson's disease: a SPECT semiquantitative study. Nucl Med Commun. 2006;27(4):381–6.
68. Perju-Dumbrava LD, Kovacs GG, Pirker S, Jellinger K, Hoffmann M, Asenbaum S, et al. Dopamine transporter imaging in autopsy-confirmed Parkinson's disease and multiple system atrophy. Mov Disord. 2012;27(1):65–71.
69. Goebel G, Seppi K, Donnemiller E, Warwitz B, Wenning GK, Virgolini I, et al. A novel computer-assisted image analysis of [123I]beta-CIT SPECT images improves the diagnostic accuracy of parkinsonian disorders. Eur J Nucl Med Mol Imaging. 2011;38(4):702–10.
70. Cilia R, Rossi C, Frosini D, Volterrani D, Siri C, Pagni C, et al. Dopamine transporter SPECT imaging in Corticobasal syndrome. PLoS One. 2011;6(5):e18301.
71. O'Sullivan SS, Burn DJ, Holton JL, Lees AJ. Normal dopamine transporter single photon-emission CT scan in corticobasal degeneration. Mov Disord. 2008;23(16):2424–6.
72. McNeill A, Wu RM, Tzen KY, Aguiar PC, Arbelo JM, Barone P, et al. Dopaminergic neuronal imaging in genetic Parkinson's disease: insights into pathogenesis. PLoS One. 2013;8(7):e69190.
73. Doepp F, Plotkin M, Siegel L, Kivi A, Gruber D, Lobsien E, et al. Brain parenchyma sonography and 123I-FP-CIT SPECT in Parkinson's disease and essential tremor. Mov Disord. 2008;23(3):405–10.
74. Wang J, Jiang YP, Liu XD, Chen ZP, Yang LQ, Liu CJ, et al. 99mTc-TRODAT-1 SPECT study in early Parkinson's disease and essential tremor. Acta Neurol Scand. 2005;112(6):380–5.
75. Breit S, Reimold M, Reischl G, Klockgether T, Wullner U. [(11)C]d-threo-methylphenidate PET in patients with Parkinson's disease and essential tremor. J Neural Transm. 2006;113(2):187–93.
76. Asenbaum S, Pirker W, Angelberger P, Bencsits G, Pruckmayer M, Brucke T. [123I]beta-CIT and SPECT in essential tremor and Parkinson's disease. J Neural Transm. 1998;105(10–12):1213–28.

77. Isaias IU, Marotta G, Hirano S, Canesi M, Benti R, Righini A, et al. Imaging essential tremor. Mov Disord. 2010;25(6):679–86.
78. Benito-Leon J, Louis ED, Bermejo-Pareja F, Neurological Disorders in Central Spain Study G. Risk of incident Parkinson's disease and parkinsonism in essential tremor: a population based study. J Neurol Neurosurg Psychiatry. 2009;80(4):423–5.
79. Minen MT, Louis ED. Emergence of Parkinson's disease in essential tremor: a study of the clinical correlates in 53 patients. Mov Disord. 2008;23(11):1602–5.
80. Marshall V, Grosset D. Role of dopamine transporter imaging in routine clinical practice. Mov Disord. 2003;18(12):1415–23.
81. Hughes AJ, Daniel SE, Kilford L, Lees AJ. Accuracy of clinical diagnosis of idiopathic Parkinson's disease: a clinico-pathological study of 100 cases. J Neurol Neurosurg Psychiatry. 1992;55(3):181–4.
82. Mark MH, Sage JI, Walters AS, Duvoisin RC, Miller DC. Binswanger's disease presenting as levodopa-responsive parkinsonism: clinicopathologic study of three cases. Mov Disord. 1995;10(4):450–4.
83. Kalra S, Grosset DG, Benamer HT. Differentiating vascular parkinsonism from idiopathic Parkinson's disease: a systematic review. Mov Disord. 2010;25(2):149–56.
84. Tzen KY, Lu CS, Yen TC, Wey SP, Ting G. Differential diagnosis of Parkinson's disease and vascular parkinsonism by (99m)Tc-TRODAT-1. J Nucl Med. 2001;42(3):408–13.
85. Gerschlager W, Bencsits G, Pirker W, Bloem BR, Asenbaum S, Prayer D, et al. [123I]beta-CIT SPECT distinguishes vascular parkinsonism from Parkinson's disease. Mov Disord. 2002;17(3):518–23.
86. Lorberboym M, Djaldetti R, Melamed E, Sadeh M, Lampl Y. 123I-FP-CIT SPECT imaging of dopamine transporters in patients with cerebrovascular disease and clinical diagnosis of vascular parkinsonism. J Nucl Med. 2004;45(10):1688–93.
87. Zijlmans J, Evans A, Fontes F, Katzenschlager R, Gacinovic S, Lees AJ, et al. [123I] FP-CIT spect study in vascular parkinsonism and Parkinson's disease. Mov Disord. 2007;22(9):1278–85.
88. Susatia F, Fernandez HH. Drug-induced parkinsonism. Curr Treat Options Neurol. 2009;11(3):162–9.
89. Marsden CD, Jenner P. The pathophysiology of extrapyramidal side-effects of neuroleptic drugs. Psychol Med. 1980;10(1):55–72.
90. Diaz-Corrales FJ, Sanz-Viedma S, Garcia-Solis D, Escobar-Delgado T, Mir P. Clinical features and 123I-FP-CIT SPECT imaging in drug-induced parkinsonism and Parkinson's disease. Eur J Nucl Med Mol Imaging. 2010;37(3):556–64.
91. Brooks DJ, Pavese N. Recent imaging advances in the diagnosis and management of Parkinson's disease. F1000 Med Rep. 2009:1.
92. Lorberboym M, Treves TA, Melamed E, Lampl Y, Hellmann M, Djaldetti R. [123I]-FP/CIT SPECT imaging for distinguishing drug-induced parkinsonism from Parkinson's disease. Mov Disord. 2006;21(4):510–4.
93. Kagi G, Bhatia KP, Tolosa E. The role of DAT-SPECT in movement disorders. J Neurol Neurosurg Psychiatry. 2010;81(1):5–12.
94. Schneider SA, Edwards MJ, Mir P, Cordivari C, Hooker J, Dickson J, et al. Patients with adult-onset dystonic tremor resembling parkinsonian tremor have scans without evidence of dopaminergic deficit (SWEDDs). Mov Disord. 2007;22(15):2210–5.
95. Gupta A, Lang AE. Psychogenic movement disorders. Curr Opin Neurol. 2009;22(4):430–6.
96. Puschmann A, Wszolek ZK. Diagnosis and treatment of common forms of tremor. Semin Neurol. 2011;31(1):65–77.
97. Benaderette S, Zanotti Fregonara P, Apartis E, Nguyen C, Trocello JM, Remy P, et al. Psychogenic parkinsonism: a combination of clinical, electrophysiological, and [(123)I]-FP-CIT SPECT scan explorations improves diagnostic accuracy. Mov Disord. 2006;21(3):310–7.
98. Gaig C, Marti MJ, Tolosa E, Valldeoriola F, Paredes P, Lomena FJ, et al. 123I-Ioflupane SPECT in the diagnosis of suspected psychogenic parkinsonism. Mov Disord. 2006;21(11):1994–8.

99. van der Flier WM, Scheltens P. Epidemiology and risk factors of dementia. J Neurol Neurosurg Psychiatry. 2005;76(Suppl 5):v2–7.
100. Ala TA, Yang KH, Sung JH, Frey WH 2nd. Hallucinations and signs of parkinsonism help distinguish patients with dementia and cortical Lewy bodies from patients with Alzheimer's disease at presentation: a clinicopathological study. J Neurol Neurosurg Psychiatry. 1997;62(1):16–21.
101. Colloby SJ, McParland S, O'Brien JT, Attems J. Neuropathological correlates of dopaminergic imaging in Alzheimer's disease and Lewy body dementias. Brain. 2012;135(Pt 9):2798–808.
102. Samuel W, Caligiuri M, Galasko D, Lacro J, Marini M, McClure FS, et al. Better cognitive and psychopathologic response to donepezil in patients prospectively diagnosed as dementia with Lewy bodies: a preliminary study. Int J Geriatr Psychiatry. 2000;15(9):794–802.
103. Emre M, Cummings JL, Lane RM. Rivastigmine in dementia associated with Parkinson's disease and Alzheimer's disease: similarities and differences. J Alzheimers Dis. 2007;11(4):509–19.
104. Rozzini L, Chilovi BV, Bertoletti E, Conti M, Delrio I, Trabucchi M, et al. Cognitive and psychopathologic response to rivastigmine in dementia with Lewy bodies compared to Alzheimer's disease: a case control study. Am J Alzheimers Dis Other Dement. 2007;22(1):42–7.
105. Nelson PT, Jicha GA, Kryscio RJ, Abner EL, Schmitt FA, Cooper G, et al. Low sensitivity in clinical diagnoses of dementia with Lewy bodies. J Neurol. 2010;257(3):359–66.
106. Kemp PM, Hoffmann SA, Tossici-Bolt L, Fleming JS, Holmes C. Limitations of the HMPAO SPECT appearances of occipital lobe perfusion in the differential diagnosis of dementia with Lewy bodies. Nucl Med Commun. 2007;28(6):451–6.
107. Koeppe RA, Gilman S, Junck L, Wernette K, Frey KA. Differentiating Alzheimer's disease from dementia with Lewy bodies and Parkinson's disease with (+)-[11C]dihydrotetrabenazine positron emission tomography. Alzheimers Dement. 2008;4(1 Suppl 1):S67–76.
108. Walker Z, Jaros E, Walker RW, Lee L, Costa DC, Livingston G, et al. Dementia with Lewy bodies: a comparison of clinical diagnosis, FP-CIT single photon emission computed tomography imaging and autopsy. J Neurol Neurosurg Psychiatry. 2007;78(11):1176–81.
109. Walker Z, Costa DC, Walker RW, Lee L, Livingston G, Jaros E, et al. Striatal dopamine transporter in dementia with Lewy bodies and Parkinson disease: a comparison. Neurology. 2004;62(9):1568–72.
110. McKeith I, O'Brien J, Walker Z, Tatsch K, Booij J, Darcourt J, et al. Sensitivity and specificity of dopamine transporter imaging with 123I-FP-CIT SPECT in dementia with Lewy bodies: a phase III, multicentre study. Lancet Neurol. 2007;6(4):305–13.
111. McKeith IG, Boeve BF, Dickson DW, Halliday G, Taylor JP, Weintraub D, et al. Diagnosis and management of dementia with Lewy bodies: fourth consensus report of the DLB consortium. Neurology. 2017;89(1):88–100.
112. Tadic V, Kasten M, Bruggemann N, Stiller S, Hagenah J, Klein C. Dopa-responsive dystonia revisited: diagnostic delay, residual signs, and nonmotor signs. Arch Neurol. 2012;69(12):1558–62.
113. Huang C-C, Yen T-C, Weng Y-H, Lu C-S. Normal dopamine transporter binding in dopa responsive dystonia. J Neurol. 2002;249(8):1016–20.
114. Hwang W-J, Yao W-J, Wey S-P, Ting G. Clinical and [99mTc]TRODAT-1/[123I]IBZM SPECT imaging findings in dopa-responsive dystonia. Eur Neurol. 2004;51(1):26–9.
115. Jeon BS, Jeong JM, Park SS, Kim JM, Chang YS, Song HC, et al. Dopamine transporter density measured by [123I]beta-CIT single-photon emission computed tomography is normal in dopa-responsive dystonia. Ann Neurol. 1998;43(6):792–800.
116. O'Sullivan JD, Costa DC, Gacinovic S, Lees AJ. SPECT imaging of the dopamine transporter in juvenile-onset dystonia. Neurology. 2001;56(2):266–7.
117. Wirdefeldt K, Burgess CE, Westerberg L, Payami H, Schalling M. A linkage study of candidate loci in familial Parkinson's disease. BMC Neurol. 2003;3:6.

118. Mencacci NE, Isaias IU, Reich MM, Ganos C, Plagnol V, Polke JM, et al. Parkinson's disease in GTP cyclohydrolase 1 mutation carriers. Brain. 2014;137(Pt 9):2480–92.
119. Erro R, Pappatà S, Amboni M, Vicidomini C, Longo K, Santangelo G, et al. Anxiety is associated with striatal dopamine transporter availability in newly diagnosed untreated Parkinson's disease patients. Parkinsonism Relat Disord. 2012;18(9):1034–8.
120. Kiferle L, Ceravolo R, Giuntini M, Linsalata G, Puccini G, Volterrani D, et al. Caudate dopaminergic denervation and visual hallucinations: evidence from a ^{123}I-FP-CIT SPECT study. Parkinsonism Relat Disord. 2014;20(7):761–5.
121. Santangelo G, Vitale C, Picillo M, Cuoco S, Moccia M, Pezzella D, et al. Apathy and striatal dopamine transporter levels in de-novo, untreated Parkinson's disease patients. Parkinsonism Relat Disord. 2015;21(5):489–93.
122. Kim R, Jun JS. Association of autonomic symptoms with presynaptic striatal dopamine depletion in drug-naive Parkinson's disease: an analysis of the PPMI data. Auton Neurosci. 2019;216:59–62.
123. Voon V, Rizos A, Chakravartty R, Mulholland N, Robinson S, Howell NA, et al. Impulse control disorders in Parkinson's disease: decreased striatal dopamine transporter levels. J Neurol Neurosurg Psychiatry. 2014;85(2):148–52.
124. Navalpotro-Gomez I, Dacosta-Aguayo R, Molinet-Dronda F, Martin-Bastida A, Botas-Peñin A, Jimenez-Urbieta H, et al. Nigrostriatal dopamine transporter availability, and its metabolic and clinical correlates in Parkinson's disease patients with impulse control disorders. Eur J Nucl Med Mol Imaging. 2019;46(10):2065–76.
125. de la Fuente-Fernandez R. Role of DaTSCAN and clinical diagnosis in Parkinson disease. Neurology. 2012;78(10):696–701.
126. Perlmutter JS, Eidelberg D. To scan or not to scan: DaT is the question. Neurology. 2012;78(10):688–9.
127. Hauser RA, Grosset DG. [123I]FP-CIT (DaTscan) SPECT brain imaging in patients with suspected parkinsonian syndromes. J Neuroimaging. 2012;22(3):225–30.
128. Bajaj N, Hauser RA, Grachev ID. Clinical utility of dopamine transporter single photon emission CT (DaT-SPECT) with (123I) ioflupane in diagnosis of parkinsonian syndromes. J Neurol Neurosurg Psychiatry. 2013;84(11):1288–95.
129. van Dyck CH, Seibyl JP, Malison RT, Laruelle M, Zoghbi SS, Baldwin RM, et al. Age-related decline in dopamine transporters: analysis of striatal subregions, nonlinear effects, and hemispheric asymmetries. Am J Geriatr Psychiatry. 2002;10(1):36–43.
130. Melikian HE, Buckley KM. Membrane trafficking regulates the activity of the human dopamine transporter. J Neurosci. 1999;19(18):7699–710.
131. Campabadal A, Segura B, Junque C, Serradell M, Abos A, Uribe C, et al. Cortical gray matter and hippocampal atrophy in idiopathic rapid eye movement sleep behavior disorder. Front Neurol. 2019;10:312.
132. Oppedal K, Ferreira D, Cavallin L, Lemstra AW, Ten Kate M, Padovani A, et al. A signature pattern of cortical atrophy in dementia with Lewy bodies: a study on 333 patients from the European DLB consortium. Alzheimers Dement. 2019;15(3):400–9.
133. Peralta C, Strafella AP, van Eimeren T, Ceravolo R, Seppi K, Kaasinen V, et al. Pragmatic approach on neuroimaging techniques for the differential diagnosis of parkinsonisms. Mov Disord Clin Pract. 2022;9(1):6–19.
134. Xu X, Han Q, Lin J, Wang L, Wu F, Shang H. Grey matter abnormalities in Parkinson's disease: a voxel-wise meta-analysis. Eur J Neurol. 2020;27(4):653–9.
135. Zeighami Y, Fereshtehnejad SM, Dadar M, Collins DL, Postuma RB, Dagher A. Assessment of a prognostic MRI biomarker in early de novo Parkinson's disease. NeuroImage Clinical. 2019;24:101986.
136. Zeighami Y, Ulla M, Iturria-Medina Y, Dadar M, Zhang Y, Larcher KM, et al. Network structure of brain atrophy in de novo Parkinson's disease. elife. 2015;4:4.
137. Rahayel S, Tremblay C, Vo A, Zheng YQ, Lehéricy S, Arnulf I, et al. Brain atrophy in prodromal synucleinopathy is shaped by structural connectivity and gene expression. Brain. 2022;145(9):3162–78.

138. Rahayel S, Gaubert M, Postuma RB, Montplaisir J, Carrier J, Monchi O, et al. Brain atrophy in Parkinson's disease with polysomnography-confirmed REM sleep behavior disorder. Sleep. 2019;42(6):zsz062.
139. Rahayel S, Postuma RB, Montplaisir J, Genier Marchand D, Escudier F, Gaubert M, et al. Cortical and subcortical gray matter bases of cognitive deficits in REM sleep behavior disorder. Neurology. 2018;90(20):e1759–e70.
140. Rahayel S, Postuma RB, Montplaisir J, Bedetti C, Brambati S, Carrier J, et al. Abnormal gray matter shape, thickness, and volume in the motor cortico-subcortical loop in idiopathic rapid eye movement sleep behavior disorder: association with clinical and motor features. Cereb Cortex. 2018;28(2):658–71.
141. Rahayel S, Montplaisir J, Monchi O, Bedetti C, Postuma RB, Brambati S, et al. Patterns of cortical thinning in idiopathic rapid eye movement sleep behavior disorder. Mov Disord. 2015;30(5):680–7.
142. Pereira JB, Weintraub D, Chahine L, Aarsland D, Hansson O, Westman E. Cortical thinning in patients with REM sleep behavior disorder is associated with clinical progression. npj Parkinson's Dis. 2019;5(1):7.
143. Chen S, Tan HY, Wu ZH, Sun CP, He JX, Li XC, et al. Imaging of olfactory bulb and gray matter volumes in brain areas associated with olfactory function in patients with Parkinson's disease and multiple system atrophy. Eur J Radiol. 2014;83(3):564–70.
144. Blair JC, Barrett MJ, Patrie J, Flanigan JL, Sperling SA, Elias WJ, et al. Brain MRI reveals ascending atrophy in Parkinson's disease across severity. Front Neurol. 2019;10:1329.
145. Lee SY, Chen MH, Chiang PL, Chen HL, Chou KH, Chen YC, et al. Reduced gray matter volume and respiratory dysfunction in Parkinson's disease: a voxel-based morphometry study. BMC Neurol. 2018;18(1):73.
146. Tessitore A, Amboni M, Cirillo G, Corbo D, Picillo M, Russo A, et al. Regional gray matter atrophy in patients with Parkinson disease and freezing of gait. AJNR Am J Neuroradiol. 2012;33(9):1804–9.
147. Damier P, Hirsch EC, Agid Y, Graybiel AM. The substantia nigra of the human brain. I. Nigrosomes and the nigral matrix, a compartmental organization based on calbindin D(28K) immunohistochemistry. Brain. 1999;122(Pt 8):1421–36.
148. Good PF, Olanow CW, Perl DP. Neuromelanin-containing neurons of the substantia nigra accumulate iron and aluminum in Parkinson's disease: a LAMMA study. Brain Res. 1992;593(2):343–6.
149. Langkammer C, Schweser F, Krebs N, Deistung A, Goessler W, Scheurer E, et al. Quantitative susceptibility mapping (QSM) as a means to measure brain iron? A post mortem validation study. Neuroimage. 2012;62(3):1593–9.
150. Biondetti E, Gaurav R, Yahia-Cherif L, Mangone G, Pyatigorskaya N, Valabregue R, et al. Spatiotemporal changes in substantia nigra neuromelanin content in Parkinson's disease. Brain. 2020;143(9):2757–70.
151. Pyatigorskaya N, Sharman M, Corvol JC, Valabregue R, Yahia-Cherif L, Poupon F, et al. High nigral iron deposition in LRRK2 and Parkin mutation carriers using R2* relaxometry. Mov Disord. 2015;30(8):1077–84.
152. Shin JH, Lee JY, Kim YK, Shin SA, Kim H, Nam H, et al. Longitudinal change in dopamine transporter availability in idiopathic REM sleep behavior disorder. Neurology. 2020;95(23):e3081–e92.
153. Meijer FJ, van Rumund A, Fasen BA, Titulaer I, Aerts M, Esselink R, et al. Susceptibility-weighted imaging improves the diagnostic accuracy of 3T brain MRI in the work-up of parkinsonism. AJNR Am J Neuroradiol. 2015;36(3):454–60.
154. Mazzucchi S, Frosini D, Costagli M, Del Prete E, Donatelli G, Cecchi P, et al. Quantitative susceptibility mapping in atypical parkinsonisms. NeuroImage Clin. 2019;24:101999.
155. Sjöström H, Granberg T, Westman E, Svenningsson P. Quantitative susceptibility mapping differentiates between parkinsonian disorders. Parkinsonism Relat Disord. 2017;44:51–7.

156. Sjostrom H, Surova Y, Nilsson M, Granberg T, Westman E, van Westen D, et al. Mapping of apparent susceptibility yields promising diagnostic separation of progressive supranuclear palsy from other causes of parkinsonism. Sci Rep. 2019;9(1):6079.
157. Bergsland N, Zivadinov R, Schweser F, Hagemeier J, Lichter D, Guttuso T Jr. Ventral posterior substantia nigra iron increases over 3 years in Parkinson's disease. Mov Disord. 2019;34(7):1006–13.
158. Hopes L, Grolez G, Moreau C, Lopes R, Ryckewaert G, Carriere N, et al. Magnetic resonance imaging features of the nigrostriatal system: biomarkers of Parkinson's disease stages? PLoS One. 2016;11(4):e0147947.
159. Li J, Yuan Y, Wang M, Zhang J, Zhang L, Jiang S, et al. Decreased interhemispheric homotopic connectivity in Parkinson's disease patients with freezing of gait: a resting state fMRI study. Parkinsonism Relat Disord. 2018;52:30–6.
160. Wieler M, Gee M, Martin WR. Longitudinal midbrain changes in early Parkinson's disease: iron content estimated from R2*/MRI. Parkinsonism Relat Disord. 2015;21(3):179–83.
161. Damier P, Hirsch EC, Agid Y, Graybiel AM. The substantia nigra of the human brain. II. Patterns of loss of dopamine-containing neurons in Parkinson's disease. Brain. 1999;122(Pt 8):1437–48.
162. Blazejewska AI, Schwarz ST, Pitiot A, Stephenson MC, Lowe J, Bajaj N, et al. Visualization of nigrosome 1 and its loss in PD: pathoanatomical correlation and in vivo 7 T MRI. Neurology. 2013;81(6):534–40.
163. Kwon DH, Kim JM, Oh SH, Jeong HJ, Park SY, Oh ES, et al. Seven-Tesla magnetic resonance images of the substantia nigra in Parkinson disease. Ann Neurol. 2012;71(2):267–77.
164. Schwarz ST, Afzal M, Morgan PS, Bajaj N, Gowland PA, Auer DP. The 'swallow tail' appearance of the healthy nigrosome – a new accurate test of Parkinson's disease: a case-control and retrospective cross-sectional MRI study at 3T. PLoS One. 2014;9(4):e93814.
165. Brammerloh M, Kirilina E, Alkemade A, Bazin PL, Jantzen C, Jager C, et al. Swallow tail sign: revisited. Radiology. 2022;305(3):674–7.
166. Lehéricy S, Bardinet E, Poupon C, Vidailhet M, François C. 7 Tesla magnetic resonance imaging: a closer look at substantia nigra anatomy in Parkinson's disease. Mov Disord. 2014;29(13):1574–81.
167. Kim HK, Kim T, Baek MS, Kim EY, Sung YH, Lee JH, et al. Nigrosome 1 visibility and its association with nigrostriatal dopaminergic loss in Parkinson's disease. Eur J Neurol. 2023;30(6):1639–47.
168. Kim HK, Lee MJ, Yoo HS, Lee JH, Ryu YH, Lyoo CH. Temporal trajectory model for dopaminergic input to the striatal subregions in Parkinson's disease. Parkinsonism Relat Disord. 2022;103:42–9.
169. Chau MT, Todd G, Wilcox R, Agzarian M, Bezak E. Diagnostic accuracy of the appearance of Nigrosome-1 on magnetic resonance imaging in Parkinson's disease: a systematic review and meta-analysis. Parkinsonism Relat Disord. 2020;78:12–20.
170. Grossauer A, Muller C, Hussl A, Krismer F, Schocke M, Gizewski E, et al. Dorsolateral Nigral Hyperintensity on 1.5 T versus 3 T susceptibility-weighted magnetic resonance imaging in neurodegenerative parkinsonism. Mov Disord Clin Pract. 2023;10(6):914–21.
171. Noh Y, Sung YH, Lee J, Kim EY. Nigrosome 1 detection at 3T MRI for the diagnosis of early-stage idiopathic Parkinson disease: assessment of diagnostic accuracy and agreement on imaging asymmetry and clinical laterality. AJNR Am J Neuroradiol. 2015;36(11):2010–6.
172. Kim PH, Lee DH, Suh CH, Kim M, Shim WH, Kim SJ. Diagnostic performance of loss of nigral hyperintensity on susceptibility-weighted imaging in parkinsonism: an updated meta-analysis. Eur Radiol. 2021;31(8):6342–52.
173. Enochs WS, Petherick P, Bogdanova A, Mohr U, Weissleder R. Paramagnetic metal scavenging by melanin: MR imaging. Radiology. 1997;204(2):417–23.
174. Cassidy CM, Zucca FA, Girgis RR, Baker SC, Weinstein JJ, Sharp ME, et al. Neuromelanin-sensitive MRI as a noninvasive proxy measure of dopamine function in the human brain. Proc Natl Acad Sci USA. 2019;116(11):5108–17.

175. Takahashi H, Watanabe Y, Tanaka H, Mihara M, Mochizuki H, Liu T, et al. Quantifying changes in nigrosomes using quantitative susceptibility mapping and neuromelanin imaging for the diagnosis of early-stage Parkinson's disease. Br J Radiol. 2018;91(1086):20180037.
176. Gaurav R, Yahia-Cherif L, Pyatigorskaya N, Mangone G, Biondetti E, Valabregue R, et al. Longitudinal changes in neuromelanin MRI signal in Parkinson's disease: a progression marker. Mov Disord. 2021;36(7):1592–602.
177. Cho SJ, Bae YJ, Kim JM, Kim D, Baik SH, Sunwoo L, et al. Diagnostic performance of neuromelanin-sensitive magnetic resonance imaging for patients with Parkinson's disease and factor analysis for its heterogeneity: a systematic review and meta-analysis. Eur Radiol. 2021;31(3):1268–80.
178. Ohtsuka C, Sasaki M, Konno K, Kato K, Takahashi J, Yamashita F, et al. Differentiation of early-stage parkinsonisms using neuromelanin-sensitive magnetic resonance imaging. Parkinsonism Relat Disord. 2014;20(7):755–60.
179. Reimao S, Pita Lobo P, Neutel D, Guedes LC, Coelho M, Rosa MM, et al. Substantia nigra neuromelanin-MR imaging differentiates essential tremor from Parkinson's disease. Mov Disord. 2015;30(7):953–9.
180. Wang M, Jiang S, Yuan Y, Zhang L, Ding J, Wang J, et al. Alterations of functional and structural connectivity of freezing of gait in Parkinson's disease. J Neurol. 2016;263(8):1583–92.
181. Soares JM, Marques P, Alves V, Sousa N. A hitchhiker's guide to diffusion tensor imaging. Front Neurosci. 2013;7:31.
182. Boska MD, Hasan KM, Kibuule D, Banerjee R, McIntyre E, Nelson JA, et al. Quantitative diffusion tensor imaging detects dopaminergic neuronal degeneration in a murine model of Parkinson's disease. Neurobiol Dis. 2007;26(3):590–6.
183. Vaillancourt DE, Spraker MB, Prodoehl J, Abraham I, Corcos DM, Zhou XJ, et al. High-resolution diffusion tensor imaging in the substantia nigra of de novo Parkinson disease. Neurology. 2009;72(16):1378–84.
184. Schwarz ST, Abaei M, Gontu V, Morgan PS, Bajaj N, Auer DP. Diffusion tensor imaging of nigral degeneration in Parkinson's disease: a region-of-interest and voxel-based study at 3 T and systematic review with meta-analysis. NeuroImage Clin. 2013;3:481–8.
185. Hirata FCC, Sato JR, Vieira G, Lucato LT, Leite CC, Bor-Seng-Shu E, et al. Substantia nigra fractional anisotropy is not a diagnostic biomarker of Parkinson's disease: a diagnostic performance study and meta-analysis. Eur Radiol. 2017;27(6):2640–8.
186. Deng XY, Wang L, Yang TT, Li R, Yu G. A meta-analysis of diffusion tensor imaging of substantia nigra in patients with Parkinson's disease. Sci Rep. 2018;8(1):2941.
187. Atkinson-Clement C, Pinto S, Eusebio A, Coulon O. Diffusion tensor imaging in Parkinson's disease: review and meta-analysis. NeuroImage Clin. 2017;16:98–110.
188. Pasternak O, Sochen N, Gur Y, Intrator N, Assaf Y. Free water elimination and mapping from diffusion MRI. Magn Reson Med. 2009;62(3):717–30.
189. Febo M, Perez PD, Ceballos-Diaz C, Colon-Perez LM, Zeng H, Ofori E, et al. Diffusion magnetic resonance imaging-derived free water detects neurodegenerative pattern induced by interferon-gamma. Brain Struct Funct. 2020;225(1):427–39.
190. Ofori E, Krismer F, Burciu RG, Pasternak O, McCracken JL, Lewis MM, et al. Free water improves detection of changes in the substantia nigra in parkinsonism: a multisite study. Mov Disord. 2017;32(10):1457–64.
191. Ofori E, Pasternak O, Planetta PJ, Burciu R, Snyder A, Febo M, et al. Increased free water in the substantia nigra of Parkinson's disease: a single-site and multi-site study. Neurobiol Aging. 2015;36(2):1097–104.
192. Planetta PJ, Ofori E, Pasternak O, Burciu RG, Shukla P, DeSimone JC, et al. Free-water imaging in Parkinson's disease and atypical parkinsonism. Brain. 2016;139(Pt 2):495–508.
193. Burciu RG, Ofori E, Archer DB, Wu SS, Pasternak O, McFarland NR, et al. Progression marker of Parkinson's disease: a 4-year multi-site imaging study. Brain. 2017;140(8):2183–92.

194. Guttuso T Jr, Bergsland N, Hagemeier J, Lichter DG, Pasternak O, Zivadinov R. Substantia Nigra free water increases longitudinally in Parkinson disease. AJNR Am J Neuroradiol. 2018;39(3):479–84.
195. Zhang D, Shi Y, Yao J, Zhou L, Wei H, Liu J, et al. Free-water imaging of the substantia Nigra in GBA pathogenic variant carriers. Mov Disord. 2023;38(5):764–73.
196. Zhang D, Zhou L, Shi Y, Liu J, Wei H, Tong Q, et al. Increased free water in the substantia Nigra in asymptomatic LRRK2 G2019S mutation carriers. Mov Disord. 2023;38(1):138–42.
197. Zhou L, Li G, Zhang Y, Zhang M, Chen Z, Zhang L, et al. Increased free water in the substantia nigra in idiopathic REM sleep behaviour disorder. Brain. 2021;144(5):1488–97.
198. Archer DB, Bricker JT, Chu WT, Burciu RG, McCracken JL, Lai S, et al. Development and validation of the automated imaging differentiation in parkinsonism (AID-P): a multicentre machine learning study. Lancet Digit Health. 2019;1(5):e222–e31.
199. Smith SM. Overview of fMRI analysis. Br J Radiol. 2004;77 Spec No 2:S167–75.
200. Hacker CD, Perlmutter JS, Criswell SR, Ances BM, Snyder AZ. Resting state functional connectivity of the striatum in Parkinson's disease. Brain. 2012;135(Pt 12):3699–711.
201. Gratton C, Koller JM, Shannon W, Greene DJ, Maiti B, Snyder AZ, et al. Emergent functional network effects in Parkinson disease. Cereb Cortex. 2019;29(4):1701.
202. Tahmasian M, Bettray LM, van Eimeren T, Drzezga A, Timmermann L, Eickhoff CR, et al. A systematic review on the applications of resting-state fMRI in Parkinson's disease: does dopamine replacement therapy play a role? Cortex. 2015;73:80–105.
203. Gratton C, Koller JM, Shannon W, Greene DJ, Maiti B, Snyder AZ, et al. Emergent functional network effects in Parkinson disease. Cereb Cortex. 2019;29(6):2509–23.
204. Maiti B, Koller JM, Snyder AZ, Tanenbaum AB, Norris SA, Campbell MC, et al. Cognitive correlates of cerebellar resting-state functional connectivity in Parkinson disease. Neurology. 2020;94(4):e384–e96.
205. Helmich RC, Janssen MJ, Oyen WJ, Bloem BR, Toni I. Pallidal dysfunction drives a cerebellothalamic circuit into Parkinson tremor. Ann Neurol. 2011;69(2):269–81.
206. Dirkx MF, den Ouden H, Aarts E, Timmer M, Bloem BR, Toni I, et al. The cerebral network of Parkinson's tremor: an effective connectivity fMRI study. J Neurosci. 2016;36(19):5362–72.
207. Cerasa A, Koch G, Donzuso G, Mangone G, Morelli M, Brusa L, et al. A network centred on the inferior frontal cortex is critically involved in levodopa-induced dyskinesias. Brain. 2015;138(Pt 2):414–27.
208. Fling BW, Cohen RG, Mancini M, Carpenter SD, Fair DA, Nutt JG, et al. Functional reorganization of the locomotor network in Parkinson patients with freezing of gait. PLoS One. 2014;9(6):e100291.
209. Lenka A, Naduthota RM, Jha M, Panda R, Prajapati A, Jhunjhunwala K, et al. Freezing of gait in Parkinson's disease is associated with altered functional brain connectivity. Parkinsonism Relat Disord. 2016;24:100–6.
210. Gilat M, Ehgoetz Martens KA, Miranda-Dominguez O, Arpan I, Shine JM, Mancini M, et al. Dysfunctional limbic circuitry underlying freezing of gait in Parkinson's disease. Neuroscience. 2018;374:119–32.
211. Chung SJ, Choi YH, Kwon H, Park YH, Yun HJ, Yoo HS, et al. Sleep disturbance may alter white matter and resting state functional connectivities in Parkinson's disease. Sleep. 2017;40(3):zsx009.
212. Dayan E, Browner N. Alterations in striato-thalamo-pallidal intrinsic functional connectivity as a prodrome of Parkinson's disease. NeuroImage Clin. 2017;16:313–8.
213. Rolinski M, Griffanti L, Piccini P, Roussakis AA, Szewczyk-Krolikowski K, Menke RA, et al. Basal ganglia dysfunction in idiopathic REM sleep behaviour disorder parallels that in early Parkinson's disease. Brain. 2016;139(Pt 8):2224–34.
214. Baggio HC, Segura B, Sala-Llonch R, Marti MJ, Valldeoriola F, Compta Y, et al. Cognitive impairment and resting-state network connectivity in Parkinson's disease. Hum Brain Mapp. 2015;36(1):199–212.

215. Diez-Cirarda M, Strafella AP, Kim J, Pena J, Ojeda N, Cabrera-Zubizarreta A, et al. Dynamic functional connectivity in Parkinson's disease patients with mild cognitive impairment and normal cognition. NeuroImage Clin. 2018;17:847–55.
216. Putcha D, Ross RS, Cronin-Golomb A, Janes AC, Stern CE. Altered intrinsic functional coupling between core neurocognitive networks in Parkinson's disease. NeuroImage Clin. 2015;7:449–55.
217. Bejr-Kasem H, Pagonabarraga J, Martinez-Horta S, Sampedro F, Marin-Lahoz J, Horta-Barba A, et al. Disruption of the default mode network and its intrinsic functional connectivity underlies minor hallucinations in Parkinson's disease. Mov Disord. 2019;34(1):78–86.
218. Bae YJ, Kim JM, Kim E, Lee KM, Kang SY, Park HS, et al. Loss of Nigral Hyperintensity on 3 Tesla MRI of parkinsonism: comparison with (123) I-FP-CIT SPECT. Mov Disord. 2016;31(5):684–92.
219. Frosini D, Ceravolo R, Tosetti M, Bonuccelli U, Cosottini M. Nigral involvement in atypical parkinsonisms: evidence from a pilot study with ultra-high field MRI. J Neural Transm (Vienna). 2016;123(5):509–13.
220. Ceravolo R, Antonini A, Frosini D, De Iuliis A, Weis L, Cecchin D, et al. Nigral anatomy and striatal denervation in genetic parkinsonism: a family report. Mov Disord. 2015;30(8): 1148–9.
221. Frosini D, Cosottini M, Donatelli G, Costagli M, Biagi L, Pacchetti C, et al. Seven tesla MRI of the substantia nigra in patients with rapid eye movement sleep behavior disorder. Parkinsonism Relat Disord. 2017;43:105–9.
222. Schirinzi T, Madeo G, Martella G, Maltese M, Picconi B, Calabresi P, et al. Early synaptic dysfunction in Parkinson's disease: insights from animal models. Mov Disord. 2016;31(6):802–13.
223. Sofic E, Paulus W, Jellinger K, Riederer P, Youdim MB. Selective increase of iron in substantia nigra zona compacta of parkinsonian brains. J Neurochem. 1991;56(3):978–82.
224. Riederer P, Monoranu C, Strobel S, Iordache T, Sian-Hulsmann J. Iron as the concert master in the pathogenic orchestra playing in sporadic Parkinson's disease. J Neural Transm (Vienna). 2021;128(10):1577–98. PMC8507512.
225. Pyatigorskaya N, Sanz-Morere CB, Gaurav R, Biondetti E, Valabregue R, Santin M, Yahia-Cherif L, Lehericy S. Iron imaging as a diagnostic tool for Parkinson's disease: a systematic review and meta-analysis. Front Neurol. 2020;11:366. PMC7270360.

Role of Molecular Imaging in Parkinson's Disease

Dongning Su and A. Jon Stoessl

Abstract Functional imaging studies utilizing single-photon emission computed tomography (SPECT) or positron emission tomography (PET) have garnered increasing attention in neurology, particularly in Parkinson's disease (PD) and other neurodegenerative disorders. These imaging techniques have proven to be important for diagnosis and understanding disease mechanisms. In this chapter, we aim to explore the principles of molecular imaging in PD, as well as its limitations. Also, it is essential to view PET and MRI as complementary rather than competing approaches in addressing these complex questions.

Keywords Positron emission tomography · Single-Photon Emission Computed Tomography · Dopamine · Prodromal · Biomarker

1 Diagnosis and Differential Diagnosis of PD and Atypical Parkinsonian Syndromes

1.1 Identification of Dopamine Deficiency

PET/SPECT imaging of the dopaminergic system includes approaches to imaging both pre- and post-synaptic function (Table 1). Pre-synaptic functional imaging encompasses imaging of the dopamine synthesis, vesicular monoamine transporter 2 (VMAT2), and dopamine transporter (DAT). Post-synaptic functional imaging refers primarily to imaging of dopamine receptors on the post-synaptic membrane.

D. Su
Beijing Tiantan Hospital, Beijing, China

A. J. Stoessl (✉)
Pacific Parkinson's Research Centre, University of British Columbia, Vancouver, BC, Canada
e-mail: jon.stoessl@ubc.ca

P. A. Broderick (ed.), *Biomedical Imaging in Neurodegeneration*,
https://doi.org/10.1007/978-3-031-95878-6_3

Table 1 Types of dopaminergic PET and currently utilized tracers

Location	PET target	^{11}C	^{18}F	^{123}I
Pre-synaptic	VMAT2	C-DTBZ	F-DTBZ	
	DAT	C-CFTC-MP	F-CFT	I-altropane
		C-PE2I	F-PE2I	I-β-CIT
				I-FP-CIT
	DA synthesis		F-DOPA	
			F-FMT	
Post-synaptic	D_1	C-NNC 112		
		C-SCH23390	F-DMFP	
	D_2/D_3	C-Raclopride	F-Fallypride	I-IBZM
		C-FLB457		Epidepride
		C-NMSP		
		C-MNPA		
		C-PHNO		
		C-NPA		
		C-NMB		

DA Synthesis Imaging

^{18}F-DOPA PET/SPECT imaging reflects the process of ^{18}F-DOPA being taken up across the blood-brain barrier by the large neutral amino acid transporter and converted by aromatic amino acid decarboxylase (AADC) in monoaminergic neurons into ^{18}F-dopamine, as well as the ultimate storage of ^{18}F-dopamine in synaptic vesicles located in the terminals of dopaminergic neurons. As ^{18}F-DOPA uptake reflects the ability of AADC to metabolize ^{18}F-DOPA in dopaminergic neurons, it can be easily influenced by various factors. For instance, reduced availability of dopamine in the early stages of PD can lead to compensatory increases in pre-synaptic AADC activity, resulting in increased ^{18}F-DOPA uptake relative to the number of surviving dopaminergic nerve terminals, thereby leading to underestimation of the degree of dopaminergic nerve terminal loss. Additionally, ^{18}F-DOPA can be taken up by serotonergic neurons and noradrenergic neurons, lowering its specificity for dopamine system imaging.

Several studies have demonstrated consistent and asymmetric reductions of striatal F-DOPA uptake in PD, with a caudal to rostral gradient, in which the posterior putamen is affected earliest and more severely than more anterior regions. This pattern, while characteristic, is not specific to PD, as it may be seen in other atypical Parkinsonian syndromes including multiple system atrophy (MSA) and progressive supranuclear palsy (PSP), although in the latter condition, a greater degree of caudate involvement is often present. In early PD, increased uptake of ^{18}F-DOPA may be seen in the anterior cingulate, dorsolateral prefrontal cortex, and medial globus pallidus, likely reflecting the compensatory upregulation of AADC and uptake of ^{18}F-DOPA into 5-HTergic and noradrenergic terminals [1].

The symptomatic threshold for PD can be estimated for different radiotracers. For ^{18}F-DOPA, this threshold ranges from 47% to 62% of normal putamen values [2]. For ^{11}C-DTBZ, the threshold is 38–48%, while for ^{11}C-MP the threshold value is 29–44%. These values provide insights into the levels at which PD symptoms become apparent when using these specific radiotracers. Indeed, reduced ^{18}F-DOPA update in bilateral striatum and substantia nigra has been shown in early stage PD [3]. Piccini and others studied putaminal ^{18}F-DOPA uptake in relatives of patients with familial PD and found a 25% reduction in asymptomatic adult family members [4]. Meanwhile, the pallidum lost over 30% of its capacity to take up ^{18}F-DOPA [5]. Reductions in F-DOPA uptake can be seen as well in other individuals with subclinical disease, including subjects exposed to the dopaminergic toxin MPTP [6] as well as carriers of pathogenic variants in *SNCA* [7] and *LRRK2* [8, 9].

F-DOPA PET data are typically acquired over 90–120 min following tracer administration, during which tracer uptake predominantly reflects conversion to F-dopamine. However, by employing longer scan times, it is possible to estimate the rate of egress of F-dopamine trapped in synaptic vesicles. The ratio of k_{loss} to uptake rate correlates well with standard neurochemical measures of dopamine turnover and is increased in PD patients with levodopa-induced dyskinesia [10].

VMAT2 Imaging

VMAT2 is a monoamine neurotransmitter transporter protein on the pre-synaptic vesicle membrane of dopaminergic neurons, which participates in the intravesicular transport of dopamine, serotonin, and norepinephrine neurotransmitters and thereby prevents their degradation. Newly synthesized dopamine in the neuron and dopamine reabsorbed from the synaptic cleft are stored in vesicles; VMAT2 is an indispensable protein on the neuroendocrine cell membrane. Although expressed in all monoaminergic neurons, VMAT imaging (mostly DTBZ PET) is a valuable tool in assessing the density of dopaminergic nerve terminals, as its striatal binding primarily mirrors these projections. The advantage of VMAT2 molecular imaging is that it reflects the function of dopaminergic neurons, and VMAT2 binding is not as readily influenced by exogenous dopaminergic drug treatment as F-DOPA or tracers for the DAT. For this reason, in early PD, the uptake of F-DOPA is increased relative to VMAT2 binding, while DAT binding is relatively decreased [2]. However, VMAT2 binding can still be influenced by competition from endogenous intravesicular dopamine. Consequently, DTBZ PET readings can be elevated in conditions of severe dopamine deficiency and can be lowered following levodopa treatment in advanced PD patients [11].

In patients with PD, a decrease in VMAT2 in the striatum and midbrain has been observed. Specifically, VMAT2 binding potential in the posterior putamen is reduced by 81%, the anterior putamen by 70%, and the caudate by 48% [12]. Over the course of the disease, the degree of asymmetry between the more and less affected striatum diminished, despite the maintenance of an antero-posterior gradient of severity for dopamine synthesis, storage, and reuptake [13]. Furthermore, the standardized uptake value ratio (SUVR) in the bilateral caudate, anterior putamen, posterior putamen, substantia nigra, and septal nucleus was significantly lower in early-stage PD patients compared to healthy participants. The most severe decline in SUVR was found in the contralateral posterior putamen (−81%), followed by the ipsilateral posterior putamen (−67%) [14].

When dividing the striatum into five regions: the caudate, anterior dorsal putamen, anterior ventral putamen, posterior dorsal putamen, and posterior ventral putamen, it was found that the contralateral posterior dorsal putamen (with an SUVR of 3.43) demonstrated the highest diagnostic accuracy for PD. This method achieved a sensitivity of 97.1% and a specificity of 100% in diagnosing PD [15].

DAT Imaging

DAT is a transmembrane transport protein on the pre-synaptic terminal membrane of dopaminergic neurons, which has a high affinity for dopamine. Its physiological function is to terminate the action of dopamine released into the synaptic cleft by taking it back up into the dopaminergic neuron. DAT imaging can reflect the integrity of the nigrostriatal pathway. The widespread access to DAT SPECT has led to its approval to differentiate PD from essential tremor. However, as is the case for all dopaminergic tracers, DAT imaging cannot reliably differentiate between PD and atypical parkinsonism. DAT imaging can detect abnormalities in the function of dopaminergic neurons before the onset of typical motor symptoms and can be used for the early diagnosis and risk assessment of PD. SPECT tropane tracers include: ^{123}I-beta-CIT, ^{123}I-FP-CIT, ^{123}I-altropane, and ^{99}mTc-TRODAT. PET tropane derivatives include: ^{11}C-RTI32, ^{11}C-CFT, and ^{18}F-FP-CIT while ^{11}C-methylphenidate and ^{11}C-nomifensine are nontropane radioligands.

The primary methods of assessment include visual and semi-quantitative analysis. Visual assessment of DAT often satisfies diagnostic needs, with good consistency generally observed between different observers. In early PD, the preferential loss of dopaminergic projections to the posterior putamen with relative preservation in the caudate results in the striatum losing its normal kidney-like shape and taking on an oval or round appearance. As the disease progresses, the background becomes brighter, and the radiotracer uptake in the striatum decreases. DAT imaging has high sensitivity in the early stages of PD and a high negative predictive value, which is a significant advantage of this method [16]. Although this anterior-posterior gradient can also be seen in the later stages of the disease, it is more prominent in the early stages [17].

For quantitative analysis, the most commonly used method, particularly for SPECT or static PET imaging is to calculate the ratio of striatal radiotracer uptake to background activity (specific binding ratio = (striatal count – background count)/background count). The cerebellum or occipital cortex is usually used as a reference region for this purpose. When PET with dynamic imaging is used, it is possible to determine the specific binding (non-displaceable binding potential or BP_{ND}). Binding will vary as a function of normal aging and the characteristics of the tracer and camera; thus, adequate quantitation requires age-corrected comparison with a cohort of healthy controls obtained at the same center. While the sensitivity for DAT imaging to detect dopamine deficiency (i.e., neurodegenerative parkinsonism) is very high, most studies conducted in early PD have reported that 10–15% of subjects have no evidence of dopamine deficiency. While some of these subjects likely have essential or dystonic tremor, a number of them have non-motor manifestations typically associated with PD [18], may show clinically important change over time [19], and nearly 10% have positive synuclein seeding amplification assays [20].

Dopamine Receptor Imaging
There are five types of dopamine receptors (D_1, D_2, D_3, D_4, and D_5). D_1 and D_5 receptors belong to the D_1 receptor family and are post-synaptic, while D_2, D_3, and D_4 receptors belong to the D_2 receptor family. D_2 and D_3 receptors may be located both pre- and post-synaptically, the former acting as auto-receptors that regulate the release of dopamine. Dopamine receptor imaging can reflect the functional status of dopamine receptors under physiological and pathological conditions. Untreated early-stage PD patients have increased post-synaptic membrane D_2 receptors due to a combination of denervation-induced upregulation of the post-synaptic receptor, as well as (depending upon the radioligand used) reduced occupancy by endogenous dopamine. As the disease progresses and exogenous dopamine medications are implemented, D_2 receptors downregulate. Because atypical forms of parkinsonism such as MSA or PSP may be associated with loss of post-synaptic dopamine receptors located on striatal projection neurons, the combined use of pre-synaptic dopamine imaging and D_2 receptor imaging may help discriminate between PD and AP, but this approach is not commonly used in practice, as AP may be more readily differentiated using MRI or FDG PET.

Radioligands such as ^{11}C-raclopride (or ^{11}C-RAC) for PET and ^{123}I-IBZM for SPECT are commonly utilized to evaluate the availability of striatal dopamine receptors. RAC and IBZM, due to their susceptibility to competition from endogenous dopamine for in vivo binding to D_2 receptors, can be employed to deduce alterations in synaptic dopamine concentration. This can be accomplished by conducting multiple scans pre- and post-intervention, or more recently, utilizing models to dynamically quantify the synaptic dopamine within a single scan [21, 22].

1.2 *Role of FDG PET*

^{18}F-fluorodeoxyglucose (^{18}F-FDG) is a radiolabeled form of 2-deoxyglucose, which is taken up by cells in a manner analogous to glucose, but unlike glucose, trapped after the first phosphorylation step, thereby allowing activity to be measured even with static scans. While it has historically been assumed that glucose uptake was in neurons and largely a reflection of synaptic activity, more recent evidence suggests that astrocytes are responsible for a large component of glucose uptake [23], but this occurs in response to neuronal release of glutamate and therefore still reflects synaptic activity [24]. In PD and in other neurodegenerative diseases, the patterns of change in FDG uptake may be characteristic and may appear earlier than structural changes such as brain atrophy shown in structural imaging, aiding in the early diagnosis and differential diagnosis. PD patients show increased metabolism in the pallidum, thalamus, brainstem, and cerebellum on ^{18}F-FDG imaging, along with decreased metabolism in the premotor cortex and posterior parietal cortical areas, known as the PD-related metabolic pattern (PDRP) [25]. MSA, PSP, and CBD each also exhibit distinct metabolic patterns in the brain. MSA typically shows reduced metabolism in the putamen and cerebellum [26], while PSP is characterized by decreased metabolism in the prefrontal cortex, caudate, medial thalamus, and upper brainstem [27]. CBD displays a decrease in metabolism in one hemisphere's cortical and basal ganglia regions [28]. A meta-analysis reported good sensitivity in discriminating idiopathic PD from atypical parkinsonism (range 83–86%). Specificity and positive prediction value (PPV) were both greater than 90% in discriminating idiopathic PD from atypical parkinsonism (range 94–97% and 96–98%, respectively) [29]. The sensitivity and specificity of PD versus non-PD were 86%/91%, MSA 77%/97%, PSP 74%/95%, and CBD 75%/92%.

1.3 Role of Cardiac Imaging

The status of cardiac sympathetic nerve function is reflected by the heart's uptake of the synthetic norepinephrine analog (^{123}I-MIBG). PD patients show reduced ^{123}I-MIBG uptake due to impairment of both pre- and post-ganglionic sympathetic fibers, whereas MSA, also a synucleinopathy, mainly shows impairment of preganglionic sympathetic neurons, while sympathetic neurons are unaffected in PSP. Thus, ^{123}I-MIBG uptake in APS patients is typically (although not always) normal and this imaging technique may be somewhat useful for differential diagnosis between PD and MSA, with reports showing sensitivity for distinguishing PD and MSA close to 100%, and specificity ranging from 77% to 100% [30]. ^{123}I-MIBG imaging has been recommended (grade A) for the differential diagnosis of Parkinsonian syndromes. Many patients with idiopathic REM behavior disorder (RBD), which is associated with a high risk of phenoconversion to synucleinopathy (PD, DLB, or MSA) show cardiac sympathetic denervation. This is in keeping with the Braak hypothesis that PD may start in the peripheral nervous system and then spread rostrally, although not all PD patients have RBD as the initial manifestation and PD associated with RBD tends to have a more aggressive course. This may explain why a cohort study (average follow-up of 3 years) found that in early-stage PD, patients with normal or only slightly reduced MIBG uptake were predominantly female, younger at onset, and showed slow progression of motor and cognitive impairments [31].

2 Identification of Prodromal Disease

2.1 REM Sleep Behavior Disorder (RBD)

In ^{123}I-FP-CIT PET studies, RBD showed reduced uptake in putamen and caudate nucleus (bilaterally asymmetrical, right side significant), compared with HC [32–34]. The ratio of putamen to caudate nucleus was higher in RBD than in PD without RBD and PD with RBD [35]. A longitudinal study in RBD observed continuous loss of pre-synaptic dopaminergic function (putamen, caudate nucleus) and 3 of 20 subjects were diagnosed with PD after 3 years [36]. Also, in ^{11}C-DTBZ PET studies, in RBD, compared to HC, uptake was reduced in the putamen and caudate nucleus [37]. RBD patients exhibited more symmetric nigrostriatal dopaminergic degeneration [38].

In noradrenergic imaging, RBD showed lower binding of ^{11}C-MeNER in the primary somatosensory cortex than healthy controls [39]. Additionally, the uptake of ^{18}F-DOPA in the caudate nucleus in RBD patients is significantly correlated with binding of ^{11}C-MeNER in the thalamus. This suggests simultaneous functional degradation of these dopaminergic and noradrenergic systems in idiopathic RBD patients [39]. RBD subjects had reduced glucose metabolism compared to healthy controls in the occipital lobe, cingulate gyrus, parietal lobe, temporal lobe, left prefrontal gyrus, and right frontal lobe [40, 41]. Decreased dopamine transporter binding in the caudate nucleus is associated with increased glucose metabolism in the pallidum. Synaptic metabolic dysfunction may lead to a lack of inhibition, compensatory mechanisms, or microglial activation, particularly in regions associated with nigrostriatal degeneration.

As noted above, RBD may be associated with cardiac sympathetic denervation that appears prior to clinical or imaging evidence of striatal dopamine denervation, but this may be associated with loss of cholinergic innervation in the GI tract [42] (Fig. 1). In RBD, 78–95% of patients showed pathological ^{123}I-MIBG uptake [42–44]. Those individuals with pathological ^{123}I-MIBG uptake experience more severe olfactory impairment and higher UPDRS-III scores [44]. Over a 3-year follow-up period, ^{123}I-MIBG uptake demonstrates a gradual decline [45]. The cardiac denervation in RBD begins at least 8–10 years prior to the onset of nigrostriatal dopaminergic denervation in the substantia nigra [43]. Moreover, this cardiac denervation could start at least 16–20 years before RBD patients convert to PD or LBD. Parasympathetic innervation of the gut–brain axis also changed in RBD. It was observed that the SUV of ^{11}C-donepezil in the colon and small intestine was decreased to a similar extent as in patients with PD [42]. This finding provides strong evidence that PD patients who experience REM sleep behavior disorder before the onset of parkinsonism undergo severe autonomic denervation several years prior to a formal PD diagnosis.

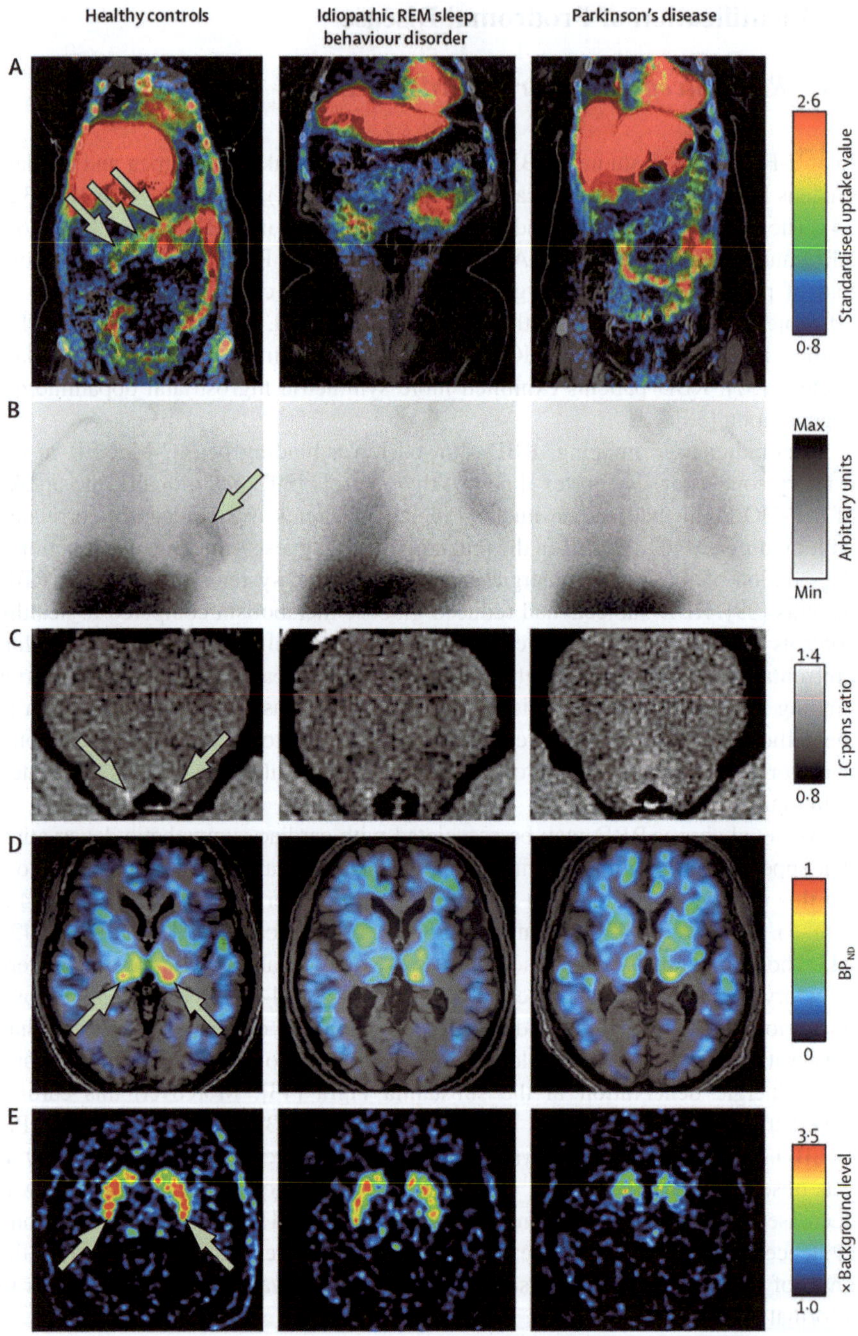

Fig. 1 Representative images from healthy controls, patients with idiopathic REM sleep behavior disorder, and patients with Parkinson's disease with the five imaging modalities. (Reproduced from Knudsen et al. [39])

2.2 Carriers of Pathogenic Genetic Variants

Compared with healthy controls, asymptomatic heterozygous *Parkin* carriers demonstrate reduced ^{18}F-FDOPA uptake in the caudate and putamen, although the interpretation is unclear [46] . Patients with *Parkin* PD show decreased uptake in the ventral striatum, locus coeruleus, midline thalamic nuclei, and globus pallidus [47]. However, there was no significant difference in striatal ^{18}F-FDOPA uptake between Young-onset Parkinson's disease (YOPD) patients with and without *Parkin* mutation [48]. No significant correlation was found between ^{18}F-FDOPA uptake, motor dysfunction, and mutation type in Parkin patients. However, in non-Parkin patients, there was a significant correlation between ^{18}F-FDOPA uptake and clinical severity [49].

Compared with HC, biallelic *PINK1* PD patients showed a 60% reduction in ^{18}F-FDOPA uptake in the caudate and putamen, while asymptomatic monoallelic carriers showed a 20% reduction in putamen uptake [50]. In PD with *GBA* mutation, the ^{123}I-FP-CIT uptake in the caudate was asymmetrically reduced, while PD with *PINK1*, *Parkin*, *LRRK2 G2019S*, and *SNCA* mutations showed relatively symmetrical reduction in caudate uptake. Asymmetrical putamen uptake reduction was seen in PD due to pathogenic *GBA* and *LRRK2* variants, while relatively symmetrical reduction of putamenal uptake was noted in PD related to *PINK1*, *Parkin*, and *SNCA* variants [51]. PD-*GBA* patients showed reduced bilateral striatal ^{18}F-FDOPA uptake [52, 53]. At baseline and follow-up, asymptomatic *GBA1* mutation carriers did not show significant differences in striatal ^{18}F-FDOPA uptake compared with controls [54].

In PD associated with pathogenic variants in *LRRK2*, reductions in striatal uptake of ^{18}F-DOPA, ^{11}C-DTBZ, and dopamine transporter binding are seen, characterized by the rostral-caudal gradient and asymmetry that are characteristic of sporadic disease [9]. Loss of striatal dopamine innervation is seen in non-manifesting carriers, although ^{18}F-DOPA and VMAT2 binding do not decline until later, whereas DAT binding may be abnormal from an early age, associated with increases in dopamine turnover [8, 55, 56]. Non-manifesting carriers of pathogenic *LRRK2* variants also demonstrate elevated serotonin transporter binding in the striatum, brainstem, and hypothalamus [56]. This might hint at compensatory modifications in serotonergic innervation prior to the emergence of motor symptoms in PD. Non-manifesting *LRRK2* carriers also displayed enhanced acetylcholinesterase hydrolysis rate in the cortex compared to healthy controls [57]. These changes in cholinergic activity could symbolize early and continuous efforts to counterbalance *LRRK2*-related dysfunction or signify alterations of acetylcholinesterase in non-neuronal cells. However, preserved cholinergic activity in *LRRK2*-PD is in keeping with the lower prevalence of RBD and cognitive impairment in these subjects [58, 59].

3 Understanding Complications of Disease and Therapy

3.1 Motor Fluctuations and LID

Using ^{11}C-DTBZ PET as an in vivo marker for dopaminergic nerve terminal density, a study found dopaminergic terminal loss plays a pathogenic role in the development of motor fluctuations in patients with asymmetric PD [60]. However, there was no significant correlation between the early decline in levodopa response and dopaminergic terminal density, indicating that dopaminergic terminal loss alone is not sufficient to explain the development of motor fluctuations. Levodopa treatment and alterations in central pharmacokinetics of levodopa, along with dopamine terminal loss, contribute to the development of motor fluctuations.

Among the subregions of the striatum, the activity of dopamine transporters in the posterior putamen is closely related to the occurrence of LID, while no significant correlation was found in the anterior putamen [61–65]. Reduced dopamine transporter activity in the posterior putamen relative to the degree of nerve terminal loss is a major risk factor for the early onset of LID [66]. This suggests that the impairment of pre-synaptic dopaminergic innervation plays a crucial role in determining the timing of LID occurrence. Meanwhile, PD patients with LID had higher radiotracer uptake in pallidal D3 receptors measured by ^{11}C-PHNO PET compared to those without LID, while tracer uptake in the lateral striatal D_2/D_3 receptors was lower in PD patients with LID compared to those without LID [67]. The interpretation of ^{11}C-PHNO PET in this setting is challenging, as the relative distribution of D_3 to D_2 receptors varies throughout the striatum and is different in PD vs. healthy controls, and is also modified by dopaminergic treatment. By studying changes in ^{11}C-RAC PET following acute therapy, it was determined that peak-dose LID patients showed a greater increase in synaptic dopamine levels 1 h after levodopa intake compared to those without LID, while both groups showed similar increases 4 h after levodopa and the degree of increase was the same at 1 and 4 h in patients with a stable response to medication [68]. This substantial increase in levodopa-induced synaptic dopamine concentration can lead to pulsatile stimulation of dopamine receptors, which may play a key role in peak-dose motor complications in PD. In 4-h ^{18}F-fluorodopa PET scans, dopamine synthesis and storage rates were lower in younger patients compared to older patients, and younger patients had dopamine turnover increased out of proportion to the degree of impairment in dopamine synthesis [10]. This finding suggests that younger onset patients have greater fluctuations in synaptic dopamine levels.

^{11}C-DASB PET imaging was used to measure serotonergic terminal function in PD-LID. Compared to PD patients without LID, those with LID showed relatively preserved serotonergic terminal function in the striatum [69]. A combined ^{11}C-DASB and ^{11}C-RAC PET study suggested that the higher the serotonergic terminal density in the pallidum and the more dopamine released, the greater the severity of LID [70]. The ratio of serotonin to dopamine terminals increases with the progression of PD and the occurrence of LID [71] and it is hypothesized that surviving serotonergic nerve terminals convert levodopa to dopamine, but without regulation by DAT or dopaminergic nerve terminal autoreceptors [72]. Another PET study employed a novel joint pattern analysis to investigate the relationship between changes in striatal dopamine and serotonin function and levodopa-induced synaptic dopamine release. It revealed a spatial pattern associated with abnormal dopamine turnover (increased dopamine release related to reduced dopaminergic and serotonergic innervation) and a significant negative correlation between dopamine release related to turnover and treatment response [73].

3.2 *Impulse Control Disorders (ICD)*

A reduction in the availability of dopamine transporters in the ventral striatum and an increase in dopamine release form the basis for the development of ICD. D_2 autoreceptors in the midbrain regulate dopamine release in the striatum through negative feedback, and the weakening of this feedback is associated with ICD in healthy individuals. In PD patients with ICD, most DAT-SPECT studies show lower availability of dopamine transporters in the ventral striatum [74–77]. Baseline D_2 receptor availability in the midbrain was negatively correlated with dopamine release in the ventral striatum. However, this relationship was not observed in individuals with ICD. ^{11}C-PHNO PET imaging suggests that the occurrence and severity of ICD may be related to decreased utilization of dopamine D_3 receptors in the ventral striatum. However, different studies have shown conflicting results, suggesting that differences in $D_{2/3}$ receptor levels alone cannot explain the development of ICD in PD [78, 79].

3.3 *Depression and Anxiety*

Although a SPECT study found no association between midbrain serotonergic lesions and depression scores in early PD patients [80], it is still widely believed that depressive symptoms in PD are influenced by changes in serotonergic function. PET studies using ^{11}C-DASB have unexpectedly shown increased tracer uptake in the prefrontal cortex, striatum, and subcortical or limbic areas of PD patients with depression. As ^{11}C-DASB binds to serotonergic nerve terminal reuptake sites, this finding seemed counter-intuitive but may indicate excessive serotonin reuptake [81, 82] or potentially reduced occupancy of the reuptake site due to low synaptic levels of 5HT. Reduced ^{11}C-DASB binding, presumably reflecting loss of 5HT terminals, has been observed in the orbitofrontal cortex, caudate nucleus, and putamen of non-depressed PD patients [83]. Additionally, reduced synaptic 5-HT1A receptor density in limbic areas has been demonstrated in depressed PD patients using PET and ^{18}F-MPPF [84]. Another study found that despite dopaminergic neuronal restoration, depressive symptoms persisted in PD patients, further supporting the influence of serotonergic alterations on depression in PD [85]. A study in newly diagnosed PD patients also confirmed a clear association between widespread limbic serotonergic degeneration and depression in PD, with severity related to anterior cingulate cortex serotonergic denervation [86].

A PET imaging study using the dopamine and noradrenaline transporter (NET) tracer ^{11}C-RTI-32 revealed a connection between the presence of depression and the extent of noradrenergic damage in the locus coeruleus and limbic areas (such as the ACC, orbitofrontal cortex, and amygdala) [87]. However, the tracer used in this study did not distinguish the respective roles of dopaminergic and noradrenergic dysfunction in depression among PD patients. Furthermore, the use of the NET reuptake inhibitor atomoxetine in treating PD did not significantly improve depressive symptoms [88]. With more selective NET tracers, it is found that sleep microstructure impairment is associated with widespread dysfunction of the noradrenergic arousal system [89]. Reduced nor-adrenergic function in PD was related to the presence of RBD, cognitive deterioration, and orthostatic hypotension [90].

3.4 Cognitive and Gait Dysfunction

Cognitive Dysfunction

For pre-synaptic cholinergic dysfunction, PET studies using ^{11}C-MP4A and ^{11}C-PMP have shown that cortical AChE is mildly to moderately reduced in PD patients without dementia, while it is severely reduced in patients with PD dementia (PDD) and dementia with Lewy body (DLB) patients [91–93]. In SPECT studies using ^{123}I-IBVM [94], it has been confirmed that vesicular acetylcholine transporter (VAChT) levels are only reduced in the parietal and occipital cortices of PD patients without dementia, while in PDD patients, the entire cortex shows significant loss, indicating a more extensive pre-synaptic cholinergic dysfunction in PDD patients. Furthermore, the loss of cortical acetylcholinesterase activity measured by ^{11}C-PMP [92] is associated with poorer performance in working memory and attention tests, but not with the severity of motor symptoms. In PDD patients, a relationship has been found between striatal ^{18}F-FDOPA uptake and cortical ^{11}C-MP4A reduction, suggesting that cognitive decline in PD occurs as the disease spreads from nigral neurons to the cortex, leading to cholinergic dysfunction in that region [91]. In PDD patients, cholinergic dysfunction is more severe than in Alzheimer's disease (AD) patients with a comparable degree of cognitive impairment, indicating that the mechanisms underlying cognitive decline in PD and AD have different etiologies [95].

For post-synaptic cholinergic dysfunction, research has shown that cortical muscarinic cholinergic receptor (mAChR) density changes in PDD and DLB samples [96, 97]. SPECT studies using ^{123}I-QNB to label mAChRs suggest that, compared to healthy controls, mAChR is significantly increased in the occipital lobes of PDD and DLB patients [98], possibly reflecting loss of cholinergic innervation. Furthermore,

PDD patients exhibit significantly lower binding in the frontal and temporal lobes compared to DLB patients. Additionally, the loss of α4β2 nicotinic acetylcholine receptor is associated with depression and cognitive decline [99].

Gait Dysfunction

Early neuropathological evidence suggests that pedunculopontine nucleus degeneration plays a role in PD and its associated gait and balance dysfunctions. Using ^{11}C-PMP PET, researchers measured thalamic AChE activity primarily originating from PPN projections and found that, regardless of the degree of nigrostriatal denervation, thalamic AChE activity was significantly reduced in patients with a history of falls compared to those without [100]. In a follow-up study, researchers employed ^{18}F-FEOBV PET (a VAChT marker) to confirm and extend these findings: cholinergic denervation in the right posterior thalamus, a relay station of the visual network, was particularly associated with a history of falls, suggesting that impaired visual information processing contributes to gait problems [101]. Thalamic cholinergic deficits were also related to impaired postural reflexes, further strengthening the pathophysiological link between PPN-thalamic cholinergic innervation and falls [102].

A significant reduction in cortical AChE activity within ^{11}C-PMP PET was associated with a decrease in walking speed, while PD patients with normal cortical cholinergic innervation had walking speeds similar to healthy controls. This effect was independent of thalamic AChE activity and the degree of nigrostriatal denervation.

The occurrence of freezing of gait has also been linked to cholinergic dysfunction, although other factors such as nigrostriatal dopaminergic dysfunction may play a more crucial role [103]. A multi-tracer PET imaging study using ^{11}C-PMP and ^{11}C-PIB showed that cortical cholinergic denervation was associated with an increased risk of freezing of gait, particularly in the presence of cortical amyloid deposition, but the study did not demonstrate an effect of PPN cholinergic thalamic projections on freezing of gait [104]. Compared to patients without freezing of gait, those with freezing exhibited significantly reduced FEOBV binding in the striatum, hippocampus, and amygdala, suggesting a role for the loss of cholinergic interneurons in the caudate nucleus in freezing of gait [101].

4 Use of Molecular Imaging for Validation of Target Engagement

4.1 Receptor Occupancy

Earlier studies applying RAC PET to investigate changes in synaptic dopamine in PD patients induced by levodopa found a decrease in RAC uptake in the caudate nucleus and putamen following levodopa administration [105]. In recent years, studies have also used RAC PET to assess the relationship between clinical improvement after a single oral dose of levodopa and drug-induced synaptic dopamine increases. The results similarly indicated a significant reduction in RAC binding potential in the caudate nucleus and putamen compared to baseline, presumed to reflect an increase in drug-induced striatal dopamine. Patients with more severe symptoms (Hoehn and Yahr stage $\geq$3) showed greater reduction in binding potential in the putamen, suggesting a relatively greater increase in synaptic dopamine levels with dopaminergic denervation due to impaired transport and storage of the neurotransmitter. RAC PET paradigm following levodopa administration produced reliable results, which were reproducible across different teams using different PET scanners [106]. Continuous dopaminergic stimulation such as levodopa-carbidopa intestinal gel could alleviate motor symptoms, controlling fluctuations, and decreasing the occurrence of sudden "off" episodes. With RAC PET imaging, a stable increase in striatal dopamine levels was increased with infusions of levodopa/carbidopa intestinal gel, correlating with improvements in motor manifestations. This provided evaluation of drug efficacy and guidance for medication treatment.

In vitro studies have shown the coexistence of D_2 receptors and adenosine A_{2A} receptors, with functional interactions between these two receptors. Blocking the function of A_{2A} receptors may prevent the downregulation of D_2 receptors. In PD, as the dopaminergic neurotransmission system in the striatum undergoes progressive dysfunctional, the expression of A_{2A} receptors may be altered [107]. A human postmortem study demonstrated higher expression of A_{2A} receptors in PD patients with levodopa-induced dyskinesias [108]. PET studies using A_{2A} receptor radioligands ^{11}C-TMSX and ^{11}C-SCH442416 have confirmed this finding in the human brain [109, 110]. These studies also found increased A_{2A} receptor expression following the initiation of anti-Parkinsonian treatment. Overall, these results provide support for the theoretical basis of A_{2A} receptors as therapeutic targets in PD, and A_{2A} receptor antagonists have shown anti-Parkinsonian effects in rodent models of PD [111]. Istradefylline, a novel A_{2A} receptor antagonist, has been used as an adjunctive medication to levodopa in Japan since 2013. It has shown to significantly reduce off-period duration, prolongs on-period duration, and has good tolerability and safety. Studies using RAC PET to evaluate the impact of the istradefylline on dopamine D_2 receptor availability and impulse control in the human brain have demonstrated that istradefylline could increase striatal D_2 binding potential and improve impulse control [112]. Recent studies have also hypothesized that istradefylline may increase striatal D_2 binding potential and alleviate neuroinflammation, and RAC PET is expected to be used for further research [113]. Another study employing ^{11}C-preladenant to evaluate the A_{2A} receptor occupancy with different doses and routes of administration found that the A_{2A} receptor occupancy in the striatum following a single dose of 20 mg and 40 mg was 39.5% and 52.1%, respectively, demonstrating a dose-dependent binding of istradefylline to A_{2A} receptors [114]. Subsequent follow-up studies found that long-term administration of istradefylline achieved greater A_{2A} receptor occupancy compared to single-dose administration, with an occupancy rate of 72.1% and 86.5% following long-term administration of 20 mg and 40 mg, respectively [115].

4.2 *Disease Pathology*

Aggregation of alpha-synuclein (aSYN) in Lewy bodies, resulting from dysfunctions in the endolysosomal system, is a salient factor in the formation of Lewy bodies [116]. Current PET/SPECT tracers approved for PD diagnosis and differential diagnosis are mostly used to evaluate the functionality of the dopaminergic system. However, 50% of substantia nigra neurons (Braak stage III) and possibly a higher proportion of striatal dopaminergic terminals have already been lost by the time motor symptoms emerge [117–119]. In contrast, aSYN deposition occurs progressively throughout the disease process, reaching Braak stage VI.

The ability to detect aSYN pathology in vivo would enable early detection of disease, currently limited to demonstration of either striatal dopaminergic denervation or cardiac sympathetic denervation, but more importantly, there are several therapies in development that target abnormal aSYN (vaccines [active immunization], monoclonal Abs [passive immunotherapy] [120], small molecules, and ASO [121]). However, so far, clinical trials of monoclonals have been disappointing [122, 123] and it is not clear whether these therapies have an impact on aSYN expression in the brain. Imaging would provide useful insights into whether therapies have engaged their intended target, though imaging of aSYN in vivo has proved challenging.

4.3 *Reminder that Target Engagement ≠ Efficacy*

Although PET imaging of aSYN holds promise for visualizing αSYN binding, it must be remembered that target engagement (i.e., an impact on tracer binding) does not necessarily indicate therapeutic efficacy. This is evident from the numerous unsuccessful trials of passive immunotherapy targeting amyloid-β (Aβ) in AD patients. In the majority of such studies, target engagement was demonstrated based on a reduction in amyloid expression determined by PET, but there was either no convincing clinical improvement or even worsening [124].

5 Prediction of Disease Trajectory and Stratification for Clinical Trials

5.1 *Dopaminergic Imaging Is Still Likely a Necessary Entry Point for Trials of Disease-Modifying Therapy Despite of Current Limited Utility of Dopaminergic Imaging to Stratify*

Clinical trials for disease-modifying therapies typically target participants who are in the early stages of PD. However, this is a time when the risk of misdiagnosis is high and approximately 15% of patients who are diagnosed with PD based on clinical features have normal DAT imaging results. These individuals, referred to as scans without evidence of dopamine deficiency (SWEDDs), represent a population of early-stage, uncertain PD-like patients [125, 126]. Longitudinal follow-up studies have shown that SWEDD PD patients exhibit slower clinical symptoms and radiological progression. Compared to patients with demonstrable dopaminergic deficits, SWEDDs show minimal changes in striatal binding ratios of β-CIT in follow-up, as well as minimal increases in UPDRS scores [127, 128]. In the long-term follow-up PRECEPT study, it was found that 44% of SWEDDs had their diagnosis changed to other diseases during the follow-up period, while the misdiagnosis rate for patients with abnormal baseline DAT scans was only 3.6% [127]. A 2-year follow-up of SWEDDs in the PPMI database revealed that out of 50 SWEDD cases, 13 were confirmed as PD in subsequent follow-up, while the remaining 37 were non-PD cases. The non-PD SWEDDs showed no significant changes in striatal specific binding ratios compared to baseline [129]. These studies suggest that DAT plays an important role in the accurate diagnosis of early-stage PD, and the likelihood of SWEDDs being diagnosed as PD during the course of the disease is relatively low.

A study that simulated clinical trials by excluding SWEDDs and including only PD participants with DAT evidence of dopaminergic abnormalities found that the required sample size to obtain positive results could be reduced by 24%, while also reducing the misdiagnosis rate and avoiding the unnecessary use of anti-Parkinsonian drugs that do not provide significant improvement in non-PD SWEDDs [19]. However, it should be noted that in this study, many of the subjects without clinically important difference over time had abnormal scans, and conversely, a significant proportion of those with SWEDD had clinically important change over time [19]. Furthermore, some patients with SWEDDs have non-motor features associated with PD [130] and in a very recent study of synuclein seeding aggregation assays (SAA), 9% of subjects with SWEDDs were SAA positive, compared to 4% of healthy controls and 8% of non-manifesting carriers of variants associated with PD [20]. Also, there is very little relationship between change in DAT binding and clinical scores over time [131, 132]. Many studies reported that various interventions can improve the DAT scan but not the clinical performance [133, 134]. Overall, despite imperfections, dopamine imaging plays a significant role in differentiating PD patients with varying degrees of dopaminergic impairment. It holds important implications for the clinical and radiological evaluation of disease-modifying treatments and is still likely a necessary entry point for trials.

5.2 *Predicting the Development of ICD and Dementia*

Dopaminergic imaging may help predict the development of ICD in PD patients. In a cohort study, 31 drug-naïve PD patients were followed for 31.56 ± 12.0 months

[76]. Twenty subjects were classified as PD without ICD while 11 later developed ICD. Compared to the PD group without ICD, the PD-ICD group showed lower DAT binding at baseline in the right ventral striatum, right anterior-dorsal striatum, and posterior part of the right caudate nucleus. The QUIP-RS scores were negatively correlated with baseline DAT binding in the right ventral striatum and right anterior-dorsal striatum. A prospective cohort study that included 320 drug-naïve PD patients showed 54 individuals being diagnosed with ICD within a 2-year period. Although the baseline DAT binding could not predict the occurrence of ICD, a decrease in time-dependent post-baseline DAT binding (assessing the relationship between absolute DAT binding at each post-baseline visit and incident ICD symptoms at that same visit) in bilateral putamen was associated with an increased incidence of ICD, even after adjusting for levodopa equivalent daily dose [75].

Studies investigating cognitive function using SPECT imaging found that reduced striatal dopamine uptake is a risk factor for cognitive decline in PD. A ^{123}I-FP-CIT SPECT imaging study showed lower uptake in the caudate nucleus on the less affected side predicted decline in executive function and visuospatial function in 4 years [135]. Another large cohort study with 249 drug-naïve PD patients and 238 healthy controls showed that the baseline DAT binding potential values in the left putamen, tail of the caudate nucleus, and right tail of the caudate nucleus were positively correlated with the Symbol Digit Modalities Test scores (a test assessing the attention and executive function) at the 2-year and 4-year follow-up visits [136]. Overall, baseline DAT binding in caudate nucleus is consistently recognized as a predictive factor for cognitive performance across different studies [137].

The application of FDG-PET to predict cognitive changes in PD patients may involve a broader range of brain regions. The glucose uptake in the angular gyrus and cerebellum predicted MMSE and MoCA scores at 18 months, even after controlling for baseline scores, age, gender, and education level [138]. In a 5-year PD-MCI follow-up study, the support vector machine (SVM) model identified a pattern with hypometabolism in the temporal and parietal lobes and hypermetabolism in the anterior cingulum, putamen, insular, mesiotemporal, and postcentral gyri. This model predicted the conversion of PD-MCI to PDD with a sensitivity of 67–95% and specificity of 80–91% [139].

Compared to tau, β-amyloid PET demonstrated a higher predictive power for cognitive changes. A small cohort study with a 2-year follow-up found no changes in the uptake rate of Flortaucipir PET in the cortical regions or substantia nigra of PD patients. The change in SUV ratio between baseline and follow-up scans was not associated with cognitive changes [140]. However, in a β-amyloid PET study, the baseline mean cortical binding potential of β-amyloid was able to predict declines in various cognitive domains, including executive function, visuospatial skills, memory, attention, and language in 4 years [141].

6 Biomarker to Study Disease Progression and Effects of Disease Modifying Therapies

6.1 *Dopaminergic Imaging*

Dopaminergic imaging plays a crucial role in the evaluation of treatment efficacy in clinical trials and is an essential component of such studies. Early studies using DAT-PET evaluated the changes in 30 clinically asymmetrical PD patients before and after 6 weeks of treatment with L-dopa (300/75 mg/d), pramipexole (1.5 mg/d), or placebo [142]. The results showed that short-term therapy with L-dopa and, to a lesser extent, pramipexole can modestly down-regulate striatal DAT in patients with early PD. By comparing the rates of loss of dopamine-terminal function in de novo patients with clinical and ^{18}F-DOPA PET evidence of early PD, a randomized 1-to-1 trial was conducted with ropinirole and levodopa. The study found that subjects receiving ropinirole had lesser reductions over time in ^{18}F-DOPA uptake in the putamen and substantia nigra [143]. Subsequently, in a large randomized, double-blind, placebo-controlled trial, the mean percent decline in ^{123}I-beta-CIT uptake was significantly greater with levodopa than with placebo [144]. While one interpretation of these studies was that dopamine agonist use was associated with slower progression of disease than treatment with levodopa, this did not translate into a clinical impact and a more likely explanation is that levodopa and dopamine agonists may differentially affect dopamine transporter binding and/or ^{18}F-DOPA uptake (L-AADC activity), rather than the actual survival of dopamine neurons.

In the field of cell transplantation, a previous double-blind, placebo-controlled surgical trial of embryonic dopamine cell implantation revealed a significant increase in FDOPA uptake in the putamen of the group receiving implants compared to the placebo surgery patients (40.3%) 1 year after surgery [145]. This indicates that embryonic dopamine cell implantation contributes to improving dopamine deficits in PD patients. Long-term clinical and PET outcomes from 33 of the original trial participants who were followed for 2 years after transplantation, and 15 of these subjects who were followed for an additional 2 years, showed significant increases in putamen ^{18}F-FDOPA uptake at all post-transplantation time points. These increases were not influenced by age or sex. Post-transplantation changes in putamen PET signal and clinical outcomes were significantly intercorrelated throughout the study [146]. However, this correlation between improved ^{18}F-DOPA uptake and clinical improvement was not seen in another double-bind trial [122], despite pathological evidence of graft survival [123].

In recent years, several new therapies for PD have emerged. In the field of antibodies, two recent phase II clinical trials, cinpanemab and prasinezumab, were completed. After 52 weeks of treatment in early PD patients, no clear clinical efficacy or DAT changes were observed [147, 148]. In a single-center, randomized, double-blind, placebo-controlled trial of intermittent intraputamenal glial cell line-derived neurotrophic factor, ^{18}F-DOPA PET imaging demonstrated a significantly increased uptake throughout the putamen in the active group only, ranging from

25% to 100%, but the trial failed to meet its primary clinical endpoint [149]. In an I/IIa study of intracerebroventricular recombinant human PDGF-BB administration in PD, secondary outcome parameters did not show dose-dependent changes in clinical rating scales, but there was a positive effect on DAT binding in the right putamen [133]. The latest multicenter, phase 2, randomized, double-blind trial on deferiprone showed no meaningful between-group difference in the change in DAT binding potentials in the posterior putamen, anterior putamen, and caudate nucleus at week 40 [150].

In conclusion, although the application of dopamine imaging in PD clinical research is still limited, it provides potentially useful information for evaluating treatment efficacy. It can verify the presence or absence of therapeutic effects and assess the source of efficacy, making it a necessary component of clinical trials, but correlation with clinical impact is often limited.

6.2 Role of FDG to Demonstrate Network Engagement

FDG PET has been employed in metabolic connectivity studies to identify disease-specific patterns and is increasingly being used in routine clinical practice. One notable application is the spatial covariance analysis of 18F-FDG PET data, which consistently reveals a stereotyped spatial covariance pattern associated with different stages of motor symptoms in PD patients [151].

The PD-related pattern (PDRP) was the first identified glucose metabolism pattern related to PD. The PDRP is characterized by relative hypermetabolism in the thalamus, putamen/pallidum, pons, cerebellum, and motor cortex, as well as decreased metabolism in the lateral premotor and parieto-occipital cortex [152]. To date, the PDRP has been validated in multiple independent patient cohorts, with consistent results regardless of patient population or imaging techniques used, demonstrating its reproducibility in assessing PD. Studies have shown a close correlation between the PDRP and motor symptoms in PD patients, particularly with the scores of bradykinesia/rigidity. Longitudinal follow-up of early-stage PD patients at baseline, 2 years, and 4 years revealed a gradual increase in the expression level of the PDRP as the disease progresses [153]. The rate of PDRP progression is similar across different cohorts, indicating an overall linear process [154]. Therefore, longitudinal measurement of the PDRP values using FDG PET can be used to assess disease progression and reflect the clinical severity of PD to varying degrees. Furthermore, effective treatment of motor symptoms in PD patients leads to a decrease in PDRP expression values. This has been observed in response to acute levodopa administration, subthalamic nucleus (STN), deep brain stimulation (DBS), and therapeutic subthalamotomy [155]. Patients scanned while on levodopa treatment (or with DBS or gene therapy) show suppressed PDRP levels compared to their OFF-state but still significantly elevated relative to healthy controls, indicating that symptomatic treatment does not completely correct the underlying network abnormality [154]. Therefore, quantifying PDRP in individual patients can aid in evaluating disease treatment.

Tremor is one of the important motor symptoms in PD, but it is not related to PDRP. To address this, researchers proposed the PD-related tremor pattern (PDTP). The PDTP is characterized by increased covariance in the cerebellum/dentate nucleus and primary motor cortex, and to a lesser extent, the caudate and putamen. PDTP expression scores are correlated with tremor but not with bradykinesia/rigidity scores [156]. In addition to motor symptoms, PD patients also experience cognitive impairments. The PD-related cognitive pattern (PDCP) is characterized by metabolic reductions in the medial prefrontal, premotor, and parietal association regions, along with relative increases in the cerebellar vermis and dentate nuclei [153]. PDCP expression is associated with executive and memory performance and increases with cognitive deterioration in PD patients. However, multiple studies have consistently shown that PDRP expression levels are consistently higher than those of PDCP, suggesting that PDRP appears earlier and progresses faster during the course of the disease. In summary, the aforementioned brain glucose metabolism patterns serve as powerful tools for the favorable diagnosis, differential diagnosis, tracking disease progression, and monitoring treatment efficacy in PD.

6.3 Assessment of Non-synuclein Pathology

6.3.1 Tau Accumulation

Tau PET imaging in PD patients has been performed using first-generation tracers such as ^{18}F-FDDNP and ^{18}F-AV-1451. ^{18}F-FDDNP was the earliest tracer used for tau PET imaging, but it lacks specificity and selectivity in vivo, as it binds to both neurofibrillary tangles and amyloid plaques in the brain. Studies using ^{18}F-FDDNP have suggested higher binding in the lateral temporal lobe in PDD patients compared to PD patients without cognitive impairment [157]. ^{18}F-AV-1451 is the most widely used tracer in patients with neurodegenerative diseases. Autopsy studies have confirmed that ^{18}F-AV-1451 strongly binds to neurofibrillary tangles and tau protein in neurites but not to Aβ or αSYN aggregates. However, there is only a limited relationship of AV-1451 binding to the degree of tau pathology, as the tracer exhibits off-target binding to monoamine oxidase (expressed in glial cells), hemosiderin and neuromelanin in the substantia nigra [158, 159]. Binding to neuromelanin results in lower uptake of ^{18}F-AV-1451 in the substantia nigra of PD patients compared to healthy controls [160, 161]. In PD patients with cognitive dysfunction, elevated ^{18}F-AV-1451 uptake has been observed in the temporal cortex and the anterior cingulate gyrus compared to healthy controls [162]. However, no abnormal ^{18}F-AV-1451 uptake was found in patients with mild cognitive impairment [163].

Second-generation tracers have addressed some of the limitations of first-generation tracers [164]. Studies using ^{18}F-THK5351 have shown lower tau tracer uptake in the midbrain and diencephalon of PD patients compared to patients with

PSP, and lower uptake in the pons compared to patients with multiple system atrophy (MSA-C) [165]. However, THK53351 still exhibits off-target binding to MAO-B and glial cells. Another study using ^{18}F-APN-1607 demonstrated lower tracer uptake in the caudate nucleus, globus pallidus, thalamus, midbrain, tegmentum, substantia nigra, red nucleus, and pons of PD patients compared to PSP patients [166].

A meta-analysis of tau PET in PD patients showed (1) PDCI patients had higher tau tracer uptake in the inferior temporal gyrus compared to healthy controls; (2) PDCI patients exhibited higher tau tracer uptake in the olfactory cortex compared to PD patients without cognitive impairment (PDNC); (3) When using first-generation tracers, PD patients showed lower tau tracer binding in the substantia nigra compared to healthy controls, likely reflecting off-target binding of ^{18}F-AV-1451 to neuromelanin; (4) When using second-generation tracers, no significant differences in tau tracer uptake in the substantia nigra were observed between PD patients and healthy controls; (5) PD patients had lower tau tracer uptake in the frontal and occipital lobes compared to AD patients [167].

6.3.2 Neuroinflammation

The translocator protein (TSPO) is a mitochondrial translocator protein that is normally present at minimal levels in the brain. However, its expression significantly increases in cases of neuroinflammation, making it a suitable biomarker for activated microglia. PET imaging techniques utilizing TSPO and its ligands are employed to measure neuroinflammation mediated by microglia in vivo.

In the first generation of TSPO ligands, such as ^{11}C-PK11195, there were limitations due to its high lipophilicity, resulting in low specific binding, a low signal-to-noise ratio, poor brain penetration, and high binding to plasma proteins. However, as this tracer avoids some of the problems encountered with second generation tracers (see below), it continues to be a first choice for many investigators.

The second generation of TSPO ligands includes ^{18}F-FEPPA, ^{11}C-DPA713, ^{11}C-PBR28, and ^{18}F-DPA714. However, these ligands are affected by a single nucleotide polymorphism (SNP) known as rs6971, which is located on exon 4 of the TSPO gene. Depending on the presence of this SNP, individuals can be categorized as high-affinity binders, low-affinity binders, or mixed-affinity binders. PET imaging using most ligands is not useful for low-affinity binders due to their low signal. In a study with 37 Caucasian, 2 Asian, 1 mixed Caucasian/Asian, and 1 Hispanic, 66% of the subjects were classified as high-affinity binders, 29% as mixed-affinity binders, and 5% as low-affinity binders [168]. Interestingly, according to the data from the 1000 Genomes Project Phase 3, the Chinese population has a significantly higher proportion of high-affinity binders, estimated at around 95% [169].

Ligands which can be used across different individuals regardless of their genetic variations would offer improved specificity and reliability in PET imaging of TSPO. The third generation of TSPO ligands, including ^{11}C-ER-176 and ^{18}F-GE-180, are insensitive to the rs6971 SNP in vitro binding assay. However, in vivo ^{18}F-GE-180 quantification is still affected [170].

While the second generation TSPO ligands have a higher signal-to-noise ratio than PK11195, variations in affinity state remain a challenge for interpretation, and all of the tracers suffer from lack of a consistently defined reference region that is free of specific binding.

Several studies have indicated higher TSPO binding in the putamen of PD patients compared to healthy controls [169, 171, 172]. Additionally, there have been reports of increased TSPO binding in the frontal cortex [171], thalamus, substantia nigra, temporal cortex, parietal cortex, and occipital cortex [172] in PD patients. However, these findings are supported by only a limited number of studies. No significant associations have been found between TSPO binding and the severity of motor symptoms or reduced dopamine transporter binding [171, 173]. Moreover, not all studies demonstrate a correlation between peripheral inflammation markers and microglial activation in the brain of PD patients [169].

An interesting observation has been made regarding the comparison between PD and MSA. There is a distinct pattern of elevated regional ^{11}C-PBR28 binding to TSPO in MSA, particularly in the lentiform nucleus and cerebellar white matter, when compared to PD. Furthermore, within the subtypes of MSA, the lentiform nucleus exhibited the most significant and largest effect on TSPO binding in MSA-P compared to PD, while the cerebellar white matter showed a similar effect in MSA-C versus PD. These findings may aid in the differential diagnosis between PD and MSA [174].

6.3.3 Synaptic Loss

Currently, there are several probes/candidate probes available for synaptic vesicle glycoprotein 2A (SV2A), as shown in Table 2.

PD patients exhibit significant loss of SV2A binding in the substantia nigra [184–186]. Some studies have also demonstrated a significant decrease in SV2A binding in the red nucleus, locus coeruleus, parahippocampal gyrus, posterior cingulate cortex, orbitofrontal cortex, and ventromedial prefrontal cortex compared to healthy controls [185, 187]. This may be somewhat unexpected, as PD is associated with loss of nigral dopamine neurons, but not typically with loss of synaptic input to the substantia nigra. It should also be remembered that SV2A is expressed ubiquitously, while the synaptic vesicular glycoprotein preferentially expressed in dopaminergic nigral neurons is SV2C [188]. However, the results for these brain regions lack consistency across different studies and require further validation. Interestingly, patients with DLB or PDD showed reduced SV2A binding in the substantia nigra and all cortical regions of interest, including the middle occipital cortex (mOC), lateral occipital-temporal cortex (lOT), lateral occipital cortex (lOC), lateral parietal cortex (lPC), superior parietal lobule (sPL), primary sensorimotor cortex (M1S1), middle frontal gyrus (mFG), and orbitofrontal cortex (OFC), except

Table 2 Synaptic vesicle glycoprotein 2A PET imaging probes

Tracer	Year	Use	Advantages and disadvantages
^{11}C-UCB-J	2016	Human (2016) [175]	Short half-life of ^{11}C (20.38 mins) has to be produced and used on-site
^{11}C-UCB-A	2016 [176]	Human (2016) [176]	Slow binding kinetics
^{18}F-UCB-H	2013 [177]	Human (2015)	Low specific binding in the brain
^{18}F-UCB-J	2019 [178]	Nonhuman (2019) [178]	Long half-life (110 mins), unsuitable for routine production
^{18}F-SynVesT-1	2019 [179]	Nonhuman (2019) [180] Human (2021) [181]	High brain uptake, fast and reversible kinetics, excellent test-retest reproducibility, and binding specificity to SV2A
^{18}F-SynVesT-2	2020 [182]	Nonhuman (2020 [182])	
^{18}F-SDM-16	2022 [183]	Nonhuman (2022) [183]	Highest affinity (Ki = 0.9 nM) to human SV2A among all reported SV2A ligands. High specific binding in monkey brains and was metabolically more stable than the other SV2A PET tracers. The fP of ^{18}F-SDM-16 was 69%, which was higher than those of ^{11}C-UCB-J (46%), ^{18}F-SynVesT-1 (43%), ^{18}F-SynVesT-2 (41%), and ^{18}F-UCB-H (43%)

for the hippocampus and amygdala [186]. In 1- or 2-year follow-ups of early-stage PD, no significant changes in SV2A were observed compared to baseline [187, 189]. There was a significant positive correlation between 11C-UCB-J binding in the caudate and ^{18}F-FE-PE2I binding in the caudate and the dorsal striatum [184], indicating a potential relationship between striatal dopaminergic degeneration and synaptic loss.

7 Understanding of Brain Function

7.1 *Reward, Including Placebo Effects*

Expectations for clinical efficacy shaped by patients' perceptions and experiences constitute a "placebo effect" that activates the reward system (including the ventral striatum, nucleus accumbens (NA), ventromedial prefrontal cortex, orbitofrontal cortex, amygdala, and anterior cingulate cortex), thus producing therapeutic effects. Notably, the activation of the reward system is associated with dopamine release from the ventral striatum (mesolimbic system). The "placebo effect" plays a crucial role in the clinical efficacy in PD patients [190]. In drug-induced placebo effects, de la Fuente-Fernandez et al. demonstrated that, compared to baseline conditions, placebo can induce dopamine release in the dorsal striatum. Moreover, placebo-induced dopamine release in the dorsal striatum is higher in patients who experience treatment benefit than in those who do not, suggesting its association with the subjective experience of benefit [191]. Further study revealed that the magnitude of placebo-evoked dopamine release varies according to the degree of expectation, but the relationship is non-linear [192], in keeping with a role for uncertainty in expectation-induced dopamine release [193]. rTMS over motor cortex can induce bilateral dopamine release in the motor striatum of healthy controls. A similar response can be evoked by sham TMS, but the response is localized to the ventral striatum [194]. In a glucose PET study investigating the neural circuitry in PD patients who randomly received either real or sham surgery for gene therapy in the subthalamic nucleus, metabolic imaging was performed before randomization and again 6 and 12 months after surgery. The placebo response was associated with increased metabolism in a specific cerebellar-limbic circuit, which involves the cerebellum, ACC, striatum, amygdala, and ventral anterior thalamus [195].

7.2 *Plasticity, Including Effect of Exercise*

Brain plasticity refers to the brain's ability to change structure and function [196]. As age advances, the plasticity of the brain decreases. However, in degenerative conditions like PD, the brain is still able to undergo some compensatory changes.

PET studies using ^{18}F-DOPA have shown that the striatum begins to lose dopaminergic neurons several years before clinical symptoms of PD appear. These losses are asymptomatic, indicating that the nigrostriatal pathway has certain compensatory functions, capable of offsetting a portion of the dopamine loss. Symptoms only appear when the uptake of ^{18}F-DOPA in the dorsal caudal putamen is reduced by 50%. Studies have suggested that PD patients exhibit compensatory increases in the uptake of ^{18}F-DOPA in the dorsal midbrain and anterior cingulate cortex, but the significance of these changes remains unclear. A 2003 study showed a decrease in the uptake of ^{18}F-DOPA in the caudal putamen and an increase in the globus pallidus internus (GPi) in early PD. In advanced PD, the uptake of ^{18}F-DOPA in the GPi also decreases, elucidating a compensatory pathway in early PD from the substantia nigra pars compacta to GPi. The loss of this pathway in the later stages is an important mechanism in PD progression [197].

In addition to these inherent compensations, exercise can improve PD symptoms and promote brain remodeling. Aerobic exercise can modify the brain's vascular system, enhance neurotrophic factor signaling, and foster neurogenesis. PET studies with ^{11}C-RAC have indicated that aerobic exercise can increase transcranial magnetic stimulation (TMS)-induced dopamine release in the caudate nucleus [198]. ^{11}C-RAC PET scans before and after acute cycling exercises showed, compared to sedentary participants, habitually exercising participants had higher dopamine release in the caudate nucleus, as well as lower apathy scores and bradykinesia scores, and increased ventral striatum activation during performance of an MRI reward task. This suggests that exercise likely results in increased striatal dopamine release acutely [199]. Research with ^{18}F-fallypride PET has also suggested that exercise may increase the level of dopamine D_2 receptor binding in the striatum of the basal ganglia [200].

8 Future Challenges and Opportunities

8.1 Focus on Challenges of Finding Ligands for aSYN, Tau, Inflammation

Ideal PET tracers need to fulfill several prerequisites: low molecular weight, sensitivity, specificity, high affinity, selectivity, moderate lipophilicity, sufficient brain uptake, reversible binding properties, favorable pharmacokinetics, and no radioactive metabolites in the brain.

8.1.1 aSYN

Aβ, tau, and αSYN form similar β-sheet structures upon aggregation [201]. Therefore, imaging probes that bind to Aβ or tau aggregates potentially have the ability to bind to αSYN aggregates [201]. Consequently, imaging probes that have been used for AD, such as ^{11}C-PIB, ^{11}C-BF227, and ^{11}C-PBB3, were tested for imaging αSYN aggregates. Although these tracers demonstrated high affinity for αSYN fibrils, their selectivity for αSYN was lower compared to Aβ and tau, which have similar β-sheet core structures. Conversely, tracers that are preferential for αSYN vs. Aβ or tau may still bind to a significant degree to the latter targets, whose concentration is much higher than that of αSYN. This problem is further compounded by the common co-localized deposition of all or more than one of these misfolded proteins. Therefore, new tracers need to be developed specifically for αSYN imaging rather than repurposing those used for AD.

Currently, several small molecule ligands with favorable pharmacokinetics and high affinity have been developed. For example, MODAG-001 possesses high affinity and good selectivity for binding to αSYN fibrils, as well as favorable pharmacokinetic and biodistribution properties. It meets several important criteria for a PET tracer targeting αSYN. However, due to its lipophilicity, it exhibits high nonspecific binding, which reduces the signal-to-noise ratio (SNR). Other small molecule ligands, such as ^{11}CSIL5 and ^{18}F-SIL26, derived from ^{125}I-SIL23, show improved selectivity. Recently, a promising new PET tracer, ^{18}F-F0502B, has been discovered. It can image aggregated αSYN in synucleinopathies. ^{18}F-F0502B selectively labels αSYN aggregates in human PD samples and demonstrates specific recognition of αSYN aggregates in PET imaging of a macaque PD model. Furthermore, in two non-human primate PD models, the levels of aggregated αSYN may increase over time, leading to an elevated PET-specific binding signal [202]. However, further optimization is required based on their pharmacokinetic profiles to achieve successful αSYN imaging in vivo. Therefore, the current development of small molecule ligands for αSYN imaging is hindered by low affinity, insufficient selectivity, and suboptimal physicochemical properties, making them unsuitable for in vivo αSYN imaging. The role of recently described tracers for αSYN including ^{18}F-ACI-12589 [203] and ^{18}F-SPAL-T-06 [204] in PD (as opposed to MSA, which is characterized by a high concentration of αSYN) is as yet unclear (Fig. 2).

There are several challenges in finding αSYN tracers (Fig. 3) [205]:

(i) The target concentration of aggregated αSYN is relatively low compared to other pathological protein aggregates, such as β-amyloid and tau, or certain enzymes like monoamine oxidases [206]. This demands a highly sensitive detection tool for adequate in vivo detection and visualization [201, 207].
(ii) The target composition poses a challenge as β-sheet binding motifs may be shared among different pathological protein aggregates, leading to unselective binding or complicating differential diagnosis in cases of co-pathologies. Enhancing the selectivity and specificity of αSYN PET tracers for the target is crucial but challenging.
(iii) The target structure of pathological αSYN is mainly oligomeric rather than fibrillar. However, most αSYN materials used in radiotracer development are recombinant fibrils, which do not accurately represent aggregated αSYN in the human brain and have poor reproducibility.

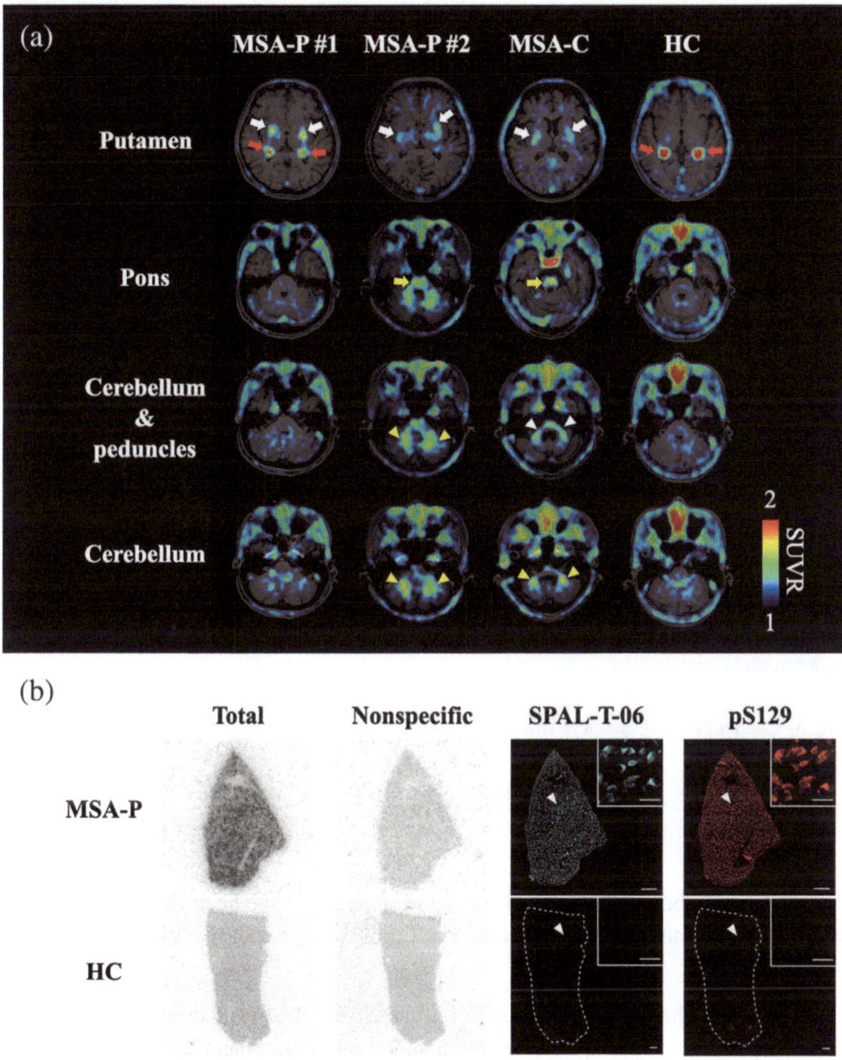

Fig. 2 In vivo and in vitro binding of SPAL-T-06 to αSYN pathologies in MSA patients. (Reproduced from Matsuoka et al. [204])

(iv) The target localization presents a challenge as αSYN aggregates are predominantly intracellular inclusion bodies, making the cell membrane an additional barrier for tracer accessibility [208].

(v) The tracer structure is important, as lipophilic compounds that effectively cross the blood-brain barrier (BBB) may exhibit high nonspecific binding, reducing the signal-to-noise ratio (SNR) [208]. Other challenges include in vivo instability, inadequate brain uptake, and pharmacokinetics [209].

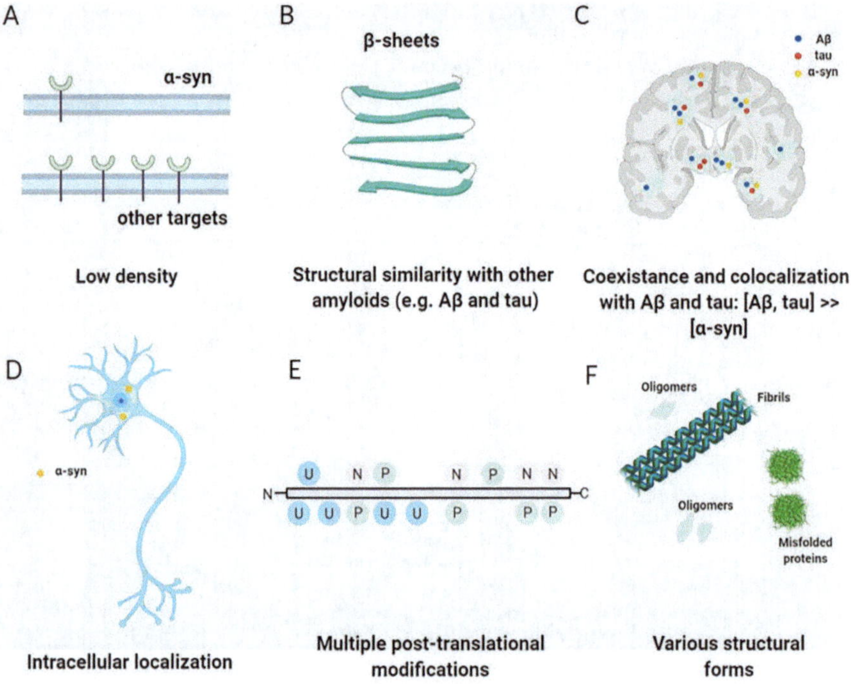

Fig. 3 Challenges and characteristics of an ideal αSYN tracer. (Reproduced from Korat et al. [205])

8.1.2 Tau

As mentioned above, although many tau PET tracers have been developed, there is still a need for more suitable tracers. The limited number of useful tracers is due to various reasons:

(i) The main challenge with tau PET imaging has been off-target binding, especially to hemosiderin, neuromelanin and monoamine oxidase (expressed in glial cells), as well as inconsistencies between ante- and postmortem results [210].
(ii) Tau PET tracers need to differentiate between Aβ plaques and neurofibrillary tangles (NFTs). However, this is challenging because the specificity of existing tau tracers is based on the folded structure of tau, which is not unique to tau alone but also present in fibrils formed by Aβ, αSYN, and other proteins [210].
(iii) Further understanding of different high-affinity binding sites on tau fibrils and their relationship with the binding patterns of different tau tracers in vivo is needed [210].
(iv) Tau has multiple molecular conformations, which can adversely affect ligand binding sites. Additionally, alternative splicing produces various isoforms of tau, some with three tandem repeat sequences (3R tau) and some with

four (4R stau). Tau pathology occurs due to hyperphosphorylation, disrupting microtubule functionality and leading to the formation of NFTs in neurons and glial cells. NFTs are aggregates of hyperphosphorylated tau with different isoforms and conformations, depending on the tauopathy [210]. Several of the tau tracers developed to date may bind to 3R, 4R tau in Alzheimer disease, but less consistently to 4R tau seen in PSP or corticobasal syndrome. In the case of ligands that bind to 4R tau, there may only a very poor relationship between tracer binding in vivo and postmortem pathology [166].

(v) The concentration of tau aggregates in the brain is low, and the forms of tau protein accumulation vary in different neurodegenerative diseases. For example, in AD, predominantly intraneuronal tau aggregates in the form of paired helical filaments are found, while in other tauopathies like corticobasal degeneration or progressive supranuclear palsy, aggregates of straight filaments are observed more frequently, also present in glial cells [206]. Therefore, ligands need to have higher sensitivity and specificity for specific forms of tau protein, which presents a major difficulty and challenge.

8.1.3 Inflammation

TSPO

Current tracers for TSPO suffers from several major limitations, such as low permeability of the blood-brain barrier and high non-specific plasma binding, resulting in a low signal-to-noise ratio in the final reconstructed PET images [211]. Additionally, the ^{11}C-PBR28 binding seems to be affected by the chromosome 1 variant rs2997325 on microglial activation [170]. These drawbacks have prompted the development of third-generation TSPO tracers. However, rs6971 polymorphisms still affect in vivo ^{18}F-GE-180 quantification, showing lower binding in patients with low-affinity binders compared to mixed- and high-affinity binders. Furthermore, it is currently uncertain whether ^{18}F-GE-180 can be used in humans due to its incomplete blood-brain barrier penetration [170]. The cellular location of the signal is another major concern for TSPO ligands. Several TSPO ligands. Further research on the next generations of TSPO tracers is needed, with selection criteria including optimal binding properties, insensitivity to TSPO polymorphism, and high glial TSPO selectivity [170, 212]. Therefore, the study of other imaging targets could be important for obtaining pathological changes in neuroinflammation.>

CSF1R

Macrophage colony-stimulating factor 1 receptor (CSF1R) is primarily expressed in microglia in the brain and shows lower expression in other cell types. There is evidence linking CSF1R to the occurrence of neuroinflammation and inhibiting CSF1R has shown promise in reducing neuroinflammation and slowing the progression of tissue diseases. Therefore, CSF1R can serve as a novel PET imaging target for neuroinflammation. Currently, representative probes for CSF1R include ^{18}F-88, ^{11}C-89 (^{11}C-AZ683), ^{11}C-90 (^{11}C-CPPC), ^{3}H-91 (^{3}H-CPPC), and ^{11}C-92 (^{11}C-GW2580).

^{18}F-88 was the first reported CSF1R-PET ligand, followed by ^{11}C-89, which exhibited high binding affinity and good selectivity for CSF1R. ^{11}C-CPPC, developed based on CSF1R inhibitors, is a radiotracer used for neuroinflammation PET studies. Evidence suggests that ^{11}C-CPPC possesses favorable biological properties for CSF1R PET imaging, but it still faces certain issues. One major concern is the low specific binding towards CSF1R. In a microglia-depleted mouse model, only a 14% reduction in radioactivity was observed in the brain, while CSF1R-knockout mice exhibited comparable brain uptake to healthy control animals. Moreover, considerable nonspecific binding of CPPC was demonstrated with ^{3}H-91 (^{3}H-CPPC) in the CNS, as evidenced by low displacement of the radioactivity by CSF1R inhibitors PLX3397, PLX5622, and BLZ945 (15%, 16%, and 25%, respectively) [213]. Additionally, ^{11}C-CPPC has a short half-life (20.4 min), limiting its synthesis for multiple doses. Increased ^{11}C-CPPC binding has recently been demonstrated in vivo in PD, but increases (widespread) were seen only in those patients with more advanced disease [214]. In comparison to ^{11}C-CPPC, ^{11}C-GW2580 is more sensitive to the proliferation of inflammatory microglia [215]. However, it only exhibits moderate affinity and requires further improvement and validation of its binding affinity.

P2X7R

The P2X7 receptor (P2X7R) is involved in the activation, proliferation, and apoptosis processes of microglia. Under normal physiological conditions, its expression in the central nervous system is very limited. However, after brain injury, P2X7R expression significantly increases, promoting the release of pro-inflammatory cytokines and chemokines. Furthermore, studies have shown that overexpression of P2X7R can enhance microglial activation and proliferation. Representative PET probes for P2X7R include ^{11}C-80 (^{11}C-JNJ-54173717), ^{18}F-81 (^{18}F-JNJ-64413739), ^{11}C-82 (^{11}C-A-740003), ^{18}F-83 (^{18}F-EFB), ^{11}C-84 (^{11}C-GSK1482160), ^{18}F-85 (^{18}F-IUR-1601), and ^{11}C-86 (^{11}C-SMW139) [213].

^{11}C-80 (^{11}C-JNJ-54173717) is one of the initial brain-penetrant ^{11}C-labeled radioligands for P2X7R imaging. In a rat model with hP2X7R overexpression, high uptake of the tracer was observed in the striatum. Pretreatment with a non-radioactive antagonist, JNJ-42253432, significantly reduced striatal tracer uptake, indicating specific binding of ^{11}C-JNJ-54173717 to hP2X7R in vivo. In a preliminary study, no increase of ^{11}C-JNJ-54173717 binding was seen in PD compared to healthy controls [216].

^{18}F-81 (^{18}F-JNJ-64413739) also exhibits high specificity, as confirmed by blocking experiments in rodent models, indicating its promising application for P2X7R imaging. However, dose-dependent inhibition of tracer binding with P2X7R antagonists was observed in both rodent and human brain tissue. Although the overall favorable characteristics support the use of ^{18}F-JNJ-64413739 for imaging P2X7R in clinical neuroinflammation, the absence of a suitable brain reference region that does not express the target hinders the full potential of kinetic modeling with this probe [213].

GSK1482160 has been found to penetrate the blood-brain barrier and enter the central nervous system, leading to the development of another class of successful

PET radioligands based on a benzamide core structure. Green et al. concluded from a clinical trial in healthy subjects that although the brain uptake of GSK1482160 was as low as 2% of the injected dose, the compound appears suitable for PET studies of P2X7R expression to monitor inflammatory processes [217]. Subsequently, ^{18}F-IUR-1601 was developed by Gao et al., and although it has been evaluated in vitro, there are no reported in vivo data to date.

Imidazoline

Research has shown that reactive astrocytes can contribute to or exacerbate neuroinflammation or neurodegenerative diseases. Therefore, it is essential to develop new astrocytic PET tracers to monitor their function and in vivo distribution.

BU99008, a novel astrocytic PET ligand targeting imidazoline-2 binding sites (I2BS) on astrocytes, might be a suitable candidate. ^{11}C-BU99008 demonstrated robust brain entry and high retention in the basal ganglia, moderate retention in the cortex, and the lowest retention in the cerebellum in porcine, rhesus monkey, and human studies. These findings align with the reported distribution of I2BS in the brain. The binding of ^{11}C-BU99008 was blocked by pretreatment with idazoxan, indicating high selectivity of this tracer to I2BS in humans. In one study in PD, ^{11}C-BU99008 binding was increased in the brainstem in early disease, with more widespread increases that correlated with cognitive decline in advanced disease [218], but this was not reproduced in another study comparing PD-dementia against healthy aging [219].

There are other neuroinflammation-targeted PET probes that have been developed so far, including tracers targeting COX-1, COX-2, ROS, MAO-B, and other targets. The main challenge in neuroinflammation PET imaging is to identify suitable targets that meet certain criteria:

(i) The target should exhibit adequate expression (high B_{max}) and high binding affinity (K_i/K_d) for successful PET visualization ($B_{max}/K_d \geq 10$).
(ii) The target should serve as a proxy for a specific aspect of the neuroinflammatory process, and significant changes in target expression/density should distinguish pathological states from healthy conditions.
(iii) The target should be druggable or traceable, with pharmacophores or pockets for binding small molecule ligands. Continuous development of PET molecules for existing and validated biological targets, which possess excellent binding affinity, high selectivity, high blood-brain barrier penetration, suitable brain kinetics, and reasonable metabolic profiles, is crucial [213] (Table 3).

8.2 *The Advantages/Disadvantages of Short (^{11}C) Versus Longer (F-18) Tracers*

^{18}F half-life is 110 min and ^{11}C half-life is 20 min. The longer half-life of ^{18}F compared to ^{11}C allows for its use after leaving the preparation center, without the

Table 3 Inflammation PET imaging probes other than TSPO

Target	Cell expression	Representative PET probe	Stage of research
P2X7R	Microglia >> astrocytes	^{11}C-JNJ-54173717	Human (PD)
		^{18}F-JNJ-64413739	Human (healthy subjects)
		^{11}C-GSK1482160	Non-human primates
		^{18}F-IUR-1601	No in vivo imaging application results reported
		^{11}C-SMW139	Human (MS)
CSF1R	Microglia	^{11}C-AZ683	Non-human primates
		^{11}C-CPPC	Human (healthy subjects and PD)
		^{11}C-GW2580	Non-human primates
Imidazoline	Astrocytes	^{11}C-BU99008	Human (PD)

need for real-time preparation like ^{11}C, or an on-site cyclotron [220, 221]. However, patients cannot undergo multiple scans in a single day [222], and they also receive a higher radiation dose.

^{11}C has a shorter half-life with strict limitations on scanning time. It requires on-site cyclotron production, which increases the cost [223]. However, it allows for multiple PET scans on the same day, enabling the use of various tracers in one session. Its shorter half-life allows for faster image acquisition and the study of faster biological processes, reducing the radiation dose received by patients, as well as the exposure time for staff and the environment. Additionally, compared to F, the C atom is widely present in various small organic molecules, and introducing a fluorinated ethyl group into a complex may modify its biological behavior. On the other hand, replacing ^{12}C with ^{11}C does not alter the substance's biological and pharmacological characteristics, making it safer [224].

8.3 Multimodal Molecular Imaging

PET is a non-invasive molecular imaging technique with high sensitivity that allows us to selectively image different targets by tracking radiolabeled molecules with extremely low concentrations in the body. However, PET imaging often lacks anatomical landmarks, which is its main drawback. The introduction of PET/CT has addressed this issue, but PET/CT is not without limitations. For example, the radiation dose in PET/CT is a major concern for patients. Additionally, PET and CT imaging in PET/CT are performed sequentially, resulting in temporal and spatial offsets [225]. Therefore, the PET/MR technology was introduced. It combines the advantages of PET and MR, offering better spatial resolution and higher diagnostic accuracy. Recent studies have provided evidence that PET/MR is superior to PET/CT [225]. Unlike PET/CT, PET/MR allows true simultaneity of PET and MRI, eliminating the misregistration effect caused by subtle motions between the two scans [225]. Furthermore, head motion is a confounder in many imaging studies [226]. Motion in PET can contaminate emission data and cause spatial misalignments, leading to significant quantitative biases in the reconstructed PET activity distributions [227]. However, MR-based motion correction of PET data has been developed, as continuous acquisition of MR sequences enables motion tracking during PET reconstruction [226].

Recently, Frost et al. utilized PET/MR to investigate the early changes in dopamine and microstructure in PD patients and their potential associations. The findings suggested that molecular degeneration plays a role in modulating microstructural disorganization and motor dysfunction in the early stages of PD [225]. This study also demonstrated the potential of PET/MR in studying dopamine deficits in PD patients. Additionally, PET/MR can be used to study other neurotransmitter systems, such as opioid, serotonin, and glutamate neurotransmitter systems [226]. Although PET/CT is unlikely to be replaced by PET/MR in the future, PET/MR holds more possibilities in future clinical applications due to various improvements. Altogether, PET/MR imaging is likely to have a significant role in unraveling in vivo neurobiology by providing novel insights into receptor pharmacology, signaling pathways, and brain function modulated by various neurotransmitters [226].

References

1. Zou J, Weng RH, Chen ZY, Wei XB, Wang R, Chen D, et al. Position emission tomography/single-photon emission tomography neuroimaging for detection of premotor Parkinson's disease. CNS Neurosci Ther. 2016;22(3):167–77.
2. Lee CS, Samii A, Sossi V, Ruth TJ, Schulzer M, Holden JE, et al. In vivo positron emission tomographic evidence for compensatory changes in presynaptic dopaminergic nerve terminals in Parkinson's disease. Ann Neurol. 2000;47(4):493–503.
3. Ito K, Morrish PK, Rakshi JS, Uema T, Ashburner J, Bailey DL, et al. Statistical parametric mapping with 18F-dopa PET shows bilaterally reduced striatal and nigral dopaminergic function in early Parkinson's disease. J Neurol Neurosurg Psychiatry. 1999;66(6):754–8.
4. Piccini P, Morrish PK, Turjanski N, Sawle GV, Burn DJ, Weeks RA, et al. Dopaminergic function in familial Parkinson's disease: a clinical and 18F-dopa positron emission tomography study. Ann Neurol. 1997;41(2):222–9.
5. Morrish PK, Sawle GV, Brooks DJ. An [18F]dopa-PET and clinical study of the rate of progression in Parkinson's disease. Brain. 1996;119(Pt 2):585–91.
6. Calne DB, Langston JW, Martin WR, Stoessl AJ, Ruth TJ, Adam MJ, et al. Positron emission tomography after MPTP: observations relating to the cause of Parkinson's disease. Nature. 1985;317(6034):246–8.

7. Samii A, Markopoulou K, Wszolek ZK, Sossi V, Dobko T, Mak E, et al. PET studies of parkinsonism associated with mutation in the alpha-synuclein gene. Neurology. 1999;53(9):2097–102.
8. Adams JR, van Netten H, Schulzer M, Mak E, McKenzie J, Strongosky A, et al. PET in LRRK2 mutations: comparison to sporadic Parkinson's disease and evidence for presymptomatic compensation. Brain. 2005;128(Pt 12):2777–85.
9. Nandhagopal R, Mak E, Schulzer M, McKenzie J, McCormick S, Sossi V, et al. Progression of dopaminergic dysfunction in a LRRK2 kindred: a multitracer PET study. Neurology. 2008;71(22):1790–5.
10. Sossi V, de la Fuente-Fernandez R, Schulzer M, Adams J, Stoessl J. Age-related differences in levodopa dynamics in Parkinson's: implications for motor complications. Brain. 2006;129(Pt 4):1050–8.
11. de la Fuente-Fernandez R, Sossi V, McCormick S, Schulzer M, Ruth TJ, Stoessl AJ. Visualizing vesicular dopamine dynamics in Parkinson's disease. Synapse. 2009;63(8):713–6.
12. Okamura N, Villemagne VL, Drago J, Pejoska S, Dhamija RK, Mulligan RS, et al. In vivo measurement of vesicular monoamine transporter type 2 density in Parkinson disease with (18)F-AV-133. J Nucl Med. 2010;51(2):223–8.
13. Nandhagopal R, Kuramoto L, Schulzer M, Mak E, Cragg J, Lee CS, et al. Longitudinal progression of sporadic Parkinson's disease: a multi-tracer positron emission tomography study. Brain. 2009;132(Pt 11):2970–9.
14. Lin SC, Lin KJ, Hsiao IT, Hsieh CJ, Lin WY, Lu CS, et al. In vivo detection of monoaminergic degeneration in early Parkinson disease by (18)F-9-fluoropropyl-(+)-dihydrotetrabenzazine PET. J Nucl Med. 2014;55(1):73–9.
15. Liu XL, Liu SY, Barret O, Tamagnan GD, Qiao HW, Song TB, et al. Diagnostic value of striatal (18)F-FP-DTBZ PET in Parkinson's disease. Front Aging Neurosci. 2022;14:931015.
16. Akdemir UO, Bora Tokcaer A, Atay LO. Dopamine transporter SPECT imaging in Parkinson's disease and parkinsonian disorders. Turk J Med Sci. 2021;51(2):400–10.
17. Biondetti E, Santin MD, Valabregue R, Mangone G, Gaurav R, Pyatigorskaya N, et al. The spatiotemporal changes in dopamine, neuromelanin and iron characterizing Parkinson's disease. Brain. 2021;144(10):3114–25.
18. Sprenger FS, Seppi K, Djamshidian A, Reiter E, Nocker M, Mair K, et al. Nonmotor symptoms in subjects without evidence of dopaminergic deficits. Mov Dis. 2015;30(7):976–81.
19. Conrado DJ, Nicholas T, Tsai K, Macha S, Sinha V, Stone J, et al. Dopamine transporter neuroimaging as an enrichment biomarker in early Parkinson's disease clinical trials: a disease progression modeling analysis. Clin Transl Sci. 2018;11(1):63–70.
20. Siderowf A, Concha-Marambio L, Lafontant DE, Farris CM, Ma Y, Urenia PA, et al. Assessment of heterogeneity among participants in the Parkinson's Progression Markers Initiative cohort using alpha-synuclein seed amplification: a cross-sectional study. Lancet Neurol. 2023;22(5):407–17.
21. Normandin MD, Schiffer WK, Morris ED. A linear model for estimation of neurotransmitter response profiles from dynamic PET data. NeuroImage. 2012;59(3):2689–99.
22. Bevington CW, Hanania JU, Ferraresso G, Cheng JK, Pavel A, Su D, et al. Novel voxel-wise residual analysis of [(11)C]raclopride PET data improves detection of low-amplitude dopamine release. J Cereb Blood Flow Metab. 2023;44:271678X231214823.
23. Zimmer ER, Parent MJ, Souza DG, Leuzy A, Lecrux C, Kim HI, et al. [(18)F]FDG PET signal is driven by astroglial glutamate transport. Nat Neurosci. 2017;20(3):393–5.
24. Stoessl AJ. Glucose utilization: still in the synapse. Nat Neurosci. 2017;20(3):382–4.
25. Peng S, Tang C, Schindlbeck K, Rydzinski Y, Dhawan V, Spetsieris PG, et al. Dynamic (18)F-FPCIT PET: quantification of Parkinson's disease metabolic networks and nigrostriatal dopaminergic dysfunction in a single imaging session. J Nucl Med. 2021;62(12):1775–82.
26. Poston KL, Eckert T, Dhawan V, Frucht S, Vonsattel J-P, Fahn S, Eidelberg D. Network correlates of disease severity in multiple system atrophy. Neurology. 2012;78:1237.

27. Ge J, Wu J, Peng S, Wu P, Wang J, Zhang H, et al. Reproducible network and regional topographies of abnormal glucose metabolism associated with progressive supranuclear palsy: multivariate and univariate analyses in American and Chinese patient cohorts. Hum Brain Mapp. 2018;39(7):2842–58.
28. Perovnik M, Tomse P, Jamsek J, Tang C, Eidelberg D, Trost M. Metabolic brain pattern in dementia with Lewy bodies: relationship to Alzheimer's disease topography. Neuroimage Clin. 2022;35:103080.
29. Walker Z, Gandolfo F, Orini S, Garibotto V, Agosta F, Arbizu J, et al. Clinical utility of FDG PET in Parkinson's disease and atypical parkinsonism associated with dementia. Eur J Nucl Med Mol Imaging. 2018;45(9):1534–45.
30. Orimo S, Suzuki M, Inaba A, Mizusawa H. 123I-MIBG myocardial scintigraphy for differentiating Parkinson's disease from other neurodegenerative parkinsonism: a systematic review and meta-analysis. Parkinsonism Relat Disord. 2012;18(5):494–500.
31. Tsujikawa K, Hasegawa Y, Yokoi S, Yasui K, Nanbu I, Yanagi T, et al. Chronological changes of123I-MIBG myocardial scintigraphy and clinical features of Parkinson's disease. J Neurol Neurosurg Psychiatry. 2015;86(9):945–51.
32. Arnaldi D, Fama F, De Carli F, Morbelli S, Ferrara M, Picco A, et al. The role of the serotonergic system in REM sleep behavior disorder. Sleep. 2015;38(9):1505–9.
33. Iranzo A, Lomena F, Stockner H, Valldeoriola F, Vilaseca I, Salamero M, et al. Decreased striatal dopamine transporter uptake and substantia nigra hyperechogenicity as risk markers of synucleinopathy in patients with idiopathic rapid-eye-movement sleep behaviour disorder: a prospective study [corrected]. Lancet Neurol. 2010;9(11):1070–7.
34. Kim YK, Yoon IY, Kim JM, Jeong SH, Kim KW, Shin YK, et al. The implication of nigrostriatal dopaminergic degeneration in the pathogenesis of REM sleep behavior disorder. Eur J Neurol. 2010;17(3):487–92.
35. Arnaldi D, De Carli F, Picco A, Ferrara M, Accardo J, Bossert I, et al. Nigro-caudate dopaminergic deafferentation: a marker of REM sleep behavior disorder? Neurobiol Aging. 2015;36(12):3300–5.
36. Iranzo A, Valldeoriola F, Lomena F, Molinuevo JL, Serradell M, Salamero M, et al. Serial dopamine transporter imaging of nigrostriatal function in patients with idiopathic rapid-eye-movement sleep behaviour disorder: a prospective study. Lancet Neurol. 2011;10(9):797–805.
37. Albin RL, Koeppe RA, Chervin RD, Consens FB, Wernette K, Frey KA, et al. Decreased striatal dopaminergic innervation in REM sleep behavior disorder. Neurology. 2000;55(9):1410–2.
38. Knudsen K, Fedorova TD, Horsager J, Andersen KB, Skjaerbaek C, Berg D, et al. Asymmetric dopaminergic dysfunction in brain-first versus body-first Parkinson's disease subtypes. J Parkinsons Dis. 2021;11(4):1677–87.
39. Andersen KB, Hansen AK, Sommerauer M, Fedorova TD, Knudsen K, Vang K, et al. Altered sensorimotor cortex noradrenergic function in idiopathic REM sleep behaviour disorder – A PET study. Parkinsonism Relat Disord. 2020;75:63–9.
40. Holtbernd F, Gagnon JF, Postuma RB, Ma Y, Tang CC, Feigin A, et al. Abnormal metabolic network activity in REM sleep behavior disorder. Neurology. 2014;82(7):620–7.
41. Wu P, Yu H, Peng S, Dauvilliers Y, Wang J, Ge J, et al. Consistent abnormalities in metabolic network activity in idiopathic rapid eye movement sleep behaviour disorder. Brain. 2014;137(Pt 12):3122–8.
42. Knudsen K, Fedorova TD, Hansen AK, Sommerauer M, Otto M, Svendsen KB, et al. In-vivo staging of pathology in REM sleep behaviour disorder: a multimodality imaging case-control study. Lancet Neurol. 2018;17(7):618–28.
43. Borghammer P. The brain-first vs. body-first model of Parkinson's disease with comparison to alternative models. J Neural Transm (Vienna). 2023;130(6):737–53.
44. Janzen A, Vadasz D, Booij J, Luster M, Librizzi D, Henrich MT, et al. Progressive olfactory impairment and cardiac sympathetic denervation in REM sleep behavior disorder. J Parkinsons Dis. 2022;12(6):1921–35.

45. Fedorova TD, Knudsen K, Andersen KB, Horsager J, Skjaerbaek C, Beier CP, et al. Imaging progressive peripheral and central dysfunction in isolated REM sleep behaviour disorder after 3 years of follow-up. Parkinsonism Relat Disord. 2022;101:99–104.
46. Binkofski F, Reetz K, Gaser C, Hilker R, Hagenah J, Hedrich K, et al. Morphometric fingerprint of asymptomatic Parkin and PINK1 mutation carriers in the basal ganglia. Neurology. 2007;69(9):842–50.
47. Pavese N, Moore RY, Scherfler C, Khan NL, Hotton G, Quinn NP, et al. In vivo assessment of brain monoamine systems in parkin gene carriers: a PET study. Exp Neurol. 2010;222(1):120–4.
48. Ribeiro MJ, Thobois S, Lohmann E, du Montcel ST, Lesage S, Pelissolo A, et al. A multitracer dopaminergic PET study of young-onset parkinsonian patients with and without parkin gene mutations. J Nucl Med. 2009;50(8):1244–50.
49. Thobois S, Ribeiro M-J, Lohmann E, Dürr A, Pollak P, Rascol O, Guillouet S, Chapoy E, Costes N, Agid Y, Remy P, Brice A, Broussolle E. Young-onset Parkinson disease with and without Parkin gene mutations. Arch Neurol. 2003;60:173.
50. Eggers C, Schmidt A, Hagenah J, Bruggemann N, Klein JC, Tadic V, et al. Progression of subtle motor signs in PINK1 mutation carriers with mild dopaminergic deficit. Neurology. 2010;74(22):1798–805.
51. McNeill A, Wu RM, Tzen KY, Aguiar PC, Arbelo JM, Barone P, et al. Dopaminergic neuronal imaging in genetic Parkinson's disease: insights into pathogenesis. PLoS One. 2013;8(7):e69190.
52. Barrett MJ, Hagenah J, Dhawan V, Peng S, Stanley K, Raymond D, et al. Transcranial sonography and functional imaging in glucocerebrosidase mutation Parkinson disease. Parkinsonism Relat Disord. 2013;19(2):186–91.
53. Saunders-Pullman R, Hagenah J, Dhawan V, Stanley K, Pastores G, Sathe S, et al. Gaucher disease ascertained through a Parkinson's center: imaging and clinical characterization. Mov Disord. 2010;25(10):1364–72.
54. Lopez G, Eisenberg DP, Gregory MD, Ianni AM, Grogans SE, Masdeu JC, et al. Longitudinal positron emission tomography of dopamine synthesis in subjects with GBA1 mutations. Ann Neurol. 2020;87(4):652–7.
55. Sossi V, de la Fuente-Fernandez R, Nandhagopal R, Schulzer M, McKenzie J, Ruth TJ, et al. Dopamine turnover increases in asymptomatic LRRK2 mutations carriers. Mov Dis. 2010;25(16):2717–23.
56. Wile DJ, Agarwal PA, Schulzer M, Mak E, Dinelle K, Shahinfard E, et al. Serotonin and dopamine transporter PET changes in the premotor phase of LRRK2 parkinsonism: cross-sectional studies. Lancet Neurol. 2017;16(5):351–9.
57. Liu SY, Wile DJ, Fu JF, Valerio J, Shahinfard E, McCormick S, et al. The effect of LRRK2 mutations on the cholinergic system in manifest and premanifest stages of Parkinson's disease: a cross-sectional PET study. Lancet Neurol. 2018;17(4):309–16.
58. Saunders-Pullman R, Alcalay RN, Mirelman A, Wang C, Luciano MS, Ortega RA, et al. REM sleep behavior disorder, as assessed by questionnaire, in G2019S LRRK2 mutation PD and carriers. Mov Dis. 2015;30(13):1834–9.
59. Alcalay RN, Mejia-Santana H, Mirelman A, Saunders-Pullman R, Raymond D, Palmese C, et al. Neuropsychological performance in LRRK2 G2019S carriers with Parkinson's disease. Parkinsonism Relat Disord. 2015;21(2):106–10.
60. Kumar A, Mann S, Sossi V, Ruth TJ, Stoessl AJ, Schulzer M, et al. [11C]DTBZ-PET correlates of levodopa responses in asymmetric Parkinson's disease. Brain. 2003;126(Pt 12):2648–55.
61. Labrador-Espinosa MA, Grothe MJ, Macias-Garcia D, Jesus S, Adarmes-Gomez A, Munoz-Delgado L, et al. Levodopa-induced dyskinesia in Parkinson disease specifically associates with dopaminergic depletion in sensorimotor-related functional subregions of the striatum. Clin Nucl Med. 2021;46(6):e296–306.
62. Jeong EH, Sunwoo MK, Song YS. Serial I-123-FP-CIT SPECT image findings of Parkinson's disease patients with levodopa-induced dyskinesia. Front Neurol. 2018;9:1133.

63. Hong JY, Oh JS, Lee I, Sunwoo MK, Ham JH, Lee JE, et al. Presynaptic dopamine depletion predicts levodopa-induced dyskinesia in de novo Parkinson disease. Neurology. 2014;82(18):1597–604.
64. Chung SJ, Lee HS, Yoo HS, Lee YH, Lee PH, Sohn YH. Patterns of striatal dopamine depletion in early Parkinson disease: prognostic relevance. Neurology. 2020;95(3):e280–e90.
65. Chung SJ, Yoo HS, Lee HS, Oh JS, Kim JS, Sohn YH, et al. The pattern of striatal dopamine depletion as a prognostic marker in De Novo Parkinson disease. Clin Nucl Med. 2018;43(11):787–92.
66. Troiano AR, de la Fuente-Fernandez R, Sossi V, Schulzer M, Mak E, Ruth TJ, et al. PET demonstrates reduced dopamine transporter expression in PD with dyskinesias. Neurology. 2009;72(14):1211–6.
67. Payer DE, Guttman M, Kish SJ, Tong J, Adams JR, Rusjan P, et al. D3 dopamine receptor-preferring [11C]PHNO PET imaging in Parkinson patients with dyskinesia. Neurology. 2016;86(3):224–30.
68. de la Fuente-Fernandez R, Sossi V, Huang Z, Furtado S, Lu JQ, Calne DB, et al. Levodopa-induced changes in synaptic dopamine levels increase with progression of Parkinson's disease: implications for dyskinesias. Brain. 2004;127(Pt 12):2747–54.
69. Politis M, Wu K, Loane C, Brooks DJ, Kiferle L, Turkheimer FE, et al. Serotonergic mechanisms responsible for levodopa-induced dyskinesias in Parkinson's disease patients. J Clin Invest. 2014;124(3):1340–9.
70. Smith R, Wu K, Hart T, Loane C, Brooks DJ, Bjorklund A, et al. The role of pallidal serotonergic function in Parkinson's disease dyskinesias: a positron emission tomography study. Neurobiol Aging. 2015;36(4):1736–42.
71. Roussakis AA, Politis M, Towey D, Piccini P. Serotonin-to-dopamine transporter ratios in Parkinson disease: relevance for dyskinesias. Neurology. 2016;86(12):1152–8.
72. Carta M, Carlsson T, Kirik D, Bjorklund A. Dopamine released from 5-HT terminals is the cause of L-DOPA-induced dyskinesia in parkinsonian rats. Brain. 2007;130(Pt 7):1819–33.
73. Fu JF, Matarazzo M, McKenzie J, Neilson N, Vafai N, Dinelle K, et al. Serotonergic system impacts levodopa response in early Parkinson's and future risk of dyskinesia. Mov Dis. 2021;36(2):389–97.
74. Navalpotro-Gomez I, Dacosta-Aguayo R, Molinet-Dronda F, Martin-Bastida A, Botas-Penin A, Jimenez-Urbieta H, et al. Nigrostriatal dopamine transporter availability, and its metabolic and clinical correlates in Parkinson's disease patients with impulse control disorders. Eur J Nucl Med Mol Imaging. 2019;46(10):2065–76.
75. Smith KM, Xie SX, Weintraub D. Incident impulse control disorder symptoms and dopamine transporter imaging in Parkinson disease. J Neurol Neurosurg Psychiatry. 2016;87(8):864–70.
76. Vriend C, Nordbeck AH, Booij J, van der Werf YD, Pattij T, Voorn P, et al. Reduced dopamine transporter binding predates impulse control disorders in Parkinson's disease. Mov Dis. 2014;29(7):904–11.
77. Cilia R, Ko JH, Cho SS, van Eimeren T, Marotta G, Pellecchia G, et al. Reduced dopamine transporter density in the ventral striatum of patients with Parkinson's disease and pathological gambling. Neurobiol Dis. 2010;39(1):98–104.
78. Payer DE, Guttman M, Kish SJ, Tong J, Strafella A, Zack M, et al. [(1)(1)C]-(+)-PHNO PET imaging of dopamine D(2/3) receptors in Parkinson's disease with impulse control disorders. Mov Dis. 2015;30(2):160–6.
79. Pagano G, Molloy S, Bain PG, Rabiner EA, Ray Chaudhuri K, Brooks DJ, et al. Impulse control disorders are associated with lower ventral striatum dopamine D3 receptor availability in Parkinson's disease: A [(11)C]-PHNO PET study. Parkinsonism Relat Disord. 2021;90:52–6.
80. Qamhawi Z, Towey D, Shah B, Pagano G, Seibyl J, Marek K, et al. Clinical correlates of raphe serotonergic dysfunction in early Parkinson's disease. Brain. 2015;138(Pt 10):2964–73.

81. Boileau I, Warsh JJ, Guttman M, Saint-Cyr JA, McCluskey T, Rusjan P, et al. Elevated serotonin transporter binding in depressed patients with Parkinson's disease: a preliminary PET study with [11C]DASB. Mov Dis. 2008;23(12):1776–80.
82. Politis M, Wu K, Loane C, Turkheimer FE, Molloy S, Brooks DJ, et al. Depressive symptoms in PD correlate with higher 5-HTT binding in raphe and limbic structures. Neurology. 2010;75(21):1920–7.
83. Guttman M, Boileau I, Warsh J, Saint-Cyr JA, Ginovart N, McCluskey T, et al. Brain serotonin transporter binding in non-depressed patients with Parkinson's disease. Eur J Neurol. 2007;14(5):523–8.
84. Ballanger B, Klinger H, Eche J, Lerond J, Vallet AE, Le Bars D, et al. Role of serotonergic 1A receptor dysfunction in depression associated with Parkinson's disease. Mov Dis. 2012;27(1):84–9.
85. Politis M, Wu K, Loane C, Quinn NP, Brooks DJ, Oertel WH, et al. Serotonin neuron loss and nonmotor symptoms continue in Parkinson's patients treated with dopamine grafts. Sci Transl Med. 2012;4(128):128ra41.
86. Maillet A, Krack P, Lhommee E, Metereau E, Klinger H, Favre E, et al. The prominent role of serotonergic degeneration in apathy, anxiety and depression in de novo Parkinson's disease. Brain. 2016;139(Pt 9):2486–502.
87. Rektorova I, Srovnalova H, Kubikova R, Prasek J. Striatal dopamine transporter imaging correlates with depressive symptoms and tower of London task performance in Parkinson's disease. Mov Dis. 2008;23(11):1580–7.
88. Weintraub D, Mavandadi S, Mamikonyan E, Siderowf AD, Duda JE, Hurtig HI, et al. Atomoxetine for depression and other neuropsychiatric symptoms in Parkinson disease. Neurology. 2010;75(5):448–55.
89. Doppler CEJ, Smit JAM, Hommelsen M, Seger A, Horsager J, Kinnerup MB, et al. Microsleep disturbances are associated with noradrenergic dysfunction in Parkinson's disease. Sleep. 2021;44(8)
90. Sommerauer M, Fedorova TD, Hansen AK, Knudsen K, Otto M, Jeppesen J, et al. Evaluation of the noradrenergic system in Parkinson's disease: an 11C-MeNER PET and neuromelanin MRI study. Brain. 2018;141(2):496–504.
91. Hilker R, Thomas AV, Klein JC, Weisenbach S, Kalbe E, Burghaus L, et al. Dementia in Parkinson disease: functional imaging of cholinergic and dopaminergic pathways. Neurology. 2005;65(11):1716–22.
92. Bohnen NI, Kaufer DI, Hendrickson R, Ivanco LS, Lopresti BJ, Constantine GM, et al. Cognitive correlates of cortical cholinergic denervation in Parkinson's disease and parkinsonian dementia. J Neurol. 2006;253(2):242–7.
93. Klein JC, Eggers C, Kalbe E, Weisenbach S, Hohmann C, Vollmar S, et al. Neurotransmitter changes in dementia with Lewy bodies and Parkinson disease dementia in vivo. Neurology. 2010;74(11):885–92.
94. Kuhl DE, Minoshima S, Fessler JA, Frey KA, Foster NL, Ficaro EP, et al. In vivo mapping of cholinergic terminals in normal aging, Alzheimer's disease, and Parkinson's disease. Ann Neurol. 1996;40(3):399–410.
95. Bohnen NI, Kaufer DI, Ivanco LS, Lopresti B, Koeppe RA, Davis JG, et al. Cortical cholinergic function is more severely affected in parkinsonian dementia than in Alzheimer disease: an in vivo positron emission tomographic study. Arch Neurol. 2003;60(12):1745–8.
96. Perry EK, Irving D, Kerwin JM, McKeith IG, Thompson P, Collerton D, et al. Cholinergic transmitter and neurotrophic activities in Lewy body dementia: similarity to Parkinson's and distinction from Alzheimer disease. Alzheimer Dis Assoc Disord. 1993;7(2):69–79.
97. Shiozaki K, Iseki E, Uchiyama H, Watanabe Y, Haga T, Kameyama K, et al. Alterations of muscarinic acetylcholine receptor subtypes in diffuse lewy body disease: relation to Alzheimer's disease. J Neurol Neurosurg Psychiatry. 1999;67(2):209–13.
98. Colloby SJ, Pakrasi S, Firbank MJ, Perry EK, Piggott MA, Owens J, et al. In vivo SPECT imaging of muscarinic acetylcholine receptors using (R,R) 123I-QNB in dementia with Lewy bodies and Parkinson's disease dementia. NeuroImage. 2006;33(2):423–9.

99. Meyer PM, Strecker K, Kendziorra K, Becker G, Hesse S, Woelpl D, et al. Reduced alpha4beta2*-nicotinic acetylcholine receptor binding and its relationship to mild cognitive and depressive symptoms in Parkinson disease. Arch Gen Psychiatry. 2009;66(8):866–77.
100. Bohnen NI, Muller ML, Koeppe RA, Studenski SA, Kilbourn MA, Frey KA, et al. History of falls in Parkinson disease is associated with reduced cholinergic activity. Neurology. 2009;73(20):1670–6.
101. Bohnen NI, Kanel P, Zhou Z, Koeppe RA, Frey KA, Dauer WT, et al. Cholinergic system changes of falls and freezing of gait in Parkinson's disease. Ann Neurol. 2019;85(4):538–49.
102. Muller ML, Albin RL, Kotagal V, Koeppe RA, Scott PJ, Frey KA, et al. Thalamic cholinergic innervation and postural sensory integration function in Parkinson's disease. Brain. 2013;136(Pt 11):3282–9.
103. Snijders AH, Takakusaki K, Debu B, Lozano AM, Krishna V, Fasano A, et al. Physiology of freezing of gait. Ann Neurol. 2016;80(5):644–59.
104. Bohnen NI, Frey KA, Studenski S, Kotagal V, Koeppe RA, Constantine GM, et al. Extra-nigral pathological conditions are common in Parkinson's disease with freezing of gait: an in vivo positron emission tomography study. Mov Dis. 2014;29(9):1118–24.
105. Tedroff J, Pedersen M, Aquilonius SM, Hartvig P, Jacobsson G, Langstrom B. Levodopa-induced changes in synaptic dopamine in patients with Parkinson's disease as measured by [11C]raclopride displacement and PET. Neurology. 1996;46(5):1430–6.
106. Pavese N, Evans AH, Tai YF, Hotton G, Brooks DJ, Lees AJ, et al. Clinical correlates of levodopa-induced dopamine release in Parkinson disease: a PET study. Neurology. 2006;67(9):1612–7.
107. Mori A. Mode of action of adenosine A2A receptor antagonists as symptomatic treatment for Parkinson's disease. Int Rev Neurobiol. 2014;119:87–116.
108. Calon F, Dridi M, Hornykiewicz O, Bedard PJ, Rajput AH, Di Paolo T. Increased adenosine A2A receptors in the brain of Parkinson's disease patients with dyskinesias. Brain. 2004;127(Pt 5):1075–84.
109. Mishina M, Ishiwata K, Naganawa M, Kimura Y, Kitamura S, Suzuki M, et al. Adenosine A(2A) receptors measured with [C]TMSX PET in the striata of Parkinson's disease patients. PLoS One. 2011;6(2):e17338.
110. Ramlackhansingh AF, Bose SK, Ahmed I, Turkheimer FE, Pavese N, Brooks DJ. Adenosine 2A receptor availability in dyskinetic and nondyskinetic patients with Parkinson disease. Neurology. 2011;76(21):1811–6.
111. Shiozaki S, Ichikawa S, Nakamura J, Kitamura S, Yamada K, Kuwana Y. Actions of adenosine A2A receptor antagonist KW-6002 on drug-induced catalepsy and hypokinesia caused by reserpine or MPTP. Psychopharmacology. 1999;147(1):90–5.
112. Okita K, Kato K, Shigemoto Y, Sato N, Matsumoto T, Matsuda H. Effects of an adenosine A(2A) receptor antagonist on striatal dopamine D2-type receptor availability: a randomized control study using positron emission tomography. Front Neurosci. 2021;15:729153.
113. Okita K, Matsumoto T, Funada D, Murakami M, Kato K, Shigemoto Y, et al. Potential treat-to-target approach for methamphetamine use disorder: a pilot study of adenosine 2A receptor antagonist with positron emission tomography. Front Pharmacol. 2022;13:820447.
114. Ishibashi K, Miura Y, Wagatsuma K, Toyohara J, Ishiwata K, Ishii K. Occupancy of adenosine A(2A) receptors by istradefylline in patients with Parkinson's disease using (11)C-preladenant PET. Neuropharmacology. 2018;143:106–12.
115. Ishibashi K, Miura Y, Wagatsuma K, Toyohara J, Ishiwata K, Ishii K. Adenosine A(2A) receptor occupancy by long-term istradefylline administration in Parkinson's disease. Mov Dis. 2021;36(1):268–9.
116. Dehay B, Bourdenx M, Gorry P, Przedborski S, Vila M, Hunot S, et al. Targeting alpha-synuclein for treatment of Parkinson's disease: mechanistic and therapeutic considerations. Lancet Neurol. 2015;14(8):855–66.
117. Ross GW, Petrovitch H, Abbott RD, Nelson J, Markesbery W, Davis D, et al. Parkinsonian signs and substantia nigra neuron density in decendents elders without PD. Ann Neurol. 2004;56(4):532–9.

118. Fearnley JM, Lees AJ. Ageing and Parkinson's disease: substantia nigra regional selectivity. Brain. 1991;114(Pt 5):2283–301.
119. Riederer P, Wuketich S. Time course of nigrostriatal degeneration in parkinson's disease. A detailed study of influential factors in human brain amine analysis. J Neural Transm. 1976;38(3–4):277–301.
120. Masliah E, Rockenstein E, Adame A, Alford M, Crews L, Hashimoto M, et al. Effects of alpha-synuclein immunization in a mouse model of Parkinson's disease. Neuron. 2005;46(6):857–68.
121. Wegrzynowicz M, Bar-On D, Calo L, Anichtchik O, Iovino M, Xia J, et al. Depopulation of dense alpha-synuclein aggregates is associated with rescue of dopamine neuron dysfunction and death in a new Parkinson's disease model. Acta Neuropathol. 2019;138(4):575–95.
122. Warren Olanow C, Goetz CG, Kordower JH, Stoessl AJ, Sossi V, Brin MF, Shannon KM, Nauert GM, Perl DP, Godbold J, Freeman TB. A double-blind controlled trial of bilateral fetal nigral transplantation in Parkinson's disease. Ann Neurol. 2003;54:403.
123. Kordower JH, Goetz CG, Chu Y, Halliday GM, Nicholson DA, Musial TF, et al. Robust graft survival and normalized dopaminergic innervation do not obligate recovery in a Parkinson disease patient. Ann Neurol. 2017;81(1):46–57.
124. Espay AJ, Sturchio A, Schneider LS, Ezzat K. Soluble amyloid-beta consumption in Alzheimer's disease. J Alzheimers Dis. 2021;82(4):1403–15.
125. Erro R, Schneider SA, Stamelou M, Quinn NP, Bhatia KP. What do patients with scans without evidence of dopaminergic deficit (SWEDD) have? New evidence and continuing controversies. J Neurol Neurosurg Psychiatry. 2016;87(3):319–23.
126. Connolly BS, Lang AE. Pharmacological treatment of Parkinson disease: a review. JAMA. 2014;311(16):1670–83.
127. Marek K, Seibyl J, Eberly S, Oakes D, Shoulson I, Lang AE, et al. Longitudinal follow-up of SWEDD subjects in the PRECEPT study. Neurology. 2014;82(20):1791–7.
128. Marshall VL, Reininger CB, Marquardt M, Patterson J, Hadley DM, Oertel WH, et al. Parkinson's disease is overdiagnosed clinically at baseline in diagnostically uncertain cases: a 3-year European multicenter study with repeat [123I]FP-CIT SPECT. Move Dis. 2009;24(4):500–8.
129. Lee JW, Song YS, Kim H, Ku BD, Lee WW. Patients with scans without evidence of dopaminergic deficit (SWEDD) do not have early Parkinson's disease: analysis of the PPMI data. PLoS One. 2021;16(2):e0246881.
130. Taylor S, Gafton J, Shah B, Pagano G, Chaudhuri KR, Brooks DJ, et al. Progression of nonmotor symptoms in subgroups of patients with non-dopamine-deficient parkinsonism. Mov Dis. 2016;31(3):344–51.
131. Rahmim A, Salimpour Y, Jain S, Blinder SA, Klyuzhin IS, Smith GS, et al. Application of texture analysis to DAT SPECT imaging: relationship to clinical assessments. Neuroimage Clin. 2016;12:e1–9.
132. Simuni T, Siderowf A, Lasch S, Coffey CS, Caspell-Garcia C, Jennings D, et al. Longitudinal change of clinical and biological measures in early Parkinson's disease: Parkinson's progression markers initiative cohort. Mov Dis. 2018;33(5):771–82.
133. Paul G, Zachrisson O, Varrone A, Almqvist P, Jerling M, Lind G, et al. Safety and tolerability of intracerebroventricular PDGF-BB in Parkinson's disease patients. J Clin Invest. 2015;125(3):1339–46.
134. Lin CH, Chang CH, Tai CH, Cheng MF, Chen YC, Chao YT, et al. A double-blind, randomized, controlled trial of lovastatin in early-stage Parkinson's disease. Mov Dis. 2021;36(5):1229–37.
135. Arnaldi D, Campus C, Ferrara M, Fama F, Picco A, De Carli F, et al. What predicts cognitive decline in de novo Parkinson's disease? Neurobiol Aging. 2012;33(6):1127 e11–20.
136. Fiorenzato E, Antonini A, Bisiacchi P, Weis L, Biundo R. Asymmetric dopamine transporter loss affects cognitive and motor progression in Parkinson's disease. Mov Dis. 2021;36(10):2303–13.
137. Yousaf T, Pagano G, Niccolini F, Politis M. Predicting cognitive decline with non-clinical markers in Parkinson's disease (PRECODE-2). J Neurol. 2019;266(5):1203–10.

138. Firbank MJ, Yarnall AJ, Lawson RA, Duncan GW, Khoo TK, Petrides GS, et al. Cerebral glucose metabolism and cognition in newly diagnosed Parkinson's disease: ICICLE-PD study. J Neurol Neurosurg Psychiatry. 2017;88(4):310–6.
139. Booth S, Park KW, Lee CS, Ko JH. Predicting cognitive decline in Parkinson's disease using FDG-PET-based supervised learning. J Clin Invest. 2022;132(20)
140. Hansen AK, Parbo P, Ismail R, Ostergaard K, Brooks DJ, Borghammer P. Tau tangles in Parkinson's disease: A 2-year follow-up flortaucipir PET study. J Parkinsons Dis. 2020;10(1):161–71.
141. Myers PS, O'Donnell JL, Jackson JJ, Lessov-Schlaggar CN, Miller RL, Foster ER, et al. Proteinopathy and longitudinal cognitive decline in Parkinson disease. Neurology. 2022;99(1):e66–76.
142. Guttman M, Stewart D, Hussey D, Wilson A, Houle S, Kish S. Influence of L-dopa and pramipexole on striatal dopamine transporter in early PD. Neurology. 2001;56(11):1559–64.
143. Whone AL, Watts RL, Stoessl AJ, Davis M, Reske S, Nahmias C, et al. Slower progression of Parkinson's disease with ropinirole versus levodopa: the REAL-PET study. Ann Neurol. 2003;54(1):93–101.
144. Fahn S, Oakes D, Shoulson I, Kieburtz K, Rudolph A, Lang A, et al. Levodopa and the progression of Parkinson's disease. N Engl J Med. 2004;351(24):2498–508.
145. Nakamura T, Dhawan V, Chaly T, Fukuda M, Ma Y, Breeze R, et al. Blinded positron emission tomography study of dopamine cell implantation for Parkinson's disease. Ann Neurol. 2001;50(2):181–7.
146. Ma Y, Tang C, Chaly T, Greene P, Breeze R, Fahn S, et al. Dopamine cell implantation in Parkinson's disease: long-term clinical and (18)F-FDOPA PET outcomes. J Nucl Med. 2010;51(1):7–15.
147. Lang AE, Siderowf AD, Macklin EA, Poewe W, Brooks DJ, Fernandez HH, et al. Trial of cinpanemab in early Parkinson's disease. N Engl J Med. 2022;387(5):408–20.
148. Pagano G, Taylor KI, Anzures-Cabrera J, Marchesi M, Simuni T, Marek K, et al. Trial of prasinezumab in early-stage Parkinson's disease. N Engl J Med. 2022;387(5):421–32.
149. Whone A, Luz M, Boca M, Woolley M, Mooney L, Dharia S, et al. Randomized trial of intermittent intraputamenal glial cell line-derived neurotrophic factor in Parkinson's disease. Brain. 2019;142(3):512–25.
150. Devos D, Labreuche J, Rascol O, Corvol JC, Duhamel A, Guyon Delannoy P, et al. Trial of deferiprone in Parkinson's disease. N Engl J Med. 2022;387(22):2045–55.
151. Liu C, Jiang J, Zhou H, Zhang H, Wang M, Jiang J, et al. Brain functional and structural signatures in Parkinson's disease. Front Aging Neurosci. 2020;12:125.
152. Meles SK, Oertel WH, Leenders KL. Circuit imaging biomarkers in preclinical and prodromal Parkinson's disease. Mol Med. 2021;27(1):111.
153. Niethammer M, Eidelberg D. Metabolic brain networks in translational neurology: concepts and applications. Ann Neurol. 2012;72(5):635–47.
154. Schindlbeck KA, Eidelberg D. Network imaging biomarkers: insights and clinical applications in Parkinson's disease. Lancet Neurol. 2018;17(7):629–40.
155. Perovnik M, Rus T, Schindlbeck KA, Eidelberg D. Functional brain networks in the evaluation of patients with neurodegenerative disorders. Nat Rev Neurol. 2023;19(2):73–90.
156. Peralta C, Biafore F, Depetris TS, Bastianello M. Recent advancement and clinical implications of 18FDG-PET in Parkinson's disease, atypical Parkinsonisms, and other movement disorders. Curr Neurol Neurosci Rep. 2019;19(8):56.
157. Buongiorno M, Antonelli F, Compta Y, Fernandez Y, Pavia J, Lomena F, et al. Cross-sectional and longitudinal cognitive correlates of FDDNP PET and CSF amyloid-beta and tau in Parkinson's disease1. J Alzheimers Dis. 2017;55(3):1261–72.
158. Marquie M, Normandin MD, Vanderburg CR, Costantino IM, Bien EA, Rycyna LG, et al. Validating novel tau positron emission tomography tracer [F-18]-AV-1451 (T807) on postmortem brain tissue. Ann Neurol. 2015;78(5):787–800.
159. Sander K, Lashley T, Gami P, Gendron T, Lythgoe MF, Rohrer JD, et al. Characterization of tau positron emission tomography tracer [(18)F]AV-1451 binding to postmortem tissue

in Alzheimer's disease, primary tauopathies, and other dementias. Alzheimers Dement. 2016;12(11):1116–24.

160. Coakeley S, Cho SS, Koshimori Y, Rusjan P, Ghadery C, Kim J, et al. [(18)F]AV-1451 binding to neuromelanin in the substantia nigra in PD and PSP. Brain Struct Funct. 2018;223(2):589–95.
161. Smith R, Scholl M, Londos E, Ohlsson T, Hansson O. (18)F-AV-1451 in Parkinson's Disease with and without dementia and in Dementia with Lewy Bodies. Sci Rep. 2018;8(1):4717.
162. Gomperts SN, Locascio JJ, Makaretz SJ, Schultz A, Caso C, Vasdev N, et al. Tau positron emission tomographic imaging in the Lewy Body diseases. JAMA Neurol. 2016;73(11):1334–41.
163. Hansen AK, Damholdt MF, Fedorova TD, Knudsen K, Parbo P, Ismail R, et al. In vivo cortical tau in Parkinson's disease using 18F-AV-1451 positron emission tomography. Mov Dis. 2017;32(6):922–7.
164. Wang YT, Edison P. Tau imaging in neurodegenerative diseases using positron emission tomography. Curr Neurol Neurosci Rep. 2019;19(7):45.
165. Schonecker S, Brendel M, Palleis C, Beyer L, Hoglinger GU, Schuh E, et al. PET imaging of astrogliosis and tau facilitates diagnosis of parkinsonian syndromes. Front Aging Neurosci. 2019;11:249.
166. Li L, Liu FT, Li M, Lu JY, Sun YM, Liang X, et al. Clinical utility of (18) F-APN-1607 tau PET imaging in patients with progressive supranuclear palsy. Mov Disord. 2021;36(10):2314–23.
167. Zhang J, Jin J, Su D, Feng T, Zhao H. Tau-PET imaging in Parkinson's disease: a systematic review and meta-analysis. Front Neurol. 2023;14:1145939.
168. Owen DR, Yeo AJ, Gunn RN, Song K, Wadsworth G, Lewis A, et al. An 18-kDa translocator protein (TSPO) polymorphism explains differences in binding affinity of the PET radioligand PBR28. J Cereb Blood Flow Metab. 2012;32(1):1–5.
169. Liu SY, Qiao HW, Song TB, Liu XL, Yao YX, Zhao CS, et al. Brain microglia activation and peripheral adaptive immunity in Parkinson's disease: a multimodal PET study. J Neuroinflammation. 2022;19(1):209.
170. Zhou R, Ji B, Kong Y, Qin L, Ren W, Guan Y, et al. PET imaging of neuroinflammation in Alzheimer's disease. Front Immunol. 2021;12:739130.
171. Lavisse S, Goutal S, Wimberley C, Tonietto M, Bottlaender M, Gervais P, et al. Increased microglial activation in patients with Parkinson disease using [(18)F]-DPA714 TSPO PET imaging. Parkinsonism Relat Disord. 2021;82:29–36.
172. Yacoubian TA, Fang YD, Gerstenecker A, Amara A, Stover N, Ruffrage L, et al. Brain and systemic inflammation in De Novo Parkinson's disease. Mov Dis. 2023;
173. Varnas K, Cselenyi Z, Jucaite A, Halldin C, Svenningsson P, Farde L, et al. PET imaging of [(11)C]PBR28 in Parkinson's disease patients does not indicate increased binding to TSPO despite reduced dopamine transporter binding. Eur J Nucl Med Mol Imaging. 2019;46(2):367–75.
174. Jucaite A, Cselenyi Z, Kreisl WC, Rabiner EA, Varrone A, Carson RE, et al. Glia imaging differentiates multiple system atrophy from Parkinson's disease: a positron emission tomography study with [(11) C]PBR28 and machine learning analysis. Mov Dis. 2022;37(1):119–29.
175. Finnema SJ, Nabulsi NB, Eid T, Detyniecki K, Lin S-F, Chen M-K, et al. Imaging synaptic density in the living human brain. Sci Transl Med. 2016;8(348):348ra96.
176. Estrada S, Lubberink M, Thibblin A, Sprycha M, Buchanan T, Mestdagh N, et al. [(11)C]UCB-A, a novel PET tracer for synaptic vesicle protein 2A. Nucl Med Biol. 2016;43(6):325–32.
177. Bretin F, Warnock G, Bahri MA, Aerts J, Mestdagh N, Buchanan T, et al. Preclinical radiation dosimetry for the novel SV2A radiotracer [18F]UCB-H. EJNMMI Res. 2013;3(1):35.
178. Li S, Cai Z, Zhang W, Holden D, Lin S-F, Finnema SJ, et al. Synthesis and in vivo evaluation of [18F]UCB-J for PET imaging of synaptic vesicle glycoprotein 2A (SV2A). Eur J Nucl Med Mol Imaging. 2019;46(9):1952–65.

179. Li S, Cai Z, Wu X, Holden D, Pracitto R, Kapinos M, et al. Synthesis and in vivo evaluation of a novel PET radiotracer for imaging of synaptic vesicle glycoprotein 2A (SV2A) in nonhuman primates. ACS Chem Neurosci. 2019;10(3):1544–54.
180. Constantinescu CC, Tresse C, Zheng M, Gouasmat A, Carroll VM, Mistico L, et al. Development and in vivo preclinical imaging of fluorine-18-labeled synaptic vesicle protein 2A (SV2A) PET tracers. Mol Imaging Biol. 2019;21(3):509–18.
181. Naganawa M, Li S, Nabulsi N, Henry S, Zheng M-Q, Pracitto R, et al. First-in-human evaluation of 18F-SynVesT-1, a radioligand for PET imaging of synaptic vesicle glycoprotein 2A. J Nucl Med. 2021;62(4):561–7.
182. Cai Z, Li S, Zhang W, Pracitto R, Wu X, Baum E, et al. Synthesis and preclinical evaluation of an 18F-labeled synaptic vesicle glycoprotein 2A PET imaging probe: [18F]SynVesT-2. ACS Chem Neurosci. 2020;11(4):592–603.
183. Zheng C, Holden D, Zheng M-Q, Pracitto R, Wilcox KC, Lindemann M, et al. A metabolically stable PET tracer for imaging synaptic vesicle protein 2A: synthesis and preclinical characterization of [18F]SDM-16. Eur J Nucl Med Mol Imaging. 2022;49(5):1482–96.
184. Delva A, Van Weehaeghe D, Koole M, Van Laere K, Vandenberghe W. Loss of presynaptic terminal integrity in the substantia Nigra in early Parkinson's disease. Mov Dis. 2020;35(11):1977–86.
185. Matuskey D, Tinaz S, Wilcox KC, Naganawa M, Toyonaga T, Dias M, et al. Synaptic changes in Parkinson disease assessed with in vivo imaging. Ann Neurol. 2020;87(3):329–38.
186. Andersen KB, Hansen AK, Damholdt MF, Horsager J, Skjaerbaek C, Gottrup H, et al. Reduced synaptic density in patients with Lewy body dementia: an [(11) C]UCB-J PET imaging study. Mov Dis. 2021;36(9):2057–65.
187. Wilson H, Pagano G, de Natale ER, Mansur A, Caminiti SP, Polychronis S, et al. Mitochondrial complex 1, sigma 1, and synaptic vesicle 2A in early drug-naive Parkinson's disease. Mov Dis. 2020;35(8):1416–27.
188. Dunn AR, Stout KA, Ozawa M, Lohr KM, Hoffman CA, Bernstein AI, et al. Synaptic vesicle glycoprotein 2C (SV2C) modulates dopamine release and is disrupted in Parkinson disease. Proc Natl Acad Sci USA. 2017;114(11):E2253–E62.
189. Delva A, Van Laere K, Vandenberghe W. Longitudinal positron emission tomography imaging of presynaptic terminals in early Parkinson's disease. Mov Dis. 2022;37(9):1883–92.
190. Christopher G, Goetz SL, Raman R, Stebbins GT. Objective changes in motor function during placebo treatment in PD. Neurology. 2000;
191. de la Fuente-Fernández R, Ruth TJ, Sossi V, Schulzer M, Calne DB, Stoessl AJ. Expectation and dopamine release: mechanism of the placebo effect in Parkinson's disease. Science (New York, NY). 2001;293(5532):1164–6.
192. Lidstone SC, Schulzer M, Dinelle K, Mak E, Sossi V, Ruth TJ, et al. Effects of expectation on placebo-induced dopamine release in Parkinson disease. Arch Gen Psychiatry. 2010;67(8):857–65.
193. Fiorillo CD, Tobler PN, Schultz W. Discrete coding of reward probability and uncertainty by dopamine neurons. Science (New York, NY). 2003;299(5614):1898–902.
194. Strafella AP, Ko JH, Monchi O. Therapeutic application of transcranial magnetic stimulation in Parkinson's disease: the contribution of expectation. NeuroImage. 2006;31(4):1666–72.
195. Ko JH, Feigin A, Mattis PJ, Tang CC, Ma Y, Dhawan V, et al. Network modulation following sham surgery in Parkinson's disease. J Clin Invest. 2014;124(8):3656–66.
196. Kolb B, Whishaw IQ. Brain plasticity and behavior. Annu Rev Psychol. 1998;49:43–64.
197. Whone AL, Moore RY, Piccini PP, Brooks DJ. Plasticity of the nigropallidal pathway in Parkinson's disease. Ann Neurol. 2003;53(2):206–13.
198. Sacheli MA, Neva JL, Lakhani B, Murray DK, Vafai N, Shahinfard E, et al. Exercise increases caudate dopamine release and ventral striatal activation in Parkinson's disease. Mov Dis. 2019;34(12):1891–900.

199. Sacheli MA, Murray DK, Vafai N, Cherkasova MV, Dinelle K, Shahinfard E, et al. Habitual exercisers versus sedentary subjects with Parkinson's disease: multimodal PET and fMRI study. Mov Dis. 2018;33(12):1945–50.
200. Fisher BE, Li Q, Nacca A, Salem GJ, Song J, Yip J, et al. Treadmill exercise elevates striatal dopamine D2 receptor binding potential in patients with early Parkinson's disease. Neuroreport. 2013;24(10):509–14.
201. Alzghool OM, van Dongen G, van de Giessen E, Schoonmade L, Beaino W. Alpha-Synuclein radiotracer development and in vivo imaging: recent advancements and new perspectives. Mov Dis. 2022;37(5):936–48.
202. Xiang J, Tao Y, Xia Y, Luo S, Zhao Q, Li B, et al. Development of an α-synuclein positron emission tomography tracer for imaging synucleinopathies. Cell.
203. Smith R, Capotosti F, Schain M, Ohlsson T, Vokali E, Molette J, et al. The alpha-synuclein PET tracer [18F] ACI-12589 distinguishes multiple system atrophy from other neurodegenerative diseases. Nat Commun. 2023;14(1):6750.
204. Matsuoka K, Ono M, Takado Y, Hirata K, Endo H, Ohfusa T, et al. High-contrast imaging of alpha-synuclein pathologies in living patients with multiple system atrophy. Mov Dis. 2022;37(10):2159–61.
205. Korat S, Bidesi NSR, Bonanno F, Di Nanni A, Hoang ANN, Herfert K, et al. Alpha-synuclein PET tracer development-an overview about current efforts. Pharmaceuticals (Basel). 2021;14(9)
206. Barthel H, Villemagne VL, Drzezga A. Future directions in molecular imaging of neurodegenerative disorders. J Nucl Med. 2022;63(Suppl 1):68S–74S.
207. Haider A, Elghazawy NH, Dawoud A, Gebhard C, Wichmann T, Sippl W, et al. Translational molecular imaging and drug development in Parkinson's disease. Mol Neurodegener. 2023;18(1):11.
208. Kuebler L, Buss S, Leonov A, Ryazanov S, Schmidt F, Maurer A, et al. [(11)C]MODAG-001-towards a PET tracer targeting alpha-synuclein aggregates. Eur J Nucl Med Mol Imaging. 2021;48(6):1759–72.
209. Ni R, Nitsch RM. Recent developments in positron emission tomography tracers for proteinopathies imaging in dementia. Front Aging Neurosci. 2021;13:751897.
210. Mena AM, Strafella AP. Imaging pathological tau in atypical parkinsonisms: a review. Clin Park Relat Disord. 2022;7:100155.
211. Wei P, Wang K, Luo C, Huang Y, Misilimu D, Wen H, et al. Cordycepin confers long-term neuroprotection via inhibiting neutrophil infiltration and neuroinflammation after traumatic brain injury. J Neuroinflammation. 2021;18(1):137.
212. Van Camp N, Lavisse S, Roost P, Gubinelli F, Hillmer A, Boutin H. TSPO imaging in animal models of brain diseases. Eur J Nucl Med Mol Imaging. 2021;49(1):77–109.
213. Chen Z, Haider A, Chen J, Xiao Z, Gobbi L, Honer M, et al. The repertoire of small-molecule PET probes for neuroinflammation imaging: challenges and opportunities beyond TSPO. J Med Chem. 2021;64(24):17656–89.
214. Mills KA, Du Y, Coughlin JM, Horti AG, Jenkins K, Spiro E, et al. Microglial CSF1R radioligand [(11)C]CPPC as a marker of disease severity in early Parkinson's disease: a pilot study. medRxiv. 2023.
215. Lee H, Park JH, Kim H, Woo SK, Choi JY, Lee KH, et al. Synthesis and evaluation of a (18)F-labeled ligand for PET imaging of colony-stimulating factor 1 receptor. Pharmaceuticals (Basel). 2022;15(3)
216. Van Weehaeghe D, Koole M, Schmidt ME, Deman S, Jacobs AH, Souche E, et al. [(11)C]JNJ54173717, a novel P2X7 receptor radioligand as marker for neuroinflammation: human biodistribution, dosimetry, brain kinetic modelling and quantification of brain P2X7 receptors in patients with Parkinson's disease and healthy volunteers. Eur J Nucl Med Mol Imaging. 2019;46(10):2051–64.
217. Schmidt S, Isaak A, Junker A. Spotlight on P2X7 receptor PET imaging: a bright target or a failing star? Int J Mol Sci. 2023;24(2)

218. Wilson H, Dervenoulas G, Pagano G, Tyacke RJ, Polychronis S, Myers J, et al. Imidazoline 2 binding sites reflecting astroglia pathology in Parkinson's disease: an in vivo11C-BU99008 PET study. Brain. 2019;142(10):3116–28.
219. Mohamed MA, Zeng Z, Gennaro M, Lao-Kaim NP, Myers JFM, Calsolaro V, et al. Astrogliosis in aging and Parkinson's disease dementia: a new clinical study with (11)C-BU99008 PET. Brain Commun. 2022;4(5):fcac199.
220. Wadsak W, Mien LK, Ettlinger DE, Eidherr H, Haeusler D, Sindelar KM, et al. 18F fluoroethylations: different strategies for the rapid translation of 11C-methylated radiotracers. Nucl Med Biol. 2007;34(8):1019–28.
221. Shao X, Hockley BG, Hoareau R, Schnau PL, Scott PJ. Fully automated preparation of [11C]choline and [18F]fluoromethylcholine using TracerLab synthesis modules and facilitated quality control using analytical HPLC. Appl Radiat Isot. 2011;69(2):403–9.
222. Li W, Lao-Kaim NP, Roussakis AA, Martin-Bastida A, Valle-Guzman N, Paul G, et al. (11) C-PE2I and (18) F-Dopa PET for assessing progression rate in Parkinson's: a longitudinal study. Mov Dis. 2018;33(1):117–27.
223. Joshi RK, Goud NS, Nagaraj C, Kumar D, Gopinath R, Rao NP, et al. Radiosynthesis challenges of (11)C and (18)F-labeled radiotracers in the FX2C/N tracerlab and their validation through PET-MR imaging. Appl Radiat Isot. 2021;168:109486.
224. Pichler V, Berroteran-Infante N, Philippe C, Vraka C, Klebermass EM, Balber T, et al. An overview of PET radiochemistry, part 1: the covalent labels (18)F, (11)C, and (13)N. J Nucl Med. 2018;59(9):1350–4.
225. Frost GR, Longo V, Li T, Jonas LA, Judenhofer M, Cherry S, et al. Hybrid PET/MRI enables high-spatial resolution, quantitative imaging of amyloid plaques in an Alzheimer's disease mouse model. Sci Rep. 2020;10(1):10379.
226. Sander CY, Hansen HD, Wey HY. Advances in simultaneous PET/MR for imaging neuroreceptor function. J Cereb Blood Flow Metab. 2020;40(6):1148–66.
227. Meikle SR, Sossi V, Roncali E, Cherry SR, Banati R, Mankoff D, et al. Quantitative PET in the 2020s: a roadmap. Phys Med Biol. 2021;66(6):06RM01.

Imaging of Status Epilepticus

Syed M. Owais, Elysia G. James, Imran I. Ali, and Ajaz A. Sheikh

Abstract In this review, we provide an overview of changes due to status epilepticus in various brain imaging modalities. These changes may be of diagnostic, therapeutic, and prognostic importance for patients with status epilepticus.

Keywords Status epilepticus · Magnetic resonance imaging

1 Introduction

1.1 Definitions

Historically, various definitions have been used to recognize status epilepticus (SE). Conceptually, the (generalized, convulsive) status epilepticus refers to a condition in which there is failure of the normal factors that serve to terminate a typical generalized tonic-clonic seizures, leading to prolonged epileptic seizures [1]. Traditionally, >30 min duration of seizure was used to define convulsive status epilepticus [2]. However, with accumulation of more evidence, generalized convulsive status epilepticus (GCSE) in adults and children older than 5 years has been operationally defined as $\geq$5 min of (1) continuous seizure or (2) two or more discrete seizures between which there is incomplete recovery of consciousness [3].

S. M. Owais
Resident Neurology, Department of Neurology, University of Toledo, Neuroscience Center, Toledo, OH, USA
e-mail: syedmuhammad.owais@utoledo.edu

E. G. James · A. A. Sheikh (✉)
Department of Neurology, University of Toledo, Neuroscience Center, Toledo, OH, USA
e-mail: elysia.james2@utoledo.edu; ajaz.sheikh@utoledo.edu

I. I. Ali
Department of Neurology, University of Toledo, College of Medicine, Neuroscience Center, Toledo, OH, USA
e-mail: imran.ali@utoledo.edu

P. A. Broderick (ed.), *Biomedical Imaging in Neurodegeneration*,
https://doi.org/10.1007/978-3-031-95878-6_4

Table 1 Table summarizing t1 and t2 times for various types of SE

Type of SE	t1	t2
Tonic-clonic SE	5 min	30 min
Focal SE with impaired consciousness	10 min	>60 min
Absence SE	10–15 min (based on limited data)	Unknown

Since the above definitions mostly address generalized convulsive status epilepticus, there was a need for a more inclusive approach to define the status epilepticus. International league against epilepsy (ILAE) Task Force on Classification of Status Epilepticus proposed a conceptual definition that addresses all types of SE, taking into consideration current knowledge regarding the pathophysiology of SE as well as the need to address clinical treatment decision-making time points. According to the ILAE definition, "SE is a condition resulting either from the failure of the mechanisms responsible for seizure termination or from the initiation of mechanisms which lead to abnormally prolonged seizures (after time point t1). It is a condition that can have long-term consequences (after time point t2), including neuronal death, neuronal injury, and alteration of neuronal networks, depending on the type and duration of seizures" [4]. This definition introduced the concept of t1 and t2 time points; t1 being defined as time point at which seizure should be considered as abnormally prolonged and t2 as a time point beyond which there is a risk of long-term consequences. The accepted thresholds for these dimensions are presented in Table 1. Although, t2 for nonconvulsive status epilepticus was unknown in 2015, accruing evidence since has suggested that this may be around 10–20% prevalence of epileptiform activity on the continuous EEG (which would equate to 6–12 min of epileptiform activity per hour of EEG recording) [5].

The American Clinical Neurophysiology Society's (ACNS) recent guidelines pertinent to critical care EEG terminology has put forth certain standard definitions for status epilepticus and its various electrographic/electroclinical variants, the definitions of which are summarized below [6]:

1. *Electrographic Status Epilepticus (ESE)*

This is defined as an electrographic seizure lasting ≥10 continuous minutes or for a total duration of ≥20% of any 60-min period of recording. The 10-min conforms to t1 for focal status epilepticus with impaired consciousness according to ILAE [4]. Electrographic seizures (ESz) are defined as presence of epileptiform discharges averaging >2.5 Hz for ≥10 s, or, an electrographic pattern with definite evolution, lasting ≥10 s.

2. *Electroclinical Status Epilepticus (ECSE)*

This is defined as an electroclinical seizure for ≥10 continuous minutes or for a total duration of ≥20% of the 60 min period of recording. Electroclinical seizures (ECSz) are considered present if any EEG pattern is associated with any clinical signs in a time locked fashion to the pattern of any duration, or, if the EEG and clinical picture improve after administration of parenteral antiseizure medication. An exception to the 10 min rule is bilateral tonic-clonic seizure, which only needs to be present for ≥5 min continuously to qualify as ECSE. This is also referred to as convulsive status epilepticus and conforms to the ILAE definition of tonic-clonic status epilepticus as mentioned above. In any other clinical situation, the minimum duration to qualify as SE is ≥10 min.

3. *Nonconvulsive Status Epilepticus (NCSE)*

Any seizures or status epilepticus without prominent motor activity is also referred to as nonconvulsive. This term can be applied to both electrographic and electroclinical seizures. All ESz and ESE alone would be nonconvulsive, however, ECSz or ECSE without prominent motor activity could also be termed nonconvulsive.

4. *Ictal-interictal Continuum (IIC)*

This is synonymous with terms "possible electrographic seizure" or "possible electrographic status epilepticus." The IIC is a purely electrographic term and does not indicate a specific diagnosis. This electrographic pattern does not fulfill the criteria for ESz or ESE, but there is a reasonable chance of it contributing to the patient's impaired awareness or other clinical symptoms, and/or contributing to the neuronal injury. Typically, this includes periodic discharges or sharp waves with a frequency of 1–2.5 Hz, lasting >10 s, or, periodic discharges or sharp waves at 0.5–1 Hz, along with a plus modifier or fluctuation in frequency lasting over 10 s, OR, any lateralized rhythmic delta activity at >1 Hz for at least 10 s, with a plus modifier. If the treatment of IIC pattern leads to both electrographic and clinical improvement, then the pattern would meet criteria for ECSz or ECSE.

2 Pathophysiology of Peri-ictal Changes

The pathogenesis of SE and brain injury can be examined and defined thanks to neuroimaging. SE can cause both cytotoxic and vasogenic cerebral edema. Vasogenic edema is noted on MRI sequences as T2 or T2 FLAIR hyperintensity without diffusion restriction and contrast enhancement [7, 8] due to the breakdown of blood-brain barrier (see Fig. 1). On the other hand, cytotoxic edema refers to the build-up of fluid inside cells due to a breakdown in cellular metabolism. Diffusion restriction on MRI is a sign of ATP-dependent sodium/potassium (Na+/K+) membrane pump failure brought on by brain injury, such as with cellular ischemia [9, 10].

In animal models, the sequence of cellular alterations after the start of a seizure has been well studied. According to this preclinical research, extended seizures may be brought on by a decrease in the inhibitory signal or by the persistence of the excitatory signal [11–14]. Oxidative stress and disruptions in cellular metabolism are

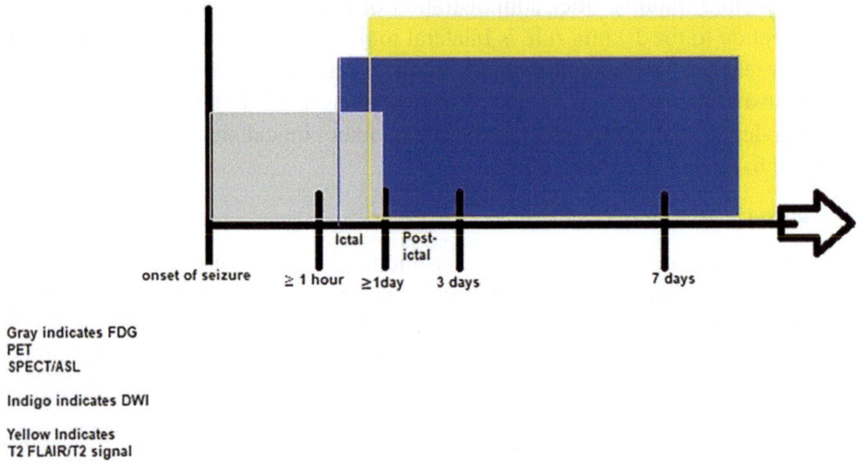

Fig. 1 Timeline of imaging changes in status epilepticus. Imaging findings that may occur during and following SE include: (1) PET hypermetabolism and SPECT or ASL measured hyperperfusion while a seizure is ongoing; (2) an increase in diffusion restriction acutely, that remains elevated for 1–7 days following seizure and gradually decreases; (3) a gradual increase in T2 FLAIR/T2 signal that rises through the post-ictal period and then typically resolves over days to weeks. Note: timing of imaging changes and relative signal increases vary. PET: positron emission tomography, SPECT: single-photon emission computed tomography, ASL: arterial spin labeling, DWI: diffusion weighted imaging, T2 FLAIR: fluid attenuated inversion recovery

caused by excess glutamate and a decrease in gamma-amino butyric acid (GABA) tone [15–17]. The intracellular, cytotoxic edema that results from mitochondrial damage, and eventually, ATP-dependent Na+/K + pump failure, is caused by calcium ion influx [18, 19]. Furthermore, clinical and preclinical investigations have demonstrated indications of blood-brain barrier disruption resulting in vasogenic edema [20–23]. Although the cause of seizure-induced injury is probably complex, research using neuroimaging is still needed to determine how much is due to intrinsic neuronal variables vs. general problems with cerebral perfusion and oxygen supply.

Understanding the etiology and location of seizure initiation may be aided by knowing the relative importance and timing of these several types of edemas in SE. Vasogenic edema is indicated by elevated T2 signal without restricted diffusion in studies with super refractory SE. This could provide additional evidence for the function of disruption of the blood-brain barrier, which have been linked to numerous instances of super refractory SE. Similar MRI abnormalities have also resolved in more common, non-super refractory cases of SE, which further corroborates the involvement of vasogenic edema [20, 21].

Diffusion restriction after seizures could be associated with postictal ischemia. Animal models of SE have corroborated this. According to these studies, ADC decreased following protracted seizures caused by chemicals, peaking 1–2 days

Table 2 Summary of imaging modalities, findings, and interpretation during and following resolution of status epilepticus

Imaging modality	Imaging findings	Interpretation
T2/FLAIR MRI	Ictal or post-ictal Hyperintensity	Cytotoxic or vasogenic edema
DWI/ADC MRI Perfusion imaging (e.g., CTP, SPECT, ASL)	Ictal: Diffusion restriction Ictal: Hyper-perfusion Post-ictal: Hypoperfusion	Cytotoxic edema Changes to Blood flow according to demand
FDG-PET	Ictal: Hypermetabolism Interictal: Hypometabolism	Changes in blood flow during and following SE

Abbreviations: *FLAIR* fluid-attenuated inversion recovery, *DWI/ADC* diffusion weight imaging/apparent diffusion coefficient, *CTP* computed tomography perfusion, *SPECT* single-photon emission computed tomography, *ASL* arterial spin labeling, *PET* positron emission tomography

later, and resolving by day 7 [24–26]. Several theories have been put up to explain the pathophysiology of these postictal imaging alterations. These possibilities include a relative increase in inhibitory transmitters and receptors, a decrease in cerebral blood flow, and a decrease in the availability or desensitization of excitatory neurotransmitters [27, 28].

Follow-up MRI, however, may show persistent abnormalities due to T2 hyperintensities along with diffusion limitation, a sign of cytotoxic edema. Hence, the existence of cytotoxic edema and these imaging results could be more concerning indicators of chronic or irreversible cellular damage (see Fig. 1).

There is abundant evidence that cerebral perfusion changes during SE. Investigations have shown enhanced perfusion during SE and hypoperfusion in the post-ictal state using CT, MRI, and, more recently, arterial spin labeling (ASL) [29, 30]. Investigating the pathogenesis and the relative contributions of blood flow and oxygen supply to cellular injury in SE and the post-ictal state may also be possible with perfusion-based methods (see Table 2).

3 Common Imaging Modalities

Neuroimaging's most important function in clinical practice is to support the diagnosis and determine the cause of status epilepticus. Neuroimaging is recommended for all patients with status epilepticus [31–34] (refer to Table 3). When it comes to detecting blood, CT scans are more easily available in emergency rooms than MRI. In patients with SE, approximately 15–33% of individuals had abnormalities found on CT scans [35–37]. This covers a wide range of acute symptomatic reasons, such as mass lesions, trauma, edema, and bleeding [31, 36, 38]. When CT and MRI are combined, the diagnostic yield is slightly higher, with abnormalities being identified in 30–36% of cases. The superiority of MRI in identifying inflammatory changes and subtle structural abnormalities accounts for a significant percentage of this improvement [36, 37]. The most frequent acute and distant symptomatic causes

Table 3 Common structural etiologies identified on neuroimaging. Acute abnormalities frequently prompt a change in management or surgical intervention. Remote causes may be amenable to further characterization

Acute symptomatic	Remote symptomatic
Ischemic stroke	Neonatal injury: stroke, HIE
Hemorrhage: IPH, SAH, SDH, EDH	Other remote injury, e.g., trauma, infection
Neoplasm	Cortical dysplasia
Cerebral edema: secondary to TBI or metabolic cause	Malformation of cortical development: e.g., polymicrogyria, heterotopias
AVM	Other genetic causes: e.g., Sturge– Weber, TSC

Abbreviations: *IPH* intraparenchymal hemorrhage, *SAH* subarachnoid hemorrhage, *SDH* subdural hemorrhage, *EDH* epidural hemorrhage, *TBI* traumatic brain injury, *AVM* arteriovenous malformation, *HIE* hypoxic ischemic encephalopathy, *TSC* tuberous sclerosis complex

of SE, according to one prospective investigation that combined CT and MRI, were cerebral dysgenesis and CNS infections, respectively [37]. Moreover, a head CT or MRI could result in a management modification. If acute symptomatic causes of SE such as bleeding, tumors, and cerebral edema are identified, adjustments in acute therapy may be necessary. On the other hand, long-standing structural abnormalities, such as neonatal or other remote brain traumas, and defects of cortical development, such as localized cortical dysplasia, are common remote symptomatic causes of epilepsy, and may not need aggressive changes in management (see Table 3).

4 Structural Imaging

4.1 Magnetic Resonance Imaging

Magnetic resonance imaging (MRI) is frequently used in status epilepticus to identify underlying structural abnormalities. In most situations, abnormalities in the T2-weighted imaging (T2WI) and diffusion-weighted imaging (DWI) signals can be observed with a cortico-subcortical distribution ipsilateral to the epileptogenic hemisphere; these abnormalities are related to the length of the SE and may be reversible [39–42]. One hypothesis is that the abnormalities on MRI result from the high metabolic demand during prolonged epileptic episodes. These changes are observed in nuclear medicine tracers used in functional neuroimaging such as positron emission tomography or single-photon emission computed tomography [43].

4.1.1 Conventional MRI

The most common MRI abnormalities linked to SE are increased T2 or fluid-attenuated inversion recovery (FLAIR) signal and/or diffusion restriction. Local edema that may be brought on by SE can be visualized using diffusion weighted imaging (DWI) and the associated apparent diffusion coefficient (ADC). ADC values in cortical and/or subcortical regions may be lowered in cases of generalized tonic-clonic seizures. Even in cases of widespread seizures, these alterations are usually more localized and often correspond with T2 or FLAIR hyperintensities [44–46] (see Fig. 2). The basal ganglia, thalamus, specifically the pulvinar, hippocampi, and cerebellum, are typical subcortical brain regions affected [47–51]. Similar to this, focal diffusion restriction and an elevated T2 or FLAIR signal that may last for days to weeks are linked to unilateral or focal SE as well as in lateralized periodic discharges seen during electroencephalograms (EEG) [52, 53].

In SE, acute hippocampal edema and reversible, transient MRI abnormalities are frequently observed. Majority of the patients had increased T2 hyperintensity in both hippocampi with MRI changes in the areas with abnormal EEG activity. Follow-up MRIs were diagnosed with mesial temporal sclerosis because of hippocampal atrophy and aberrant T2 signal intensity in the hippocampal region. Increased T2 signal intensity has been seen in the brain stem and cerebellum of the individuals who did not exhibit hippocampal abnormalities on MRI. The most frequent finding is an abnormality in the ipsilateral hippocampus (Fig. 3) [54].

There are some reversible seizure-induced MRI abnormality patterns associated with SE. The most common ones are predominant cortical (with or without subcortical, leptomeningeal or thalamic involvement), hippocampal (with or without cortical, subcortical, leptomeningeal or thalamic involvement), claustrum, predominant subcortical, and splenium involvement [55] (Fig. 4).

Summary of Changes in SE from Presentation to Follow-up (Fig. 5)
This section provides a practical example of temporal changes in the various changes in MRI sequences after an ictal event.

Correlation of MRI and EEG Changes (Figs. 6, 7, 8, and 9)
The following examples highlight how MRI changes often correlate with epileptiform findings on EEG, typically preserving laterality and maintaining topographic correspondence.

Rare Findings on MRI Imaging (Figs. 10 and 11)
Neurologists and neuroradiologists should be aware of a rare pitfall of MRI FLAIR imaging due to an interesting phenomenon called hyperintense acute reperfusion marker (HARM) recently renamed as FLAIR subarachnoid hyperintensity (FLASH) [58]. HARM most likely reflects a disruption in the blood-CSF barrier, allowing the contrast to leak into CSF spaces and making them appear hyperintense on post-contrast T2 FLAIR images, often mistaken as subarachnoid hemorrhage (SAH) or other CSF pathologies. It is a nonspecific finding and may be seen in SAH,

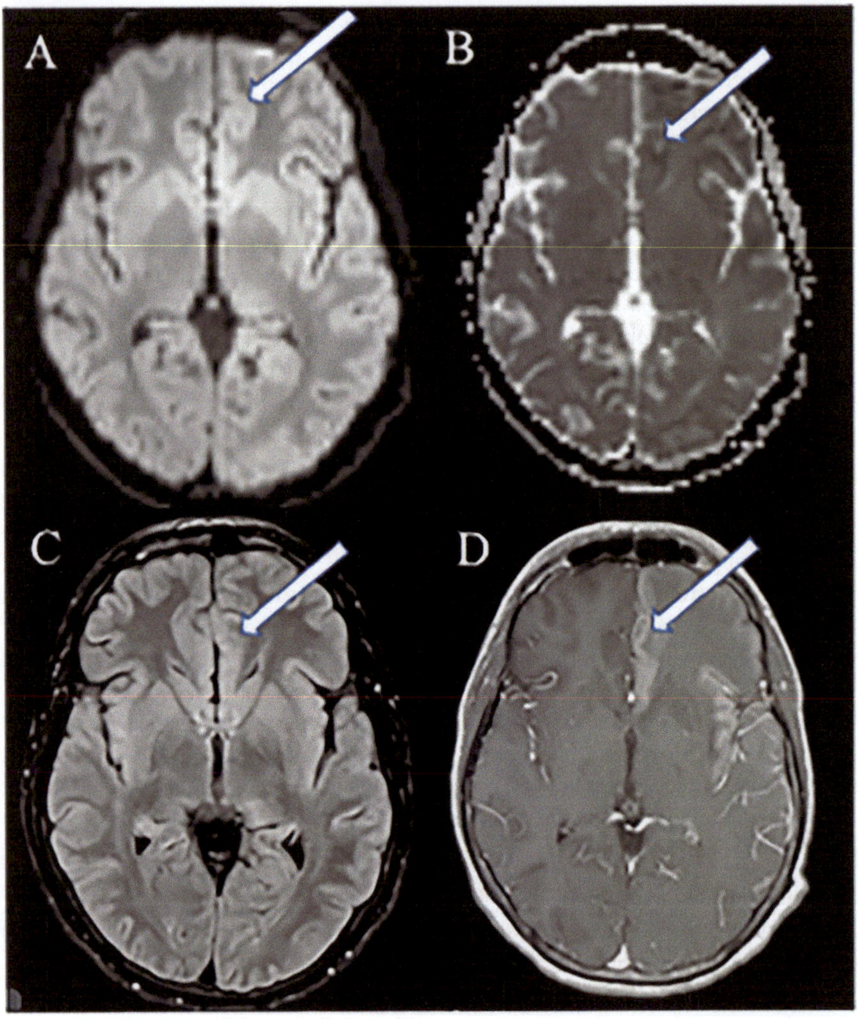

Fig. 2 The most common acute MRI findings due to SE (**a**). High intensity signal in the left frontal and insular cortices is seen in the DWI sequence with an concomitant reduction in the apparent diffusion coefficient (ADC) (**b**). Identical areas of higher intensity on FLAIR images (**c**) and contrast amplification representing vasogenic edema (**d**)

meningitis, stroke, transient ischemic attack, meningeal carcinomatosis, subdural hematomas, adjacent neoplasms, dural venous thrombosis, and status epilepticus [59–61].

The images below (Figs. 10 and 11) illustrate a case of HARM where no abnormality is seen in the initial post-contrast MRI sequence but the hyperintensity is seen in the subsequent T2 FLAIR sequences, with eventual complete resolution [62].

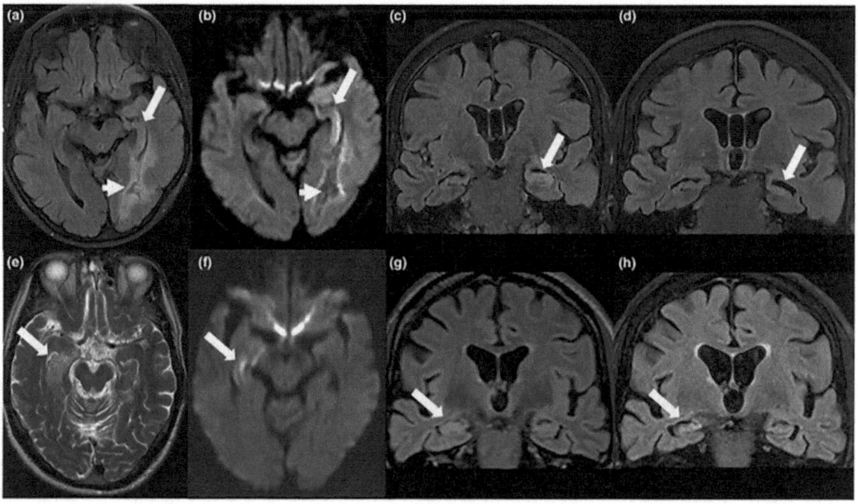

Fig. 3 Temporal onset SE. (**a**)–(**d**) Case 1: A 76-year-old woman with an SE secondary to a subacute left posterior cerebral artery infarction (**a**, **b**, arrowheads). In baseline MRI, FLAIR hyperintensity (**a**, arrow) and DWI restriction (**b**, arrow) were observed in the left hippocampus. Moreover, a coronal FLAIR image showed hippocampal enlargement and hyperintensity (**c**, arrow). The 6-month follow-up MRI showed left mesial temporal sclerosis (**d**, arrow). (**e**)–(**h**) Case 2: A 65-year-old man with nonconvulsive SE of unknown etiology. In baseline MRI, right hippocampus hyperintensity and enlargement on axial T2WI (**e**, arrow) and coronal FLAIR (**g**, arrow), and restriction on DWI (**f**, arrow) were observed. The 9-month follow-up MRI showed right mesial temporal sclerosis (**h**, arrow) [54]. (Image and caption taken from Sarria-Estrada et al. [54])

The pathophysiology of HARM/FLASH is multifactorial. The vasogenic and cytotoxic edema caused by the seizure along with the possible leptomeningeal and parenchymal uptake of gadolinium, following SE induced blood-brain barrier breakdown are hypothesized to be involved [63, 64]. Another hypothesis is that the increased gadolinium concentration in the setting of renal insufficiency can lead to gadolinium's movement across the osmotic gradient [59, 65]. This is further supported by the fact that findings resolve overtime when gadolinium concentration falls.

4.2 Perfusion Imaging

4.2.1 CT Perfusion

CT perfusion (CTP) is another modality used in SE specially in conditions where a differentiation between post-ictal state and stroke is needed. Post-ictal state can present as a stroke mimic, and it is important to differentiate because of the management plan. Perfusion CT also helps in detecting non convulsive SE.

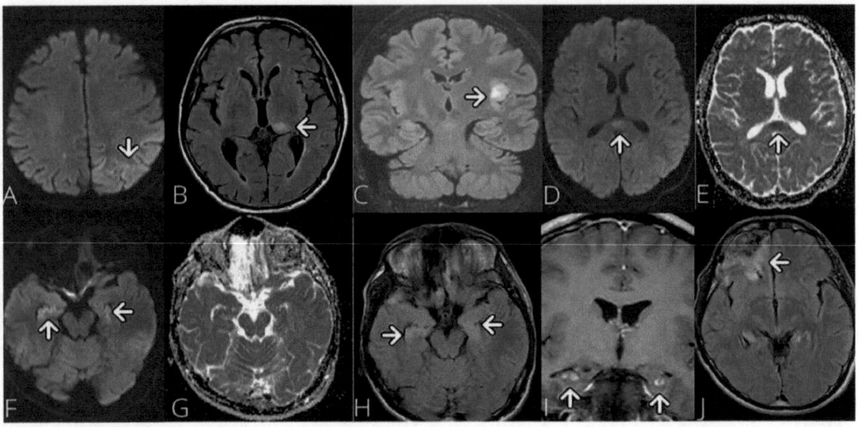

Fig. 4 Patterns of MRI signal abnormalities seen in status epilepticus. Cortical pattern—(**a**): focal diffusion restriction involving cortical gyri; (**b**): ipsilateral pulvinar signal change on fluid-attenuated inversion recovery imaging; 2. Subcortical pattern—(**c**): hyperintensity in the subcortical left insular region on fluid-attenuated inversion recovery imaging; 3. Splenial pattern—(**d**): diffusion restriction of the splenium, (**e**): corresponding decreased signal intensity on the apparent diffusion coefficient image; 4. Hippocampal pattern—(**f**): diffusion restriction in hippocampi bilaterally; (**g**): corresponding decreased signal intensity on the apparent diffusion coefficient image; (**h**): hyperintensity of hippocampi bilaterally on fluid-attenuated inversion recovery imaging; (**i**): contrast enhancement of hippocampi bilaterally on T1-weighted imaging (post-contrast); (**j**): gliosis and encephalomalacia in the right fontal region due to previous head trauma. This lesion is the etiology of status epilepticus. Note bilateral signal changes in hippocampi due to status epilepticus, away from the lesion [55]. (Image and caption from Mariajoseph et al. [55])

Perfusion CT accurately detects hyper-perfusion changes which are not seen in stroke [2]. Cortical hyperperfusion is the most common finding on CTP during ictal phase, and is specified by increased cerebral blood flow and cerebral blood volume and decreased mean transit time [66–70] (Fig. 12). Cerebral blood flow and volume eventually return to normal.

5 Nuclear Medicine Functional Brain Imaging

Nuclear imaging methods—specifically, positron emission tomography (PET) and single-photon emission computed tomography (SPECT) assist to locate epileptogenic zones—the parts of the brain that cause seizures. In order to detect regions of aberrant perfusion or hypometabolism linked to continuous seizure activity, nuclear imaging can offer information on regional cerebral blood flow and metabolic activity. To better understand the pathophysiology of status epilepticus and assess new treatment methods, nuclear imaging techniques are frequently used in research studies and clinical trials.

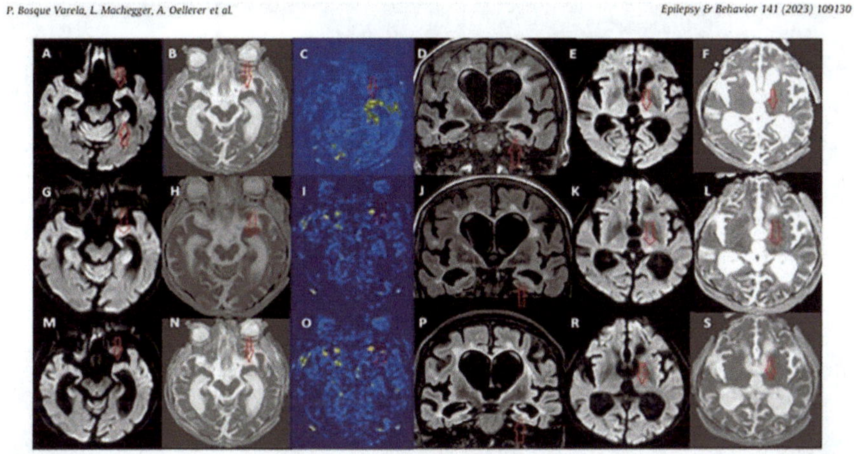

Fig. 5 An illustration of peri-ictal MRI abnormalities in a patient with SE in acute and follow-up MRIs [56] A 65-year-old woman was admitted to the hospital with focal motor SE (clonic movements of the right upper extremity) which evolved into a non-convulsive SE. Lateralized periodic discharge of varying range between 0.5 Hz and 1 Hz in left temporal region. SE lasted 48 h. After 16 h of onset, MRI with a SE protocol was performed The first row shows images of an initial MRI taken in less than 24 h after the onset of SE: DWI with diffusion restriction (**a**), normal ADC (**b**), hyperperfusion in ASL (**c**), and signal increase in FLAIR (**d**) in left hippocampus. Diffusion restriction (**e**) and signal decrease in ADC map (**f**) in left pulvinar of thalamus Second row shows images of the first follow-up MRI taken 7 days after the onset of SE: Persisting diffusion restriction (**g**), normal ADC (**h**), normal perfusion (**i**), and persisting signal increase in FLAIR (**j**) in the left hippocampus. Persisting diffusion restriction (**k**) signal decrease in ADC (**l**) in left pulvinar of thalamus Third row shows images of the second follow-up MRI taken 4 weeks after the onset of SE: Resolved diffusion restriction (**m**), normal ADC (**n**), normal perfusion (**o**), and normalized signal in FLAIR (**p**) in left hippocampus. Volume loss in the left hippocampus as compared to the first two images (**d** & **j**). Resolved diffusion restriction (**r**) and ADC map (**s**) in the left pulvinar of thalamus. Red arrows point to the affected brain areas. (Image taken from Bosque Varela et al. [56])

Besides SE, nuclear imaging techniques are utilized in pre-surgical evaluation to pinpoint the precise brain areas that trigger seizures; the epileptogenic zone. These modalities assist in evaluating the suitability of a patient for a surgical intervention, such as resection or the implantation of neurostimulation devices, such as responsive neurostimulation (RNS) systems or vagus nerve stimulation (VNS).

5.1 PET Imaging

Positron emission tomography (PET) gives higher quality and resolution images than single photon emission computed tomography (SPECT) and enables quantitative measures. PET tracers can monitor glucose, serotonin, and oxygen metabolism, cerebral blood flow, and receptor binds in epilepsy. ^{18}Fluoro-2-deoxyglucose (FDG)

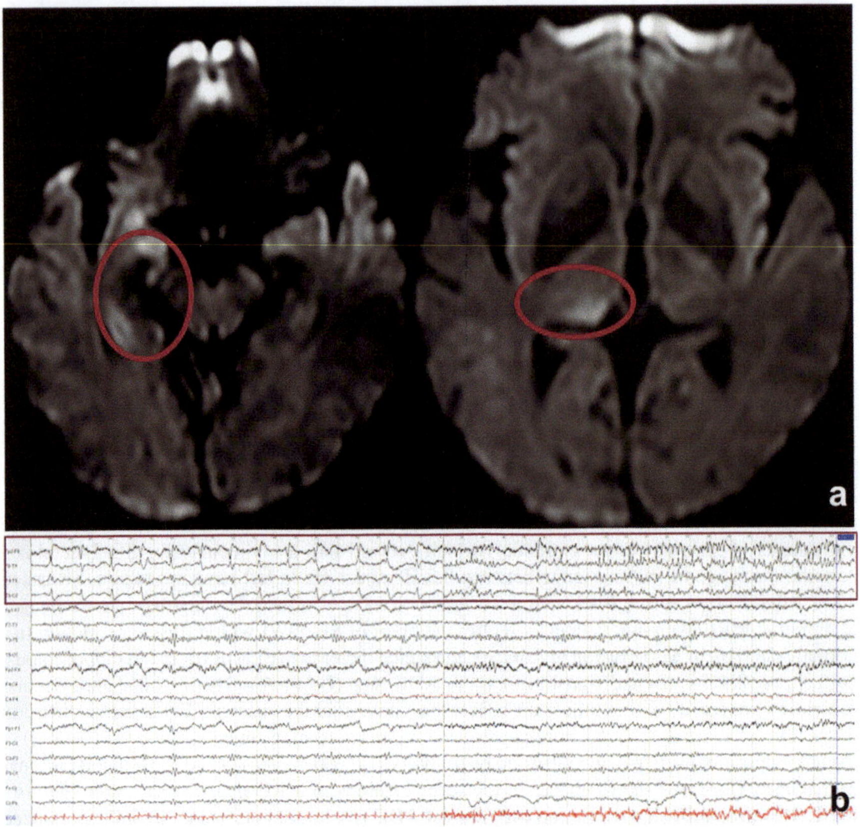

Fig. 6 (**a**) DWI signal hyperintensity in the right hippocampus and pulvinar (encircled red). (**b**) Continuous quasi-periodic spike and waves at frequency of 0.7–1 Hz, showing an evolving pattern into a poly-spikes at 4–5 Hz in the right frontotemporal leads [57]

PET, which measures brain glucose metabolism, is the most extensively used technology in clinical practice. Other PET tests require expert workers and on-site equipment, including a cyclotron. Table 4 presents PET data related to seizure focus.

PET scans are typically taken during the interictal phase. Focal hypometabolism is typically seen on interictal FDG-PET scans. Because most seizures are unpredictable and have a brief duration, doing ictal PET investigations is challenging. The lengthy 30- to 45-min brain uptake time of FDG following injection, which results in a complex pattern of elevated and decreased metabolism, is another drawback of ictal PET. Depending on when the injection was given after the seizure, postictal PET can show a complex pattern of both increased and decreased metabolism, or just increased or decreased metabolism (Fig. 13).

A postictal PET study conducted at various times after the most recent seizure revealed either hyper- or hypometabolism, depending on when the injection was given [73, 74]. PET showed focal hypermetabolism in the early postictal period

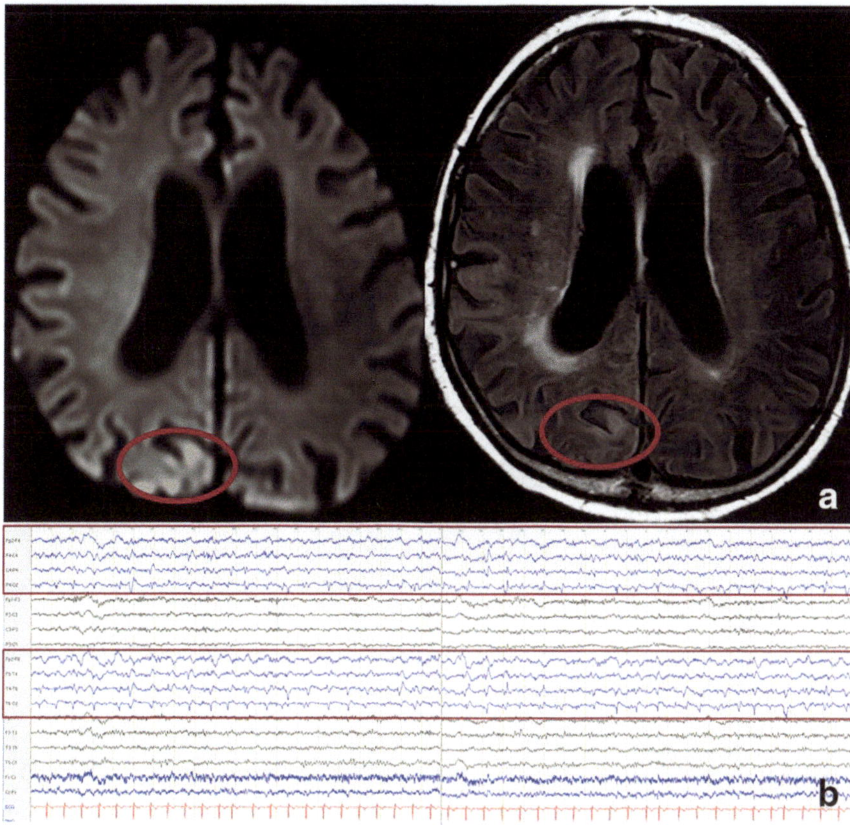

Fig. 7 (**a**) DWI and FLAIR signal hyperintensity in the right parietal cortex. (**b**) Continuous quasi-periodic sharp waves at frequency of 1 Hz on the right temporoparietal leads [57]

(seizures that occurred within 15 min of the FDG injection). The least severe regional hypometabolism happened between 24 and 48 h after the seizure, while the most severe hypometabolism happened more than 48 h after the seizure.

6 Imaging of Nonconvulsive Status Epilepticus

Treatment and prognostication in non-convulsive status epilepticus is extremely challenging. This disease paradigm, especially in refractory or super-refractory states has been referred to as "an evidence-free area" [75]. Imaging modalities to correlate affected brain regions with clinical outcomes is a continued area of study. The literature provides key concepts to consider when utilizing imaging in the care of the patient with non-convulsive status epilepticus syndromes: timing [76], patient comorbidities [75, 76], implications of involved anatomy [77–80], and choice of

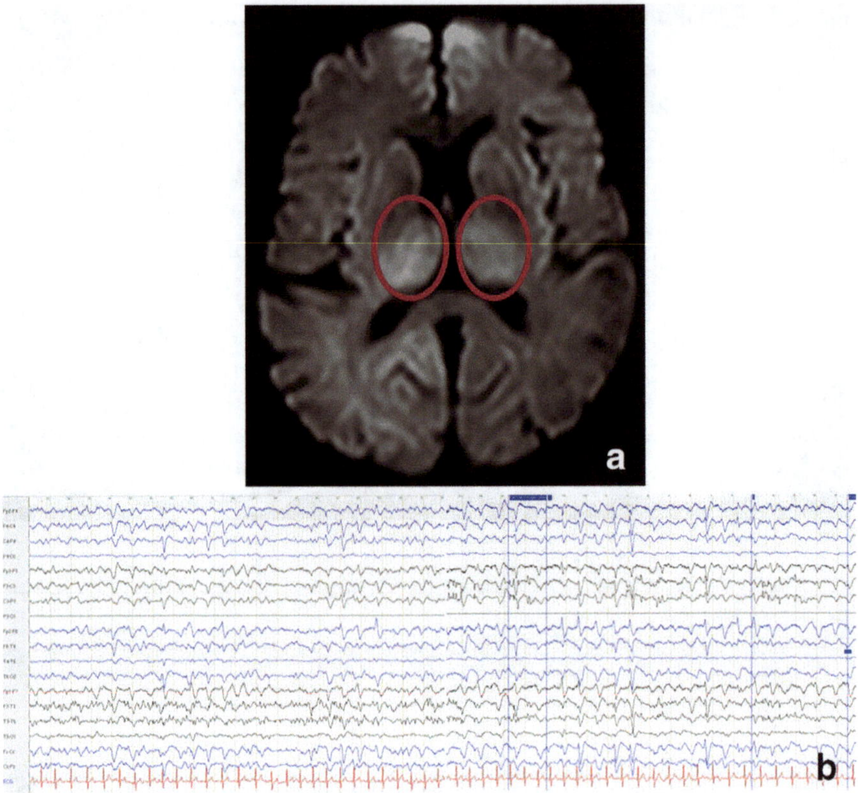

Fig. 8 (**a**) Bilateral thalamic DWI hyperintensity. (**b**) Diffuse triphasic sharp waves at frequency of 1 Hz [57]

imaging modality [75, 76]. Imaging in refractory nonconvulsive status epilepticus can be particularly challenging as these patients are critically ill requiring advanced medical care, monitoring, and limited hemodynamic stability [81]. Also, due to the potential reversibility of imaging findings that may occur over the clinical course [76], optimal timing of imaging with correlation to degree of seizure severity has yet to be determined [76, 80].

In clinical suspicion of nonconvulsive frontal status epilepticus, FDG-PET detected hypermetabolism in the left orbitofrontal area [82]. After 5 days of oxcarbazepine medication, a follow-up FDG-PET scan showed that the hypermetabolic emphasis had resolved. After exposure to seizure eliciting music, PET scan revealed hypometabolism in the right lateral temporal lobe and an area of enhanced metabolism in the right anteromesial temporal lobe [83].

Further challenges in the treatment of refractory nonconvulsive status epilepticus include classification of disease entities, further delineation of abnormal electrographic discharges, and anatomical correlation [76]. The classification of the clinical spectrum is far reaching encompassing multiple disease states [84, 85]. Secondly,

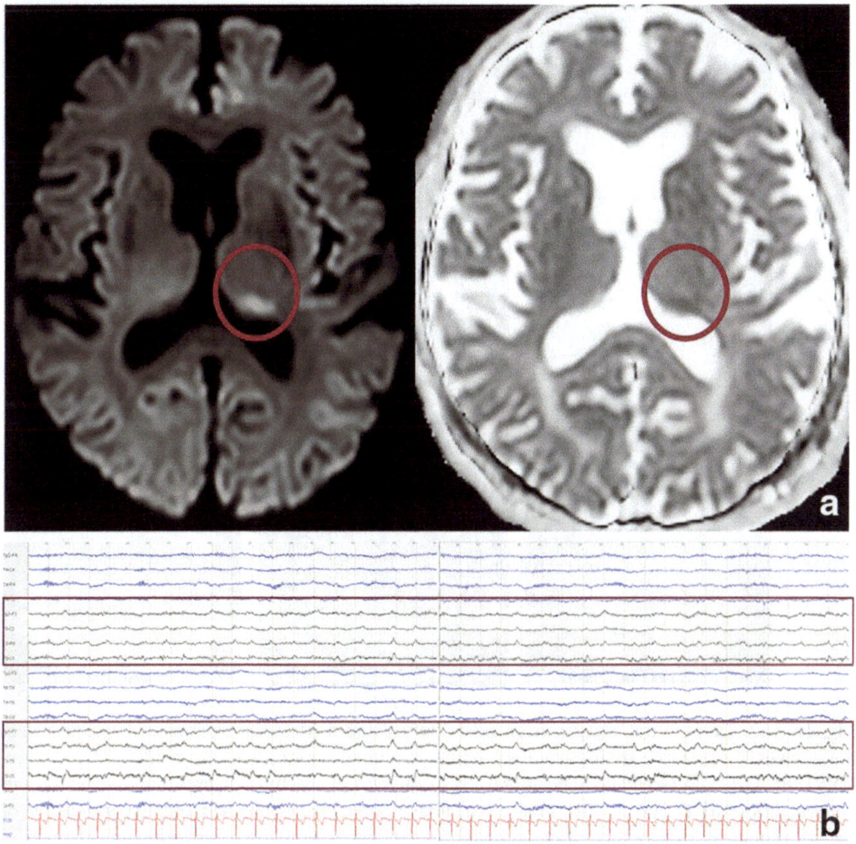

Fig. 9 (**a**) DWI signal hyperintensity and corresponding decreased ADC signal located in the left thalamus. (**b**) Continuous quasi-periodic left hemispheric sharp waves at frequency of 0.8–1 Hz [57]. (Giovannini et al. [57])

there is the question of linking super refractory status epilepticus with anatomical correlates on imaging. One region that continues to gain traction is the claustrum. The claustrum is in the insular region, present bilaterally [86] and implicated in many cortical functions with a particular predilection to inhibition. This area is implicated in modulating function in both the awake and resting states, with widespread neurotransmitter projections [80, 86]. Although a full review is beyond the scope of this chapter, a continued source of interest is the wide connectivity of this anatomical region to multi-integrated tertiary cortical function and the bilateral input [86]. There is the hypothesis that, given imaging alterations in the claustrum are seen in multiple conditions [87], this may be a marker of "cytokine-mediated neuroinflammation" [81]. In both NORSE (new onset refractory status epilepticus) and FIRES (Febrile infection-related epilepsy syndrome), abnormalities on MRI imaging have been noted in prior studies to include hyperintense

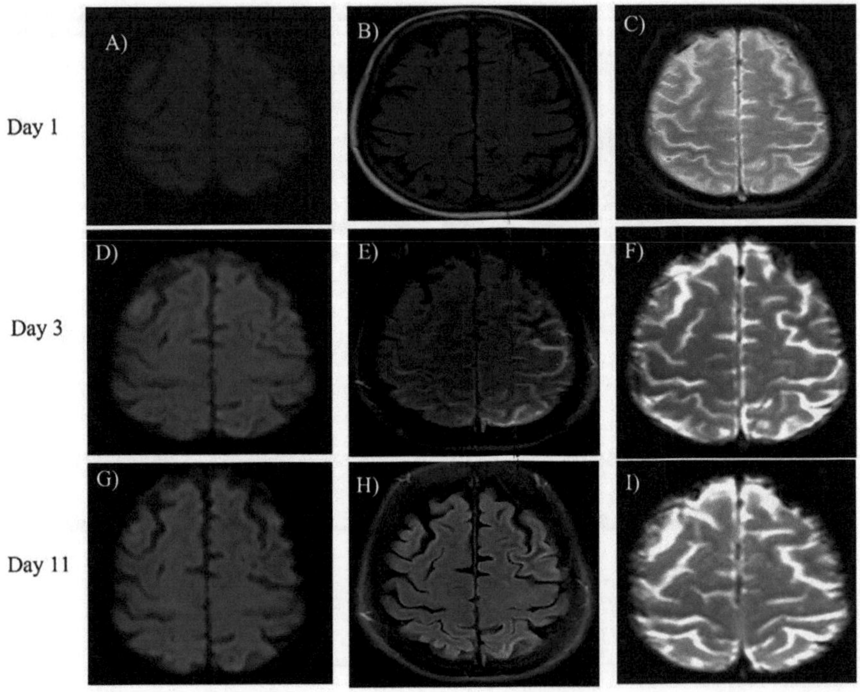

Fig. 10 [62] (**a–c**) Axial images from post-contrast brain MRI, in a patient who experienced non-convuslive seizures arising from left parieto-occipital region, with three different sequences (DWI, FLAIR, and Gradient Echo [GRE], respectively) at the time of presentation showing no abnormalities. (**d–f**) Axial MRI images (in above sequence): Subarachnoid hyperintensity on FLAIR sequence (**e**) over the left parietal and frontal convexity with no corresponding changes on DWI (**d**) or GRE (**f**). (**g–i**) Axial MRI images (in above sequence), showing complete resolution of the subarachnoid FLAIR hyperintensity (**h**) along the left parietal and frontal convexity with no abnormalities on DWI (**g**) or GRE

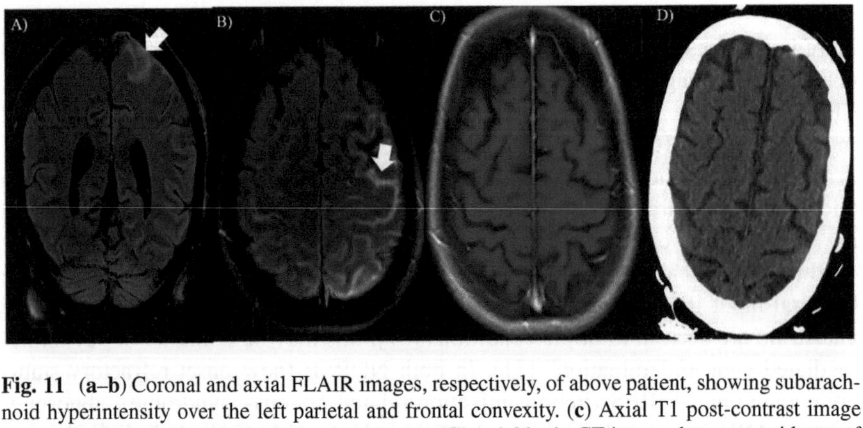

Fig. 11 (**a–b**) Coronal and axial FLAIR images, respectively, of above patient, showing subarachnoid hyperintensity over the left parietal and frontal convexity. (**c**) Axial T1 post-contrast image shows absence of any pathological enhancement. (**d**) Axial brain CT image shows no evidence of subarachnoid blood [62]. (Images and caption taken from Al-Chalabi et al. [62])

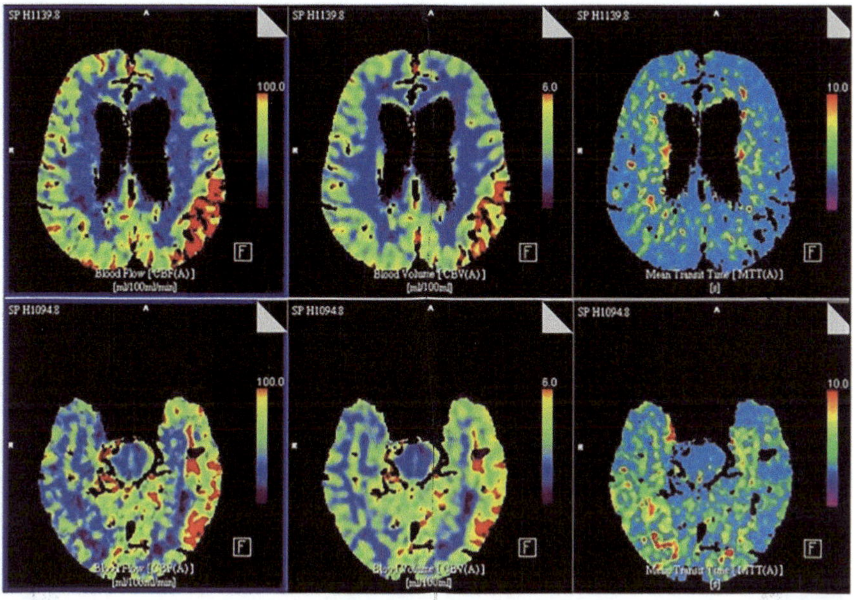

Fig. 12 80-year-old woman with remote history of subarachnoid hemorrhage presented with sensory aphasia. EEG showed epileptiform discharges in the left temporal region, but no definite ictal activity. CTP was performed in this presumed post-ictal state. Top (**a**) and bottom (**b**) slices of brain perfusion CT shows extensive perfusion alterations in the entire left parietotemporal region, consisting of a normal to slightly diminished MTT (right column), and an increased CBF (left column) and CBV (middle column) [67]

Table 4 PET findings in the area of seizure focus in patients with epilepsy [71]

PET study	Findings
Interictal 18F-FDG	Usually reduced metabolism
Ictal 18F-FDG	Increased and decreased metabolism (complex pattern)[a]
Post-ictal 18F-FDG metabolism[b]	Complex pattern, increased or decreased
GABAA-cBZR receptor	Reduced binding
Opioid receptor	Increased mu and delta receptor bindings
Serotonin receptor	Reduced binding
Dopamine receptor	Reduced binding
11C-alpha-methyl-L-tryptophan	Increased uptake
Interictal ^{15}O-H_2O	Usually reduced perfusion
Ictal ^{15}O-H_2O	Increased perfusion

[a]Due to long brain uptake period of FDG, [b]Depending on the time of injection after seizure

T2-weighted imaging in the claustrum [71, 74]. Work by Meletti et al. [77, 80] has demonstrated the severity of new onset refractory status epilepticus and alterations in the claustrum [80]. Their work also further explores the concept of the connection of timing with imaging acquisition [80]. Future themes for study have already been

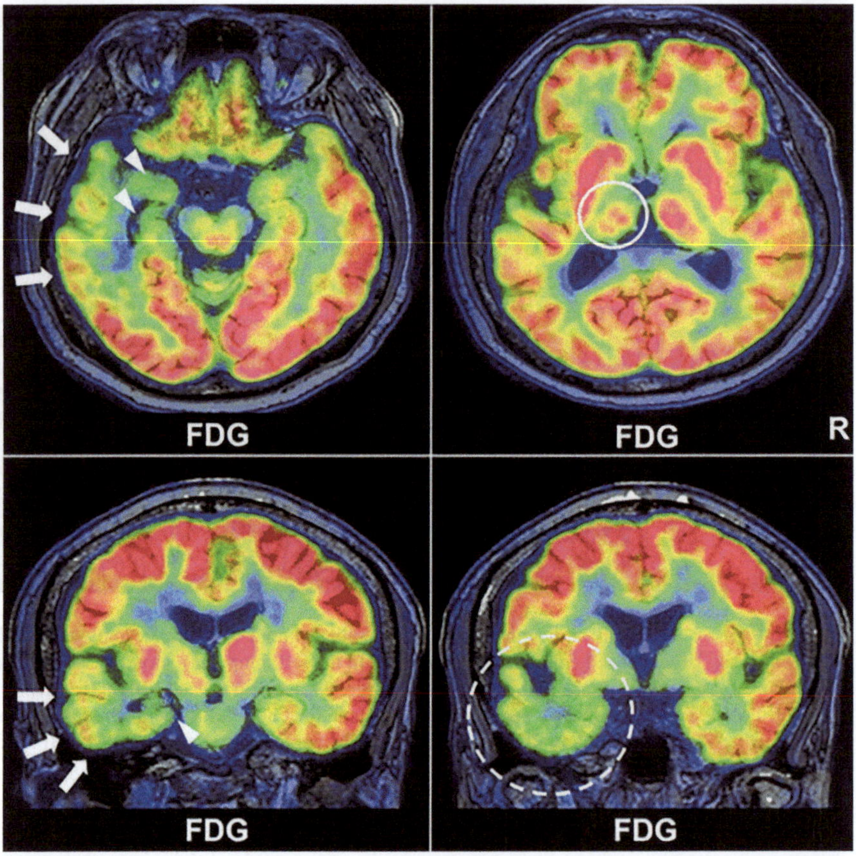

Fig. 13 Interictal PET scan in a patient with right mesial temporal epilepsy. [72]: Ipsilateral hypoperfusion in the temporal neocortex (arrows) and in the medial temporal lobe (arrowhead) are present in FDG-PET Reduced glucose metabolism can be found in the ipsilateral thalamus (circle) and temporal pole (broken circle). (Image from Sone et al. [72])

proposed in the literature including imaging use for prognostication, further defining separate subgroups within refractory nonconvulsive status epilepticus, delineation of timing intervals, and correlation between radiographic and neurophysiological findings [55, 78, 88, 89].

7 Imaging of Ictal-interictal Continuum

Besides patients with status epilepticus, neuroimaging changes have also been described in patients with high-risk epileptiform patterns on EEG, including lateralized periodic discharges (LPD) with plus modifiers (fast activity, +F, and/or rhythmic activity, +R). As mentioned earlier, in the ACNS' standardized critical care EEG terminology, LPDs with fluctuation or plus feature, as well as LRDA with plus modifier can also be classified as IIC. Since the electrographic patterns may sometime be equivocal, neuroimaging findings may potentially become relevant as an added tool to determine the need for more aggressive treatment and/or more prolonged EEG monitoring. Recently, criteria have been suggested to help distinguish interictal from peri-ictal LPDs, with the notion that the latter pose a high risk of seizures and status epilepticus [90].

The frequency of periodic epileptiform discharges is understood to be associated with likelihood of seizures; higher the frequency more likely the LPDs are to be ictal or peri-ictal. Based on FDG-PET scan studies LPDs at 1 Hz or greater were demonstrated to exhibit metabolic demands that were respectively 100% and 309% higher than patients with LPDs at 0.5 Hz [91].

Using various ligand for PET imaging, increased regional cerebral blood flow, local oxygen metabolism, and local glucose utilization have been associated with LPDs, suggesting that these discharges may be potentially ictal [92, 93].

Another study on 18 hospitalized patients, who underwent FDG-PET imaging during contemporaneous IIC patterns on EEG showed that hypermetabolism was most common (61%) and predicted electrographic or electroclinical status epilepticus with high specificity and sensitivity [94].

Studies using MR imaging have demonstrated that LPDs with concurrent clinical seizures had hyperintensities on diffusion-weighted imaging [95, 96] and hyperperfusion on perfusion MRIs [95].

A case study found an association between IIC pattern and diffusion-weighted imaging changes on MRI, in association with dynamic susceptibility contrast-MRI changes (increased cerebral blood flow and cerebral blood volume, along with decreased mean transit time) as well as hyperperfusion on Tc-99 m-HMAPO SPECT scan, which reversed with the resolution of IIC pattern [97].

Single photon emission CT (SPECT) scan studies have shown increased regional cerebral perfusion in the LPD focus [98–102]. Resolution of the hyperperfusion [99, 102], or even hypoperfusion [100, 101], was seen often after LPDs terminated. Interestingly, some cases have also demonstrated relevant-regional hypoperfusion on SPECT with the appearance of LPDs, which has been proposed to indicate brain functional impairment [103, 104]. A small retrospective study found that patients who had experienced status epilepticus and had LPD + pattern on EEG at the time of MRI had hyperperfusion pattern on ASL. The median time from seizure detection to MRI was 50 min. All of them exhibited hyperperfusion in the cortex and ipsilateral thalamus. Five patients had a bilateral hyperperfusion pattern, and in 13/15, who had a lateralized pattern, the site of hyperperfusion always matched with the side of LPD. Five patients with laterality and cerebellar blood flow changes showed hyperperfusion in the cerebellar hemisphere contralateral to the LPD laterality, a phenomenon referred to as crossed cerebellar diaschisis. Pertinently, none of the three patients with LPD without plus signs (who had also experienced status epilepticus prior to the EEG) demonstrated hyperperfusion on ASL (Fig. 14) [105].

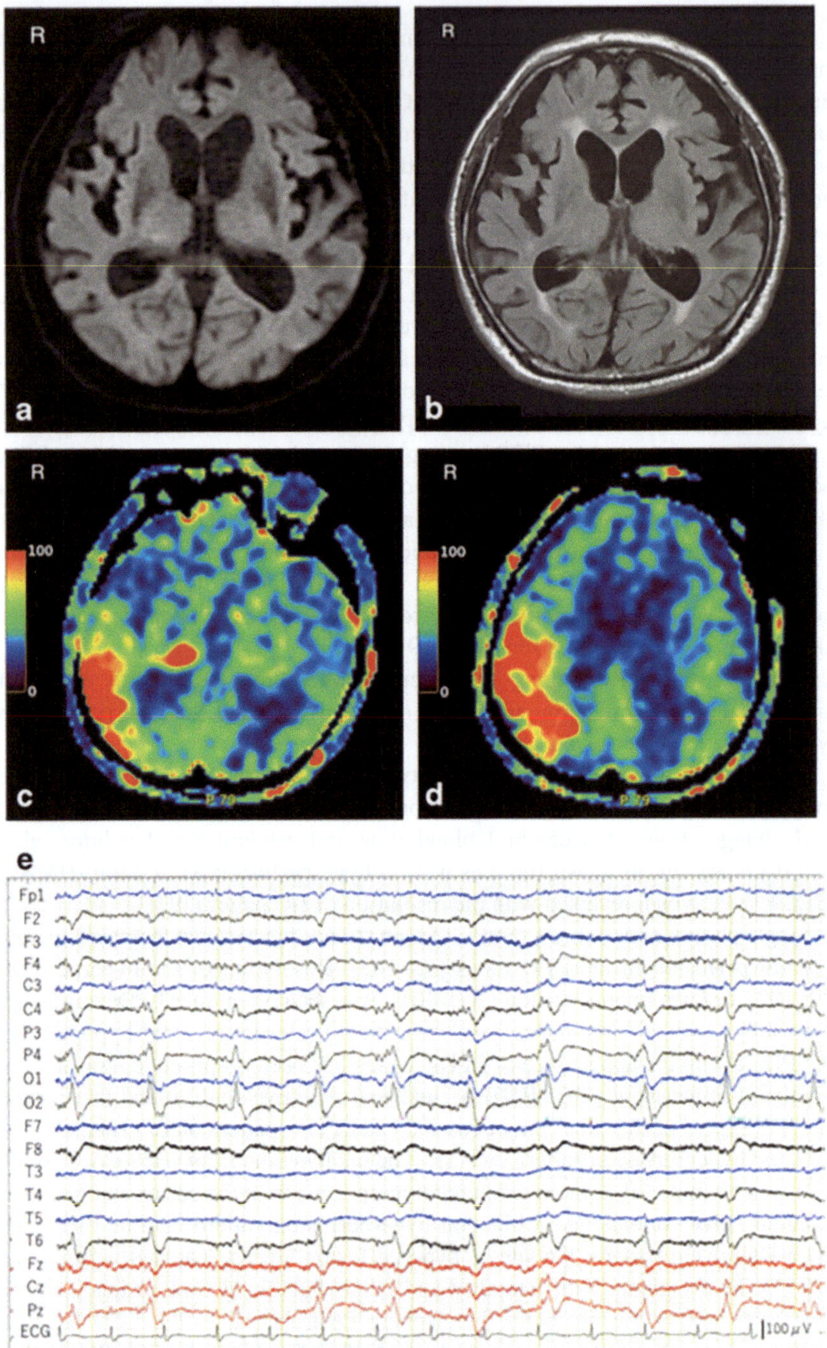

Fig. 14 MRI and EEG findings in a patient after a seizure. (**a**) No hyperintensity observed on DWI; (**b**) No hyperintensity observed on T2-FLAIR sequence. (**c**, **d**) Hyperperfusion was observed in the right temporo-parieto-occipital cortex and the ipsilateral thalamus on ASL perfusion imaging (post-labeling delay: 2.5 s); (**e**) Lateralized periodic discharges with fast activities (LPD + F) were observed in the right posterior quadrant with the monopolar montage. High-cut filter: 120 Hz; time constant: 0.3 s [105]

Another study by Eisermann et al. reported an association between periodic discharges and ASL hyperperfusion in the cortex and thalamus. They showed an increase in focal cerebral blood flow in the cortical-striatal-thalamic subdivisions, corresponding to LPDs. This increase was found topographically related with the EEG focus [106]. These findings potentially indicate a high risk of seizures in patients with LPD + pattern.

In summary, in the backdrop of above information, it is clear that many of the electrographic patterns, which were traditionally considered to be interictal or unequivocal, may either be strongly associated with seizures (peri-ictal) or may actually be ictal themselves, and neuroimaging findings may provide information complementary to the electrographic data to help make that determination.

References

1. Engel J Jr. Report of the ILAE classification core group. Epilepsia. 2006;47:1558–68.
2. Glauser TA. Designing practical evidence-based treatment plans for children with prolonged seizures and status epilepticus. J Child Neurol. 2007;22(suppl 5):38S–46S.
3. Lowenstein DH, Bleck T, Macdonald RL. It's time to revise the definition of status epilepticus. Epilepsia. 1999;40:120–2.
4. Trinka E, Cocl H, Hesdorffer D, et al. Epilepsia. 2015;56(10):1515–23.
5. Husain AM. "T" times: revisiting the timing of neuronal injury in status epilepticus. Epilepsy Curr. 2023;24(1):16–8. Published 2023 Dec 8. https://doi.org/10.1177/15357597231216003.
6. Hirsch LJ, Fong MWK, Leitinger M, et al. American clinical neurophysiology society's standardized critical care EEG terminology: 2021 version. J Clin Neurophysiol. 2021;38(1):1–29. https://doi.org/10.1097/WNP.0000000000000806.
7. Schaefer PW, Buonanno FS, Gonzalez RG, Schwamm LH. Diffusion-weighted imaging discriminates between cytotoxic and vasogenic edema in a patient with eclampsia. Stroke. 1997;28(5):1082–5.
8. Barzó P, Marmarou A, Fatouros P, Hayasaki K, Corwin F. Contribution of vasogenic and cellular edema to traumatic brain swelling measured by diffusion-weighted imaging. J Neurosurg. 1997;87(6):900–7.
9. Simard JM, Kent TA, Chen M, Tarasov KV, Gerzanich V. Brain oedema in focal ischaemia: molecular pathophysiology and theoretical implications. Lancet Neurol. 2007;6(3):258–68.
10. Stadnik TW, Demaerel P, Luypaert RR, Chaskis C, Van Rompaey KL, et al. Imaging tutorial: differential diagnosis of bright lesions on diffusion-weighted MR images. Radiographics. 2003;23(1):e7.
11. Goodkin HP, Joshi S, Mtchedlishvili Z, Brar J, Kapur J. Subunit-specific trafficking of GABAA receptors during status epilepticus. J Neurosci. 2008;28(10):2527–38.
12. Seinfeld S, Goodkin HP, Shinnar S. Status Epilepticus. Cold Spring Harb Perspect Med. 2016;6(3):a022830–13.
13. Goodkin HP. Status epilepticus increases the intracellular accumulation of GABAA receptors. J Neurosci Soc Neurosci. 2005;25(23):5511–20.
14. Rajasekaran K, Todorovic M, Kapur J. Calcium-permeable AMPA receptors are expressed in a rodent model of status epilepticus. Ann Neurol. 2012;72(1):91–102. Wiley Subscription Services, Inc., A Wiley Company.
15. Schubert D, Piasecki D. Oxidative glutamate toxicity can be a component of the excitotoxicity cascade. J Neurosci. 2001;21(19):7455–62.
16. Murphy TH, Miyamoto M, Sastre A, Schnaar RL, Coyle JT. Glutamate toxicity in a neuronal cell line involves inhibition of cystine transport leading to oxidative stress. Neuron. 1989;2(6):1547–58.

17. Aoyama K, Watabe M, Nakaki T. Regulation of neuronal glutathione synthesis. J Pharmacol Sci. 2008;108(3):227–38.
18. Griffiths T, Evans MC, Meldrum BS. Intracellular sites of early calcium accumulation in the rat hippocampus during status epilepticus. Neurosci Lett. 1982;30(3):329–34.
19. Griffiths T, Evans MC, Meldrum BS. Status epilepticus: the reversibility of calcium loading and acute neuronal pathological changes in the rat hippocampus. Neuroscience. 1984;12(2):557–67.
20. Yaffe K, Ferriero D, Barkovich AJ, Rowley H. Reversible MRI abnormalities following seizures. Neurology. 1995;45(1):104–8.
21. Hong K-S, Cho Y-J, Lee SK, Jeong S-W, Kim WK, Oh EJ. Diffusion changes suggesting predominant vasogenic oedema during partial status epilepticus. Seizure. 2004;13(5):317–21.
22. Nitsch C, Klatzo I. Regional patterns of blood-brain barrier breakdown during epileptiform seizures induced by various convulsive agents. J Neurol Sci. 1983;59(3):305–22.
23. Sammaritano M, Andermann F, Melanson D, Pappius HM, Camfield P, Aicardi J, et al. Prolonged focal cerebral edema associated with partial status epilepticus. Epilepsia. 1985;26(4):334–9. Blackwell Publishing Ltd.
24. Choy M, Cheung KK, Thomas DL, Gadian DG, Lythgoe MF, Scott RC. Quantitative MRI predicts status epilepticus-induced hippocampal injury in the lithium–pilocarpine rat model. Epilepsy Res. 2010;88(2-3):221–30.
25. Fabene PF, Marzola P, Sbarbati A, Bentivoglio M. Magnetic resonance imaging of changes elicited by status epilepticus in the rat brain: diffusion-weighted and T2- weighted images, regional blood volume maps, and direct correlation with tissue and cell damage. NeuroImage. 2003;18(2):375–89.
26. Wang Y, Majors A, Najm I, Xue M, Comair Y, Modic M, et al. Postictal alteration of sodium content and apparent diffusion coefficient in epileptic rat brain induced by Kainic acid. Epilepsia. 1996;37(10):1000–6. Blackwell Publishing Ltd.
27. Fisher RS, Schachter SC. The postictal state: a neglected entity in the management of epilepsy. Epilepsy Behav. 2000;1(1):52–9.
28. Koepp MJ, Diehl B, Woermann FG. Functional neuroimaging in the postictal state. Epilepsy Behav. 2010;19(2):127–30.
29. Mathews MS, Smith WS, Wintermark M, Dillon WP, Binder DK. Local cortical hypoperfusion imaged with CT perfusion during postictal Todd's paresis. Neuroradiology. 2008;50(5):397–401. 2nd ed. Springer-Verlag.
30. Rupprecht S, Schwab M, Fitzek C, Witte OW, Terborg C, Hagemann G. Hemispheric hypoperfusion in postictal paresis mimics early brain ischemia. Epilepsy Res. 2010;89(2-3):355–9.
31. Riviello JJ, Ashwal S, Hirtz D, Glauser T, Ballaban-Gil K, et al. Practice parameter: diagnostic assessment of the child with status epilepticus (an evidence-based review): report of the Quality Standards Subcommittee of the American Academy of Neurology and the Practice Committee of the Child Neurology Society. Neurology. 2006;67:1542–50. Lippincott Williams & Wilkins.
32. Gaillard WD, Chiron C, Helen Cross J, Simon Harvey A, Kuzniecky R, Hertz-Pannier L, et al. Guidelines for imaging infants and children with recent-onset epilepsy. Epilepsia. 2009;50(9):2147–53.
33. Brophy GM, Bell R, Claassen J, Alldredge B, Bleck TP, Glauser T, et al. Neurocritical Care Society Status Epilepticus Guideline Writing Committee. Guidelines for the evaluation and management of status epilepticus. Neurocrit Care. 2012;17(1):3–23. https://doi.org/10.1007/s12028-012-9695-z. PMID: 22528274.
34. Freilich ER, Schreiber JM, Zelleke T, Gaillard WD. Pediatric status epilepticus. Curr Opin Pediatr. 2014;26(6):655–61.
35. Maytal J, Krauss JM, Novak G, Nagelberg J, Patel M. The role of brain computed tomography in evaluating children with new onset of seizures in the emergency department. Epilepsia. 2000;41(8):950–4.

36. Lyons TW, Johnson KB, Michelson KA, Nigrovic LE, Loddenkemper T, Prabhu SP, et al. Yield of emergent neuroimaging in children with new-onset seizure and status epilepticus. Seizure: Eur J Epilepsy. 2016;1(35):4–10. BEA Trading Ltd.
37. Singh RK, Stephens S, Berl MM, Chang T, Brown K, Vezina LG, et al. Prospective study of new-onset seizures presenting as status epilepticus in childhood. Neurology. 2010;74(8):636–42. Lippincott Williams & Wilkins.
38. Earnest MP, Feldman H, Marx JA, Harris JA, Biletch M, Sullivan LP. Intracranial lesions shown by CT scans in 259 cases of first alcohol-related seizures. Neurology. 1988;38(10):1561–5.
39. Mendes A, Sampaio L. Brain magnetic resonance in status epilepticus: a focused review. Seizure. 2016;38:63–7.
40. Goyal MK, Sinha S, Ravishankar S, Shivshankar JJ. Peri-ictal signal changes in seven patients with status epilepticus: interesting MRI observations. Neuroradiology. 2009;51:151–61.
41. Eidt S, Kendall EJ, Obenaus A. Neuronal and glial cell populations in the piriform cortex distinguished by using an approximation of q-space imaging after status epilepticus. AJNR Am J Neuroradiol. 2004;25:1225–33.
42. Toledo M, Munuera J, Sueiras M, Rovira R, Álvarez-Sabín J, Rovira A. MRI findings in aphasic status epilepticus. Epilepsia. 2008;49:1465–9.
43. Von Oertzen TJ. PET and ictal SPECT can be helpful for localizing epileptic foci. Curr Opin Neurol. 2018;31:184–91.
44. Canas N, Breia P, Soares P, Saraiva P, Calado S, Jordão C, et al. The electroclinical imagiological spectrum and long-term outcome of transient periictal MR abnormalities. Epilepsy Res Elsevier BV. 2010;91(2-3):240–52.
45. Lansberg MG, O'Brien MW, Norbash AM, Moseley ME, Morrell M, Albers GW. MRI abnormalities associated with partial status epilepticus. Neurology. 1999;52(5):1021–7.
46. Kim JA, Chung JI, Yoon PH, Kim DI, Chung TS, Kim EJ, et al. Transient MR signal changes in patients with generalized tonicoclonic seizure or status epilepticus: peri-ictal diffusion-weighted imaging. AJNR Am J Neuroradiol. 2001;22(6):1149–60.
47. Nakae Y, Kudo Y, Yamamoto R, Dobashi Y, Kawabata Y, Ikeda S, et al. Relationship between cortex and pulvinar abnormalities on diffusion-weighted imaging in status epilepticus. J Neurol. 2015;263(1):127–32. Springer Berlin Heidelberg.
48. Chatzikonstantinou A, Gass A, Förster A, Hennerici MG, Szabo K. Features of acute DWI abnormalities related to status epilepticus. Epilepsy Res Elsevier BV. 2011;97(1-2):45–51.
49. Katramados AM, Burdette D, Patel SC, Schultz LR, Gaddam S, Mitsias PD. Periictal diffusion abnormalities of the thalamus in partial status epilepticus. Epilepsia. 2009;50(2):265–75.
50. Samaniego EA, Stuckert E, Fischbein N, Wijman CAC. Crossed cerebellar diaschisis in status epilepticus. Neurocrit Care. 2009;12(1):88–90.
51. Massaro AM. Teaching neuroimages:crossed cerebellar diaschisis in hemispheric status epilepticus. Neurol Am Acad Neurol. 2012;79(20):e182.
52. Gaxiola-Valdez I, Singh S, Perera T, Sandy S, Li E, et al. Seizure onset zone localization using postictal hypoperfusion detected by arterial spin labelling MRI. Brain. 2017;140(11): 2895–911.
53. Huang Y-C, Weng H-H, Tsai Y-T, Huang Y-C, Hsiao M-C, Wu C-Y, et al. Periictal magnetic resonance imaging in status epilepticus. Epilepsy Res. 2009;86(1):72–81.
54. Sarria-Estrada S, Santamarina E, Quintana M, Pareto D, Sueiras M, Auger C, Toledo M, Rovira A. Magnetic resonance imaging findings in focal-onset status epilepticus. Eur J Neurol. 2022;29(1):3–11. https://doi.org/10.1111/ene.15065. Epub 2021 Sept 3. PMID: 34390102.
55. Mariajoseph FP, Sagar P, Muthusamy S, Amukotuwa S, Seneviratne U. Seizure-induced reversible MRI abnormalities in status epilepticus: a systematic review. Seizure. 2021;92:166–73. https://doi.org/10.1016/j.seizure.2021.09.002. Epub 2021 Sept 7. PMID: 34525432.

56. Bosque Varela P, Machegger L, Oellerer A, Steinbacher J, McCoy M, Pfaff J, Trinka E, Kuchukhidze G. Imaging of status epilepticus: making the invisible visible. A prospective study on 206 patients. Epilepsy Behav. 2023;141:109130. https://doi.org/10.1016/j.yebeh.2023.109130. Epub 2023 Feb 18. PMID: 36803874.
57. Giovannini G, Kuchukhidze G, McCoy MR, Meletti S, Trinka E. Neuroimaging alterations related to status epilepticus in an adult population: definition of MRI findings and clinical-EEG correlation. Epilepsia. 2018;59(Suppl 2):120–7. https://doi.org/10.1111/epi.14493. Epub 2018 Aug 20. PMID: 30129213.
58. Althaus K, Kasel M, Ludolph AC, Kassubek J, Kassubek R. HARM revisited: etiology of subarachnoid hyperintensities in brain FLAIR MRI. Int J Stroke. 2022; https://doi.org/10.1177/17474930211064754. Epub ahead of print. PMID: 34983275.
59. Morris JM, Miller GM. Increased signal in the subarachnoid space on fluid attenuated inversion recovery imaging associated with the clearance dynamics of gadolinium chelate: a potential diagnostic pitfall. Am J Neuroradiol. 2007;28(10):1964–7.
60. Forster A, et al. Hyperintense acute reperfusion marker on FLAIR in a patient with transient ischemic attack. Case Rep Radiol. 2016:1–4. EBSCOhost. https://doi.org/10.1155/2016/9829823.
61. Stuckey SL, et al. Hyperintensity in the subarachnoid space on FLAIR MRI. Am J Roentgenol. 2007;189(4):913–21.
62. Al-Chalabi M, Bajrami S, Karim N, Sheikh A. Rare pitfall in the magnetic resonance imaging of status epilepticus. eNeurologicalSci. 2022;27:100405. https://doi.org/10.1016/j.ensci.2022.100405. PMID: 35647328; PMCID: PMC9136252.
63. Villalobos-Chavez F, et al. Sequential changes in magnetic resonance in a limbic status epilepticus. Rev Neurol. 2005;40(6):354–7.
64. Kim DW, et al. Blood-brain barrier disruption is involved in seizure and hemianopsia in nonketotic hyperglycemia. Neurologist. 2011;17(3):164–6.
65. Bozzao A, et al. Cerebrospinal fluid changes after intravenous injection of gadolinium chelate: assessment by FLAIR MR imaging. Eur Radiol. 2003;13(3):592–7.
66. Masterson K, Vargas MI, Delavelle J. Postictal deficit mimicking stroke: role of perfusion CT. J Neuroradiol. 2009;36(1):48–51. https://doi.org/10.1016/j.neurad.2008.08.006. Epub 2008 Oct 5. PMID: 18835645.
67. Dekeyzer S, Vanden Bossche S, Keereman V, Hemelsoet D, Van Driessche V. Stroke versus seizure - perfusion computerized tomography in a patient with aphasia. J Belg Soc Radiol. 2015;99(2):85–9. https://doi.org/10.5334/jbr-btr.880. PMID: 30039114; PMCID: PMC6032479.
68. Wiest R, von Bredow F, Schindler K, et al. Detection of regional blood perfusion changes in epileptic seizures with dynamic brain perfusion CT—a pilot study. Epilepsy Res. 2006;72:102–10. https://doi.org/10.1016/j.eplepsyres.2006.07.017. PMid: 17118628.
69. Gelfand JM, Wintermark M, Josephson SA. Cerebral perfusion-CT patterns following seizure. Eur J Neurol. 2010;17:594–601. https://doi.org/10.1111/j.14681331.2009.02869.x.
70. Hauf M, Slotboom J, Nirkko A, von Bredow F, Ozdoba C, Wiest R. Cortical regional hyperperfusion in nonconvulsive status epilepticus measured by dynamic brain perfusion CT. AJNR Am J Neuroradiol. 2009;30:693–8. https://doi.org/10.3174/ajnr.A1456.
71. Sarikaya I. PET studies in epilepsy. Am J Nucl Med Mol Imaging. 2015;5(5):416–30. PMID: 26550535; PMCID: PMC4620171.
72. Sone D, Maikusa N, Sato N, Kimura Y, Ota M, Matsuda H. Similar and differing distributions between ^{18}F-FDG-PET and arterial spin labeling imaging in temporal lobe epilepsy. Front Neurol. 2019;10:318. https://doi.org/10.3389/fneur.2019.00318. PMID: 31001198; PMCID: PMC6456651.
73. Chugani HT, Shewmon DA, Khanna S, Phelps ME. Interictal and postictal focal hypermetabolism on positron emission tomography. Pediatr Neurol. 1993;9:10–5.
74. Leiderman DB, Albert P, Balish M, Bromfield E, Theodore WH. The dynamics of metabolic change following seizures as measured by positron emission tomography with fluorodeoxyglucose F 18. Arch Neurol. 1994;51:932–6.

75. Trinka E, Leitinger M. Management of Status Epilepticus, refractory status epilepticus, and super-refractory status epilepticus. Continuum (Minneap Minn). 2022;28(2):559–602. https://doi.org/10.1212/CON.0000000000001103. PMID: 35393970.
76. Meletti S, Monti G, Mirandola L, Vaudano AE, Giovannini G. Neuroimaging of status epilepticus. Epilepsia. 2018;59(Suppl 2):113–9. https://doi.org/10.1111/epi.14499. Epub 2018 Aug 29. PMID: 30160066.
77. Meletti S, Slonkova J, Mareckova I, Monti G, Specchio N, Hon P, Giovannini G, Marcian V, Chiari A, Krupa P, Pietrafusa N, Berankova D, Bar M. Claustrum damage and refractory status epilepticus following febrile illness. Neurology. 2015;85(14):1224–32. https://doi.org/10.1212/WNL.0000000000001996. Epub 2015 Sept 4. PMID: 26341869; PMCID: PMC4607596.
78. Crick FC, Koch C. What is the function of the claustrum? Philos Trans R Soc Lond Ser B Biol Sci. 2005;360(1458):1271–9. https://doi.org/10.1098/rstb.2005.1661. PMID: 16147522; PMCID: PMC1569501.
79. Wong KLL, Nair A, Augustine GJ. Changing the cortical conductor's tempo: neuromodulation of the claustrum. Front Neural Circuits. 2021;15:658228. https://doi.org/10.3389/fncir.2021.658228. PMID: 34054437; PMCID: PMC8155375.
80. Meletti S, Giovannini G, d'Orsi G, Toran L, Monti G, Guha R, Kiryttopoulos A, Pascarella MG, Martino T, Alexopoulos H, Spilioti M, Slonkova J. New-onset refractory status epilepticus with claustrum damage: definition of the clinical and neuroimaging features. Front Neurol. 2017;8:111. https://doi.org/10.3389/fneur.2017.00111. PMID: 28396650; PMCID: PMC5366956.
81. Gaspard N, Foreman BP, Alvarez V, Cabrera Kang C, Probasco JC, Jongeling AC, Meyers E, Espinera A, Haas KF, Schmitt SE, Gerard EE, Gofton T, Kaplan PW, Lee JW, Legros B, Szaflarski JP, Westover BM, LaRoche SM, Hirsch LJ, Critical Care EEG Monitoring Research Consortium (CCEMRC). New-onset refractory status epilepticus: etiology, clinical features, and outcome. Neurology. 2015;85(18):1604–13. https://doi.org/10.1212/WNL.0000000000001940. Epub 2015 Aug 21. PMID: 26296517; PMCID: PMC4642147.
82. Stayman A, Abou-Khalil B. FDG-PET in the diagnosis of complex partial status epilepticus originating from the frontal lobe. Epilepsy Behav. 2011;20:721–4.
83. Mehta AD, Ettinger AB, Perrine K, Dhawan V, Patil A, Jain SK, Klein G, Schneider SJ, Eidelberg D. Seizure propagation in a patient with musicogenic epilepsy. Epilepsy Behav. 2009;14:421–4.
84. Hirsch LJ, Gaspard N, van Baalen A, Nabbout R, Demeret S, Loddenkemper T, Navarro V, Specchio N, Lagae L, Rossetti AO, Hocker S, Gofton TE, Abend NS, Gilmore EJ, Hahn C, Khosravani H, Rosenow F, Trinka E. Proposed consensus definitions for new-onset refractory status epilepticus (NORSE), febrile infection-related epilepsy syndrome (FIRES), and related conditions. Epilepsia. 2018;59(4):739–44. https://doi.org/10.1111/epi.14016. Epub 2018 Feb 5. PMID: 29399791.
85. Wickstrom R, Taraschenko O, Dilena R, Payne ET, Specchio N, Nabbout R, Koh S, Gaspard N, Hirsch LJ, International NORSE Consensus Group. International consensus recommendations for management of new onset refractory status epilepticus (NORSE) incl. Febrile infection-related epilepsy syndrome (FIRES): statements and supporting evidence. Epilepsia. 2022;63(11):2840–64. https://doi.org/10.1111/epi.17397. Epub ahead of print. PMID: 35997591; PMCID: PMC9828002.
86. Smith JB, Lee AK, Jackson J. The claustrum. Curr Biol. 2020;30(23):R1401–6. https://doi.org/10.1016/j.cub.2020.09.069. PMID: 33290700.
87. Di Dier K, Dekesel L, Dekeyzer S. The claustrum sign in febrile infection-related epilepsy syndrome (FIRES). J Belg Soc Radiol. 2023;107(1):45. https://doi.org/10.5334/jbsr.3142. PMID: 37396965; PMCID: PMC10312252.
88. Sculier C, Gaspard N. New onset refractory status epilepticus (NORSE). Seizure. 2019;68:72–8. https://doi.org/10.1016/j.seizure.2018.09.018. Epub 2018 Sept 29. PMID: 30482654.

89. Kim HJ, Lee SA, Kim HW, Kim SJ, Jeon SB, Koo YS. The timelines of MRI findings related to outcomes in adult patients with new-onset refractory status epilepticus. Epilepsia. 2020;61(8):1735–48. https://doi.org/10.1111/epi.16620. Epub 2020 Jul 27. PMID: 32715470.
90. Gelisse P, Crespel A, Genton P, Jallon P, Kaplan PW. Lateralized periodic discharges: which patterns are interictal, ictal, or peri-ictal? Clin Neurophysiol. 2021;132(7):1593–603. https://doi.org/10.1016/j.clinph.2021.04.003.
91. Subramaniam T, Jain A, Hall LT, Cole AJ, Westover MB, Rosenthal ES, et al. Lateralized periodic discharges frequency correlates with glucose metabolism. Neurology. 2019;92:e670–4. https://doi.org/10.1212/wnl.0000000000006903.
92. Franck G, Sadzot B, Salmon E, et al. Regional cerebral blood flow and metabolic rates in human focal epilepsy and status epilepticus. Adv Neurol. 1986;44:935–48.
93. Handforth A, Cheng JT, Mandelkern MA, Treiman DM. Markedly increased mesiotemporal lobe metabolism in a case with PLEDs: further evidence that PLEDs are a manifestation of partial status epilepticus. Epilepsia. 1994;35:876–81.
94. Struck AF, Westover MB, Hall LT, Deck GM, Cole AJ, Rosenthal ES. Metabolic correlates of the ictal-interictal continuum: FDG-PET during continuous EEG. Neurocrit Care. 2016;24(3):324–31. https://doi.org/10.1007/s12028-016-0245-y.
95. Szabo K, Poepel A, Pohlmann-Eden B, et al. Diffusion-weighted and perfusion MRI demonstrates parenchymal changes in complex partial status epilepticus. Brain. 2005;128:1369–76.
96. Narayanan J. Can diffusion-weighted imaging be used as a tool to predict seizures in patients with PLEDS? Epileptic Disord. 2016;18:440–6.
97. Claassen J. How I Treat Patients with EEG Patterns on the Ictal–Interictal Continuum in the Neuro ICU. Neurocrit Care. 2009;11:437–44. https://doi.org/10.1007/s12028-009-9295-8.
98. Lee BI, Schauwecker DS. Regional cerebral perfusion in PLEDs: a case report. Epilepsia. 1988;29:607–11.
99. Assal F, Papazyan JP, Slosman DO, Jallon P, Goerres GW. SPECT in periodic lateralized epileptiform discharges (PLEDs): a form of partial status epilepticus? Seizure. 2001;10:260–5.
100. Bozkurt MF, Saygi S, Erbas B. SPECT in a patient with postictal PLEDs: is hyperperfusion evidence of electrical seizure? Clin Electroencephalogr. 2002;33:171–3.
101. Ergun EL, Salanci BV, Erbas B, Saygi S. SPECT in periodic lateralized epileptiform discharges (PLEDs): a case report on PLEDs. Ann Nucl Med. 2006;20:227–31.
102. Ali II, Pirzada NA, Vaughn BV. Periodic lateralized epileptiform discharges after complex partial status epilepticus associated with increased focal cerebral blood flow. J Clin Neurophysiol. 2001;18:565–9.
103. Hisada K, Morioka T, Nishio S, et al. Magnetoencephalographic analysis of periodic lateralized epileptiform discharges (PLEDs). Clin Neurophysiol. 2000;111:122–7.
104. Aye SM, Lim KS, Ramli NM, Tan CT. Periodic lateralized epileptiform discharges (PLEDs) in cerebral lupus correlated with white-matter lesions in brain MRI and reduced cerebral blood flow in SPECT. Lupus. 2013;22:510–4.
105. Sato T, Katagiri M, Terasawa Y. Correlation between lateralized periodic discharges and arterial spin labeling perfusion imaging in patients with status epilepticus. Neurol Sci. 2024;45:5547. https://doi.org/10.1007/s10072-024-07658-7.
106. Eisermann M, Fillon L, Saitovitch A, et al. Periodic electroencephalographic discharges and epileptic spasms involve cortico-striatal-thalamic loops on Arterial Spin Labeling Magnetic Resonance Imaging. Brain Commun. 2022;4(5):fcac250. Published 2022 Oct 6. https://doi.org/10.1093/braincomms/fcac250.

Intraoperative Magnetic Resonance Imaging in the Surgical Treatment of Epilepsy

Graham Winston and Theodore H. Schwartz

Abstract Epilepsy is the only chronic neurodegenerative disorder that can be cured with surgery. Epilepsy surgery encompasses both curative and palliative procedures. Curative surgeries are based on the removal of a focal area of the brain from which the epilepsy arises. Often, this region can be visualized on magnetic resonance imaging (MRI) but is not apparent macroscopically in the operating room. Furthermore, stereotactic navigation in the operating room that is based on a preoperative acquired dataset does not account for intraoperative "brain shift" because of gravity, loss of cerebrospinal fluid, and retraction. MRI can be a useful adjunct for intraoperative neuronavigation in these situations to assure complete removal of the lesion. Surgeries such as amygdalohippocampectomy, functional hemispherectomy, and corpus callosotomy require a complete removal or disconnection of an anatomic structure in the brain. Intraoperative judgment can sometimes be inaccurate in assuring that all operative goals are met. The incorporation of intraoperative MRI scans, combined with neuronavigation, plays a pivotal role in validating or guiding the surgeon's assessment of various anatomical boundaries. Yet, it remains to be determined whether the adoption of intraoperative MRI is cost-effective and whether it significantly impacts patient outcomes, making it a subject of ongoing investigation.

Keywords Epilepsy · Seizure · Magnetic resonance imaging · Laser interstitial thermal therapy · Temporal lobe · Neurologic surgery

G. Winston · T. H. Schwartz (✉)
Weill Cornell Medical College, New York, NY, USA
e-mail: gmw2002@med.cornell.edu; schwarh@med.cornell.edu

P. A. Broderick (ed.), *Biomedical Imaging in Neurodegeneration*,
https://doi.org/10.1007/978-3-031-95878-6_5

1 Introduction

Epilepsy remains the only chronic neurodegenerative disease of the brain that can be cured with surgery. Although surgical options exist for the symptoms of other neurodegenerative diseases, such as Parkinson's disease [1], Alzheimer's disease [2], they are merely palliative at best. In contrast, surgical resection of a well-defined epileptogenic focus may be curative in as many as 70–80% of cases [3, 4]. Remarkably, the cognitive decline associated with persistent epilepsy can be reversed after cessation of seizures, and these gains can persist for up to 10 years after resection, particularly if surgery is performed in the non-dominant hemisphere [5]. The effectiveness of resection is proportional to the focality of the epileptic zone in the brain. Patients with localized pathology have a higher cure rate than those with diffuse brain abnormalities or those with normal MRI scans [6, 7]. While epilepsy is primarily a disease of brain function and not structure, association with a structural abnormality is quite common, and precise localization of the structural abnormality is often critical to successful epilepsy surgery. Not uncommonly, however, structural abnormalities that can be seen on MRI scan are not grossly apparent to the surgeon in the operating room. The ability to identify abnormal MRI signals in real-time and correlate these signals with visualized brain structures in the OR is a powerful advance in our ability to surgically treat chronic seizure disorders that is offered by intraoperative MRI (iMRI).

Despite the advantage provided by the ability to localize abnormal MRI findings in "surgical space" with intraoperative neuronavigation tools, epilepsy is first-and-foremost a disease of brain function and not structure. Accordingly, surgeons often define the margins of their resections based on electrographic data, which oftentimes have no MRI correlate. What use, then, is iMRI in this situation? Because epilepsy neurosurgeons, unlike vascular, oncologic, or spine neurosurgeons, often find themselves operating on normal appearing anatomy with no pathologic substrate, resections are frequently based on known anatomical boundaries. As will be discussed below, surgeries such as selective amygdalohippocampectomy, corpus callosotomy, and functional hemispherectomy are based on aggressive removal or disconnections of anatomical structures, such as deep gray matter nuclei or white matter tracts. The margins of these landmarks can be quite difficult to differentiate from neighboring structures in the OR, even under microscopic illumination and magnification. Real-time anatomical feedback from an iMRI can be extremely useful in anatomic-based resections for epilepsy.

Finally, we will discuss other uses for iMRI in epilepsy surgery, such as for stereotactic placement of depth electrodes and laser interstitial thermal therapy (LITT). While iMRI has theoretical advantages in epilepsy surgery, its actual impact on patient outcomes is unclear. It is an expensive technology, and it is uncertain whether it offers advantages over more affordable technology. Therefore, the author does not suggest that iMRI is necessary for successful epilepsy surgery, but rather that it may improve outcomes in specific cases.

2 Intraoperative MRI

2.1 Justification for the Technology

The implementation of iMRI into neurosurgical practice and its combination with frameless stereotaxy, also known as "neuronavigation," is the result of decades of work in the fields of image-guided, computer-guided, and stereotactic neurosurgery. The goal of this technology is to localize a point in three-dimensional space, as defined by computed tomography (CT) or MRI imaging, within a patient's head to guide a neurosurgical procedure. The use of computers with increasing processor speeds and developments in graphical interfaces has been instrumental in the growth of this field. It should be noted that with the recent advent and increase in the use of robotic stereo-EEG, iMRI is not being used to place electrodes, but rather as an aid in identifying and resecting pathologic, epileptogenic tissue.

The need for neuronavigation is based on the following facts: (1) the boundaries of a focal lesion in the brain are unknown when preparations are made to incise the skin and perform the craniotomy to expose these pathologies; (2) certain anatomical boundaries in the brain that can be viewed on MRI may not be appreciable with the naked eye macroscopically in the OR; (3) the location and trajectory to a focal lesion in the brain that does not present to the cortical surface is unknown when the brain is exposed in the OR; (4) the ability to see around corners and beyond anatomical surfaces is impossible with a conventional operating microscope but critical to successful neurosurgical procedures; (5) certain brain pathology that can be visualized on MRI or CT may not be distinguishable from normal brain by the naked eye macroscopically in the OR. Hence, an accurate system for image-guided stereotactic neuronavigation is essential for determining (1) where to make a skin incision and the extent of bone removal required; (2) where to enter the cortex to reach a lesion without damaging a normal structure; (3) the trajectory required to reach the lesion; (4) the boundaries required for a safe and successful biopsy or excision; (5) the absence or presence of residual tissue that should be resected.

Before the advent of iMRI, most surgeons relied on frame-based or frameless stereotactic systems, which are based exclusively on preoperative imaging. These systems are excellent if there are no changes in the structure of the brain between the time the data are acquired and the time they are needed for intraoperative decision-making. However, once the brain is exposed and cerebrospinal fluid is lost, or some tissue is resected, certain brain structures can shift by as much as 1–1.5 cm [8–14]. Therefore, a truly accurate image-guided system for neuronavigation requires real-time imaging to compensate for this known inaccuracy. One such possible modality is intraoperative ultrasound, which has proven to be helpful specifically in achieving gross-total resection of echogenic masses such as gliomas [15]. Intraoperative CT is another solution; however, soft tissue resolution is not comparable with MRI. Ultimately, the best solution was to integrate MRI technology with the OR environment.

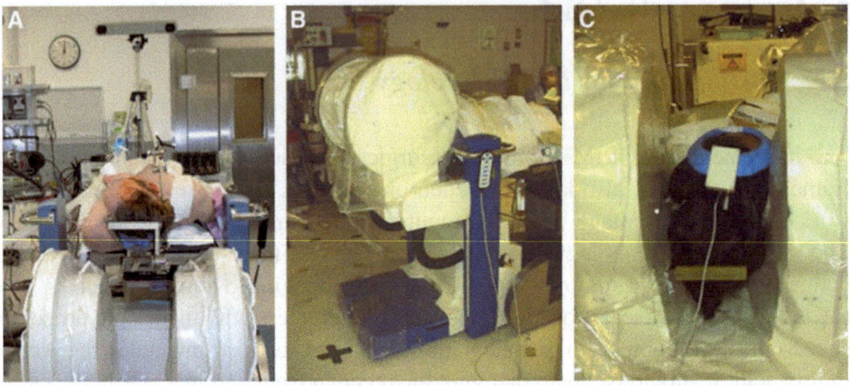

Fig. 1 The ODIN Polestar N-10 0.12 T intraoperative magnet with integrated neuronavigation. (**a**) The two poles of the magnet sit on the floor under the patient's head. The head is fixed in a MR-compatible head-holder. There is a reference frame attached to the head-holder on the right side of the patient's neck with three reflective spheres. At the foot of the bed is an infrared camera, which notes the position of the patient, the magnet, and a reflecting wand for neuronavigation (not shown). (**b**) The magnets can be robotically elevated with a push of a button into "scanning position," and the position is remembered by the system for later comparison images taken during the surgery. (**c**) A surface coil (blue) must be placed in the center of the magnet over the surface of the head for each image. This can be placed in a sterile bag for use during surgery

2.2 *Technology*

iMRI has continued to advance since its introduction into clinical practice. The first system, which was introduced at Brigham and Women's Hospital in Boston, used a 0.5-T "double donut" with two vertical magnets in which the surgeons operated using non-ferromagnetic equipment [16]. In an effort to make this technology more accessible, the next iteration of devices were made with the intention of being more compact, albeit with lower field strength magnets ranging from 0.12 to 0.20-T (Figs. 1 and 2) [17–19]. Introduction of high-field magnets, between 1.5- and 3.0-T, has allowed for use of high-resolution iMRI. These devices either require the use of non-ferromagnetic instruments or are designed with the ability to move in and out of the operative field on mounted tracts (Fig. 3) [20–23]. The imaging capabilities of the system extend to include MR spectroscopy, fMRI, and MR angiography.

The purchase and installation of an iMRI can be expensive and, in addition to the cost of the magnet, the OR must have radiofrequency shielding and dedicated anesthesia equipment, including ventilator, infusion pumps, and monitoring screens. Several modifications in anesthesia technique must be made to accommodate the high magnetic field and limit radiofrequency noise (Fig. 4) [24]. Cables cannot come in contact with either the scanner or the patient's skin to prevent burns, and

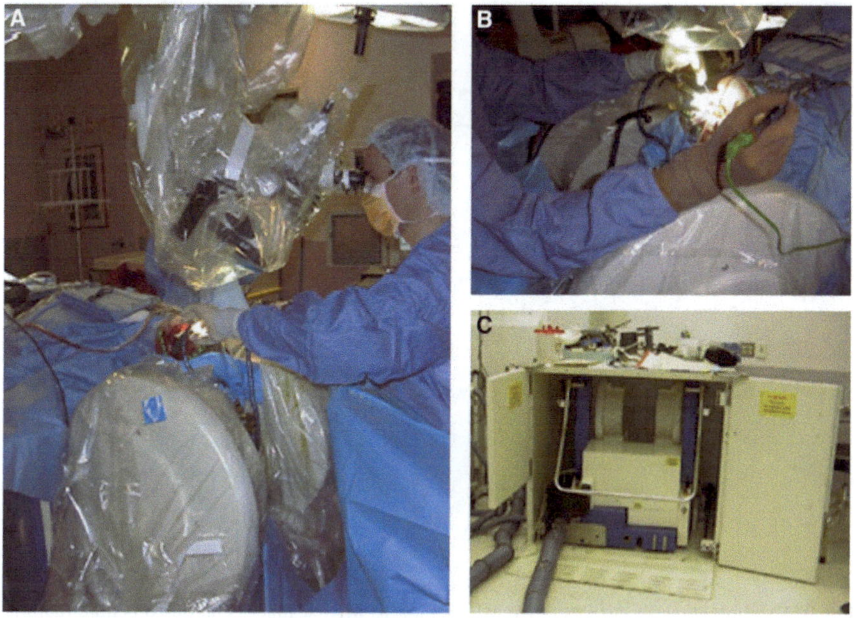

Fig. 2 The magnets can be placed in an intermediate position during surgery. (**a**) Clear sterile drapes are placed over the magnets, which remain sterile with the magnet lowered slightly. (**b**) In this position, the surgeon can use the magnets as armrests and normal instruments can still be used. (**c**) When not in use, the magnets are stored in a lead-shielded box and the OR can be used for non-MRI cases

the pressure transducers must be kept outside the scanner to prevent artifacts. The increase in OR time has been reported to be approximately 30 min, however, this is highly variable.

3 Epilepsy Surgery

Epilepsy is a clinical term referring to a disease involving recurrent seizures and effects between 1% and 2% of the United States population [25]. Epilepsies can be divided into "focal," which start in a focal area of the brain and are sub-classified into "focal-aware" and "focal with impaired awareness," and "generalized," which involve large areas of the brain simultaneously. Initial therapy is always medical, which is variably successful at reducing the frequency and severity of seizures, depending on the etiology. Patients who are refractory to medical treatment are considered for surgical therapy. Surgical therapy for epilepsy can be divided into "curative" or "palliative." Curative surgeries aim at removing the region of brain that

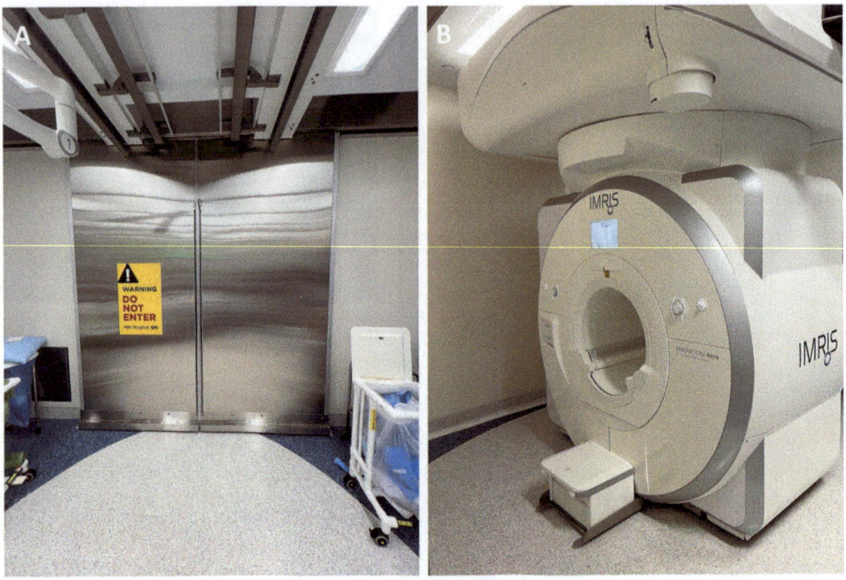

Fig. 3 Example modern intraoperative MRI setup. (**a**) MRI scanner is harbored behind radiofrequency shielded doors during the operation, allowing for use of ferromagnetic, standard operating equipment. (**b**) At time of intraoperative MRI, operating room equipment must be moved out of magnet zone, at which time the MRI scanner is brought into the operating theater on ceiling-mounted tracks. (Images courtesy of Dr. Jacob Goldberg)

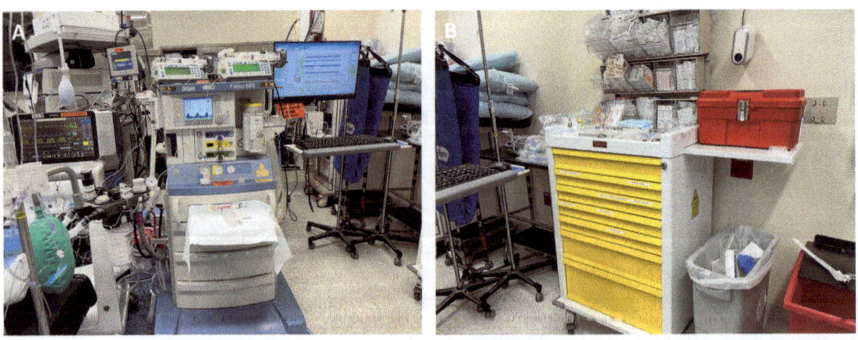

Fig. 4 iMRI requires extensive operating room radiofrequency shielding. In addition to shielding the OR walls, (**a**) anesthesia equipment must be MRI compatible, as these machines cannot be switched off during imaging. All other equipment must be shielded, such as computers. (**b**) Other supplies, including emergency patient safety items, must also be carefully shielded. (Images courtesy of Dr. John Chae)

is responsible for the onset of the seizure, the so-called "ictal onset zone." Palliative surgeries either prevent the spread of epileptic events by disrupting pathways of propagation or partially interrupt seizure initiation by sub-totally removing or disabling the ictal onset zone.

The most common focal-onset epilepsy treated surgically is medial temporal lobe epilepsy. Seizures arise from a local network involving the hippocampus, amygdala, entorhinal cortex, and parahippocampal gyrus. The removal of these structures is curative in as many as 70–80% of well-selected cases [4]. Neocortical epilepsy, in which the seizures arise from a specific abnormal region in the cerebral cortex, is also potentially curable with surgery. This disease can be divided into lesional epilepsy, in which there is an abnormality found on MRI scan, and nonlesional epilepsy, in which the MRI scan appears normal. Lesional epilepsy may be caused either by a macrostructural abnormality, such as a low-grade tumor, vascular malformation, or gliotic scar, or a microstructural abnormality, such as focal cortical dysplasia. If the epileptogenic area is focal, then surgical cure may be expected in a high percentage of patients (50–95%) [26, 27] whereas a more diffuse epileptic area may be more difficult to resect and the rate of cure will be accordingly lower. Nonlesional neocortical epilepsy in a patient with a normal MRI scan is usually caused by more diffuse, subtle cortical dysplasia or an idiopathic etiology, and the cure rates with resective surgery are lower (25–40%), again depending on the focality of the epileptic onset zone [6, 28, 29].

Palliative surgical procedures are also extremely effective in preventing the initiation and spread of seizures and can decrease both the severity and frequency of the ictal events. Corpus callosotomy is one such procedure, in which the fibers connecting the hemispheres are disrupted to prevent secondary generalization of partial onset seizures. Likewise, partial resection of an epileptic focus that overlaps with functional brain, such as motor or language cortex, may provide palliation rather than cure. Techniques such as multiple subpial transections, vagal nerve stimulation, deep brain stimulation, direct cortical stimulation, chronic drug infusion, and gene therapy are currently being investigated as further palliative approach for patients who are not suitable for resective surgery of their epileptic focus.

4 Intraoperative MRI and Epilepsy Surgery

4.1 History

The first report of epilepsy surgery performed in an iMRI was by Steinmeier et al. in 1998 [17]. They reported six patients who underwent temporal lobe resections for pharmacoresistant epilepsy in the 0.2-T Magnetom Open Scanner as part of a larger series of tumor patients. Utility in tailoring the extent of the mesial and neocortical resection to fit the "preoperative findings of morphological and electrophysiological alterations" was described. Seven additional temporal lobectomies were mentioned in a larger series of tumor patients presented by Kaibara et al. in 2000, however, there were no specific discussion of these cases [20]. In the first epilepsy-specific surgical cohort published, Buchfelder et al. reported an additional 61 cases including temporal lobectomy, corpus callosotomy, and lesionectomy [30]. This cohort of patients, who were treated at University Hospital Erlangen, has since been expanded to nearly 500 surgical patients as of reporting in 2016 (Table 1) [31].

Table 1 Reported epilepsy surgeries using iMRI

Author	Year	No. of patients	Surgery	MRI scanner
Steinmeyer et al. [17]	1998	6	Temporal lobectomy	0.2 T
Kaibara et al. [20]	2000	7	Temporal lobectomy	1.5 T
Buchfelder et al. [30]	2000	57	Temporal lobectomy/lesionectomy	0.2 T
		4	Corpus callosotomy	
Nimsky et al. [19]	2001	1	Corpus callosotomy	0.2 T
Schwartz et al. [44]	2002	5	Amygdalohippocampectomy	0.12 T
Kaibara et al. [45]	2002	14	Amygdalohippocampectomy	1.5 T
Walker et al. [49]	2002	12	Neocortical lesionectomy	0.5 T
		1	Hypothalamic hamartoma	
Kelly et al. [43]	2005	31	Selective amygdalohippocampectomy	1.5 T
		20	Temporal lobectomy	
		16	Neocortical lesionectomy	
		3	Corpus callosotomy	
Levy et al. [46]	2009	12	Neocortical lesionectomy	1.5 T
		7	Selective amygdalohippocampectomy	
		3	Corpus callosotomy	
		2	Functional hemispherectomy	
		2	Endoscopic ventricular exploration	
		1	Depth electrode	
Sun et al. [50]	2011	36	Neocortical lesionectomy	1.5 T
Sommer et al. [51]	2013	25	Neocortical lesionectomy	1.5 T
Cui et al. [52]	2014	31	Neocortical lesionectomy	1.5 T
Kurwale et al. [48]	2015	15	Temporal surgery	1.5 T
		15	"Extra-temporal" surgery	
		9	Disconnection surgery	
Roessler et al. [31]	2016	329	Temporal surgery	1.5 T
		63	"Extra-temporal" surgery	
		23	Depth electrode	
Roessler et al. [53]	2016	6	Depth electrode	3.0 T
Sacino et al. [54, 55]	2016	12	Neocortical lesionectomy	1.5 T
Sommer et al. [51]	2016	28	Neocortical lesionectomy	1.5 T
Warsi et al. [56]	2016	23	Lesionectomy	3.0 T
		16	Disconnection surgery	
van Tonder et al. [57]	2018	10	Hypothalamic hamartoma	1.5 T

4.2 Mesial Temporal Lobe Epilepsy

Advances in operative techniques and imaging technologies have made a dramatic impact on our understanding of the etiology and surgical treatment for medial temporal lobe epilepsy. The overall goal of any focal-onset epilepsy surgery is to maximize resection of epileptogenic tissue while minimizing chance of injury to critical structures. Therefore, determining the seizure outcomes and safety profiles of varying extents of surgical resection was of utmost importance for epilepsy surgeons, and the optimal approach in temporal lobe epilepsy remains highly controversial. Studies have shown that extent of resection of medial temporal structures are correlated with improved seizure outcome [32–35]. This is corroborated by studies demonstrating that residual hippocampus and parahippocampal gyrus may be responsible for recurrent seizures after initial medial temporal lobe resections [36, 37]. Therefore, many surgeons now perform a radical amygdalohippocampectomy as an integral part of their temporal lobe resections. Whether this approach confers a higher rate of visual field or neuropsychologic impairment than electrographically tailored resections has never been tested in a randomized prospective fashion [38]. Although there is clearly a subset of patients who will achieve excellent seizure control or seizure freedom after sub-total hippocampectomy [35, 39], it is not always apparent how to reliably identify these patients and whether partial hippocampectomy minimizes postoperative neuropsychological deficits. Hence, in many epilepsy centers, the goal of surgery in the treatment of medial temporal lobe epilepsy is an extensive resection of the hippocampus, amygdala, and parahippocampal gyrus.

Despite the visibility provided by the operative microscope, careful evaluation of postoperative MRIs has indicated that the extent of hippocampal resection is often overestimated by the operating surgeon [40, 41]. Additionally, there has been shown to be considerable variability in extent of "standard" procedures (selective amygdalohippocampectomy, for example), even when performed at the same center [41, 42]. While most surgeons now utilize frameless stereotaxy systems to minimize craniotomy size, locate the temporal horn of the lateral ventricle, and maximize removal of the targeted structures, the major flaw of this technique is its reliance on preoperative data sets. Brain shift caused by loss of cerebrospinal fluid, gravity, retraction, or resection can render surgical planning inaccurate [9–12]. When combined with neuronavigation, intraoperative MRI is an ideal technique for ensuring a complete but not overly aggressive resection of the medial temporal lobe structures, therefore optimizing outcome in the surgical treatment of temporal lobe epilepsy (Fig. 5).

514 temporal lobe resections for medially intractable epilepsy have been reported in the literature thus far (Table 1). These surgeries range from traditional temporal lobectomy [17, 20, 43] to amygdalohippocampectomy [43–46]. Advantages afforded by use of iMRI include (1) smaller craniotomies; (2) ease of finding the temporal horn of the lateral ventricle; (3) assuring complete and safe resection of the amygdala, hippocampus, and parahippocampal gyrus; (4) assuring complete removal of any associated lesions; (5) visualizing postoperative hematomas prior

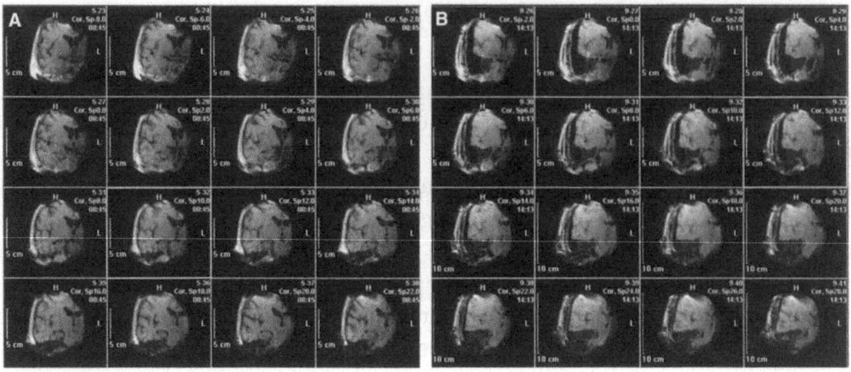

Fig. 5 Coronal T1-weighted scans through the temporal lobe, amygdala, and hippocampus obtained with the Polestar N-10 before resection (**a**) and after resection (**b**) before closure of craniotomy. If residual structures are seen, they can be removed prior to closing skin

to extubating the patient; and (6) minimizing morbidity during reoperations when anatomy is distorted. There is a growing amount of data comparing patients operated on using iMRI vs. standard image guidance, though no randomized studies to date. In the study by Kaibara et al., half of patients (7 of 14) had inadequately resected mesial temporal lobe structures at time of intraoperative MRI, requiring re-exploration and further surgery [45]. At a mean of 17 months, 93% of patients were seizure-free. The study by Kelly et al. also showed that iMRI was useful in discovering inadequate initial resection, with re-exploration surgery necessary in 27.4% of temporal lobe surgery cases (14/51) [43]. 84.3% of their temporal lobe surgery cohort achieved either Engel I or Engel II outcomes, defined as free of disabling seizures and rare disabling seizure respectively [47]. While Kurwale et al. claim that iMRI guided management decisions in focal cortical dysplasia cases, they state that there were no alterations in surgical plan based on iMRI for mesial temporal sclerosis (MTS) cases [48].

4.3 Neocortical Epilepsy

Although the mesial temporal structures are the most common location for focal seizures, epilepsy can potentially arise from any neocortical location [29, 58]. In contrast to mesial temporal lobe epilepsy, in which resections can be guided uniquely by anatomical boundaries, neocortical epilepsy can be lesional or non-lesional. Because non-lesional epilepsy is characterized by a normal MRI scan, the limits of the resection are generally defined by electrophysiology, that is, the boundaries of ictal and interictal events, which can be recorded acutely in the operating room or with chronically implanted electrodes. Lesional neocortical epilepsy, however, can be treated with a lesionectomy, although improved outcomes are generally achieved from additional electrophysiological recordings to identify adjacent epileptogenic brain from which seizures arise [26, 27, 36].

Most lesions causing chronic epilepsy are extremely difficult to visualize in the OR under microscopy. Examples include gangliogliomas, dysembryoplastic neuroepithelial tumors, and focal cortical dysplasia [27, 28]. Focal cortical dysplasia, first described by Taylor et al. in 1971 [59], is divided into two pathologic subcategories: type 1 and type 2 [60]. Type 1 is a milder form without balloon cells, and type 2 is more severe and characterized pathologically by balloon cells.

Although invisible to the naked eye on the cortical surface, several distinct MRI signal abnormalities are associated with these lesions [60]. The advantage of iMRI is the ability to localize the anatomical boundaries of the underlying lesion during real time in the OR. This information is critical not only if a lesionectomy is performed but also in identifying the area where electrode grids must be placed for intraoperative electrocorticography (ECoG) or chronic recording.

There is an increasing number of studies reporting on use of iMRI for neocortical epilepsy. Buchfelder et al. reported on a series of 29 cases of lesional temporal lobe epilepsy in which iMRI increased the completeness of resection from 73% to 87% [30]. After 15.5 months, 24 of 29 (83%) patients were free of seizures. Walker et al. reported 12 patients, of whom six had extratemporal neocortical lesions, who underwent a lesionectomy with iMRI [49]. Five patients required further resection after imaging, all of which had complete radiographic removal of lesion on follow-up imaging. Roessler et al. expands on prior studies by reporting 88 lesional temporal lobe epilepsy cases, of which 22 (25%) required further resection after imaging [61]. Of these 22 patients requiring further resection, 19 (86.3%) had complete radiographic resection and 16 (72.7) achieved complete seizure freedom.

iMRI for neocortical epilepsy has also proven to be beneficial for patients with extratemporal lesions. Kurwale et al. reported that iMRI detected residual cortical lesions not detectable microscopically in four patients, two of whom underwent further resection, and two of whom were found to have dysplasia near eloquent cortex, preventing possible neurologic deficit [48]. Cui et al. compared groups of iMRI to non-iMRI patients who underwent surgical excision of lesions near primary motor or sensory cortices [52]. While there was no significant difference in seizure outcome (as measured by Engel score), there were fewer patients in the iMRI cohort who experienced postoperative hemiparesis. Sun et al. described 20 surgeries for cavernous malformation causing epilepsy using iMRI combined with functional neuronavigation [50]. While not quantified explicitly, the authors state that iMRI was helpful for lesionectomy in terms of avoiding white matter tracts and ensuring safe total resection, especially in cases with multiple lesions. Sommer et al. reports on 25 patients with lesional epilepsy near eloquent (speech or motor) cortex who underwent surgery with iMRI, with five patients (20%) requiring additional resection after iMRI scan [51]. 84% of patients achieved Engel I or Engel II outcomes. Finally, Sacino et al. reports a retrospective analysis of pediatric patients undergoing lesionectomy for cortical dysplasia or heterotopia with or

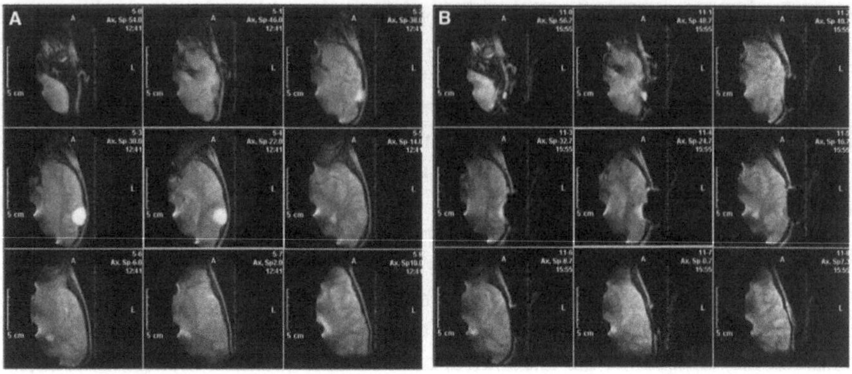

Fig. 6 Axial T2-weighted images (3.75 min scan) taken with the Polestar N-10 reveal a neocortical ganglioglioma (**a**). The lesion is clearly seen at the surface and neuronavigation can be used to locate the skin incision as well as the margins of the resection at the cortical surface. (**b**) Post-resection scan obtained before bone flap was replaced reveals that the lesion has been removed. (Images courtesy of Dr. Mark Souweidane)

without iMRI [54]. They found that seizure freedom rates (82% vs. 39%) and gross total resection rates were higher (91% vs. 44%), while postoperative neurologic deficits were lower (36% vs. 83%) in the iMRI group (Fig. 6).

4.4 Corpus Callosotomy

Division of the corpus callosum for the treatment of patients with seizure disorders dates to the observations of Van Wagenan and Herren in the 1930s [62] when they noticed that patients with strokes affecting the corpus callosum often had improvement in seizure frequency. This was supported by experimental evidence demonstrating the importance of commissural fibers in the spread of epilepsy in primates by Erickson, lending further support for this therapeutic approach [63]. Additional reports by Bogan and Vogel in adults [64] and Luessenhop in children [65] established corpus callosotomy (CC) as a standard technique in the surgical treatment of epilepsy.

Candidates for CC include patients with medically intractable primary generalized epilepsy or focal epilepsy with rapid secondary generalization that is either unlocalized or localized to unresectable cortex. The most common generalized epilepsies treated with CC are characterized by atonic or akinetic seizures, often involving sudden drop attacks [66]. In particular, children with Lennox–Gastaut syndrome show a significant improvement in seizure control and quality of life [67, 68]. While complete section is approximately twice as effective as partial section in controlling seizure frequency [69, 70], most centers perform a partial section initially to avoid a potential acute disconnection syndrome. Completion of CC can subsequently be performed as a second operation if seizure control is inadequate. For all patients, secondarily generalized seizures are controlled in 70–80% of patients whereas only 25–50% of patients find relief from focal seizures [66]. It has been shown that the percentage of patients obtaining a greater than 80% reduction in generalized seizures improves from 29% after anterior section to 62% after completion of callosal section [70].

Image guidance and stereotaxy have been shown to be useful in performing an anterior two-thirds callosotomy [71–73]. Although most surgeons prefer to approach the brain from the right side in right-handed patients to minimize retraction injury to the dominant hemisphere, the identification of draining veins that might limit exposure may lead the surgeon to approach the callosum from the dominant side. During the approach, image guidance may be helpful in distinguishing the cingulate gyrus from the callosum to avoid cortical injury. Finally, the exact extent of the callosum section can be determined intraoperatively to ensure that only the anterior two thirds is divided. However, because entry into the ventricle often occurs in this surgery, the degree of brain shift can be significant. Hence, achieving the stated objectives would be theoretically aided by iMRI.

Buchfelder et al. reported four cases in which the 0.2 T Magnetom Open MRI was used for CC with good results [30]. The case series by Kelly et al. included three CC patients, one of whom iMRI revealed incomplete callosotomy and required additional surgical resection [43]. While additional reports of using iMRI for corpus callosotomy exist in the literature [19, 46], no specific results for these patients are discussed.

4.5 *Hemispherectomy*

Complete removal, or disconnection of one cerebral hemisphere, is indicated in patients with unilateral ictal onsets that are either multifocal or nonlocalizable to a single lobe associated with severe hemispheric dysfunction involving movement. Ideal candidates for hemispherectomy are near hemiplegic or have at least lost any functional hand or finger movements. The absence of a visual field deficit is not a contraindication to hemispherectomy as long as the patient and/or primary caregivers are well-informed of the inevitable deficit. Language function will relocate to the opposite hemisphere if the surgery is performed before the age of 10, ideally before the age of 6. Outcomes can be excellent, with seizure freedom rates of 60–70% in well-selected candidates [74, 75]. Surgical procedures range from complete physical removal of the entire hemisphere to decortication or functional hemispherectomy.

Functional hemispherectomy involves anatomic disconnection of the gray matter of the hemisphere without removal of significant amounts of tissue [76]. The surgeon enters the ventricle circumferentially around the Sylvian fissure, performs a CC from within the ventricle, removes the medial temporal lobe structures, and then disconnects the basal frontal lobe and insula. At each of these steps, complete disconnection must be achieved, otherwise the procedure will fail. Assessment of disconnection is difficult intraoperatively because the remaining tissue obstructs the surgeon's view. Additionally, intraoperative electrophysiologic recordings may be misleading because the residual tissue often exhibits increases in interictal activity [77, 78]. However, the few reports in the literature using iMRI for disconnection hemispherectomy indicate that it may have limited utility. Kurwale et al. reports the results of nine hemispherectomy patients for which iMRI was used, none of which required additional resection after iMRI was complete [48]. Similarly, Warsi et al. includes three patients with hemispherectomy in their case series, none of whom had additional resection after intraoperative imaging [56].

4.6 Hypothalamic Hamartomas

Hamartomas arising from the hypothalamus are a rare case of chronic epilepsy in children characterized by precocious puberty and gelastic seizures (paroxysmal laughter). Confirmation of the etiology of the seizures has been demonstrated both with direct electrical recordings as well as functional imaging [79, 80]. Reasonable relief of seizures can be obtained either with direct surgical resection of the hamartoma or stereotactic radiosurgery (SRS), if the lesion is of the appropriate size [81, 82]. Recent advances in LITT technology have opened the possibility for its use as an alternative approach to open surgery or SRS [83–85]. Any surgery can be quite risky based on the deep location of the lesion and the indistinct margin between hamartoma and normal hypothalamus.

In theory, iMRI should be quite helpful for hypothalamic hamartoma (HH) surgery because the margin between the hamartoma and the normal hypothalamus is challenging to determine intraoperatively but is readily apparent on MRI imaging. The largest series describing use of iMRI for resection of HH, by van Tonder et al. [57], found that four of ten patients required additional resection of residual tissue after iMRI scan. While all patients achieved either Engel I or Engel II outcomes, gross total resection was only achieved in 50% of patients.

4.7 Depth Electrodes

Localization of epileptogenic foci often requires placement of chronically implanted electrodes to record seizure onsets. Subdural grid and strip electrodes provide a means for localizing foci on the cortical surface. However, potential foci also include deep structures such as the hippocampus, amygdala, and subcortical heterotopias [86–88]. Volume conduction of the electric fields generated by deep foci interferes with accurate localization by recordings at the surface of the brain [89, 90]. Thus, accurate assessment of deep foci often requires placement of depth recording electrodes.

Depth electrodes were originally placed using a double-grid system in conjunction with angiography [91]. This technique, although accurate and safe, was time consuming and imposed limitations upon the working space at the implantation site. The development of the Leksell frame enabled the use of multiple entry sites and more working space [92]. Although frame-based methodologies are highly accurate, they suffer from a number of drawbacks, including patient discomfort, time involved in frame placement, restricted access to the surgical field with limitations on craniotomy size, and, most importantly, a limited ability to rapidly define new targets and trajectories in real-time during surgery. The ability to plan trajectories in real-time has been partially alleviated with the use of robotic stereotactic trajectory guidance systems, which provide a safe and accurate alternative to manual frame-based electrode placement [93]. While frameless stereotaxy using preoperative imaging has become commonplace since it was initially reported [94–96], the problem of brain shift, field inhomogeneities, warping, and reliance on a preoperatively acquired data set may render the placement inaccurate. This may prove especially problematic when the lateral temporal approach is used, which is often utilized when depth electrodes are placed in conjunction with large arrays of subdural grids and strip electrodes. In these cases, the mesial structures may shift away from the tentorium [10–12], causing the target to migrate in a superior direction.

This loss of accuracy from shift phenomenon has been supported by the largest series of iMRI use in depth electrode placement, in which Roessler et al. described 23 patients undergoing iMRI-guided placement of temporal grids, strips, and depth electrodes [31]. Thirteen percent (3 of 23) of patients required re-positioning of electrode after iMRI, making a case that the changes from preoperative to intraoperative imaging data are significant enough to be clinically relevant. This, however, is in contrast to a later study by the same group utilizing iMRI for placement of depth electrodes alone (patients not necessitating craniotomy), in which all electrodes (6 patients, 58 electrodes) were in satisfactory position after iMRI [53].

4.8 Laser Interstitial Thermal Therapy (LITT)

Perhaps the most common current use for intraoperative MRI in epilepsy surgery is for performing LITT procedures. LITT employs a laser probe, which is passed to the target under MRI guidance, to heat brain tissue under real-time MR thermometry and achieve focused ablation [97]. This technique has been utilized for treatment of epilepsy with varying degrees of success. In mesial temporal lobe epilepsy patients, Jermakowicz et al. found that 15 of 23 (65%) were free of disabling seizures at 1-year follow up [98]. Youngerman et al. published results of 12 non-MTS patients, seven of whom (58%) were completely seizure-free [99]. Similarly, Kang et al. reported that 8 of 15 (53%) of temporal lobe epilepsy patients were free of consciousness-impairing seizures [100]. Four of these patients eventually underwent subsequent "open" surgical resection, resulting in seizure freedom in two-of-four patients. The results of Willie et al. also corroborate this approximate seizure freedom rate, as 54% (7/13) of their temporal lobe epilepsy patients achieved freedom from disabling seizures [101].

LITT has also been utilized for treatment of hypothalamic hamartomas with success. In the largest published cohort of 71 HH patients, Curry et al. reports 93% of patients were free of gelastic seizures at 1 year [83]. The study by Gagdil et al. includes 58 pediatric patients who underwent LITT for HH, 81% were free from gelastic seizures [84]. Yao et al. described the results of 47 patients who underwent LITT for HH, 72% of whom were free of gelastic seizures, and 68% of patients were completely seizure free [85]. While temporal lobe epilepsy and HH are the dominant indications for LITT for epilepsy in the literature, it should be noted that it has also been used for corpus callosotomy [102], epileptogenic cavernous malformations [103], tuberous sclerosis lesions [104, 105], and periventricular nodular heterotopia lesions [106].

4.9 Note on Electrocorticography, Electroencephalography, and Awake Mapping

Although high field strength magnets and intraoperative ECoG will interfere with each other, there should be no interference if they are not performed simultaneously. Thus, although Walker et al. report difficulty performing ECoG within a 0.5-T double donut MR [49], Buchfelder et al. had no difficulties adjacent to a 0.2-T scanner [30]. Our personal experience is that using the 0.12-T magnet at a distance of 2–3 ft. from the brain and 5 ft. from the electroencephalography (EEG) machine, ECoG can be performed successfully, even in patients who are awake. Once the EEG machine is powered off, the imaging can proceed smoothly without artifact. If a local shield is used, then the machine can remain powered on. The use of platinum electrodes is preferred because they do not conduct heat nor do they emit MRI artifact. The ability to operate on patients while they are awake is merely based on an individual patient's tolerance and comfort, and is not, in and of itself, contraindicated with iMRI. Surgeries can be performed either with the patient experiencing rigid fixation and receiving generous dose of local anesthesia or with the patient resting on a soft headrest. The author has successfully performed language mapping in the Polestar N-10 without any significant problems.

5 Cost–Benefit Analysis

iMRI is expensive for a hospital to purchase. In addition to the cost of the hardware and software, there are costs associated with shielding the OR, training personnel, and lost revenue due to increased surgery length. The majority of cost–benefit analyses for iMRI have been conducted in the tumor population, though some studies have attempted to translate this analysis into epilepsy surgery. Makary et al. found no significant difference in cost when comparing iMRI with conventional MRI for glioma patients [107]. Abraham et al. found that iMRI for high-grade glioma patients yielded an incremental benefit of 0.18 quality-adjusted life years (QALYs) at a cost of $13,447, or an incremental cost-effectiveness ratio of $76,442 per QALY (deemed to be cost-effective at a threshold of $100,000 per QALY) [108]. Eljamel et al. performed a similar study and calculated a ratio of $32,924 per QALY, which was the highest of the adjuncts studied: 5-ALA ($16,218/QALY), fluorescein ($3181/QALY), and intraoperative ultrasound ($6049/QALY) [109]. In the only published study to date attempting to quantify cost-effectiveness of iMRI in epilepsy patients, Sacino et al. compared ten pediatric patients who underwent epilepsy surgery with iMRI, seven patients who underwent conventional surgery followed by re-resection with iMRI, and 39 patients who underwent epilepsy surgery without iMRI [110]. They discovered that patients who underwent surgery with upfront iMRI were the most cost-effective, at $16,179/QALY, compared to $28,514/QALY in iMRI re-operation group and $49,960/QALY in the conventional surgery group. This small study suggests that iMRI may be fiscally practical, though this certainly warrants further evaluation before making definitive conclusions about cost-effectiveness.

Since increased extent of resection of the pathologic substrate of epilepsy improves seizure outcome, iMRI should, in theory, maximize extent of resection. However, the presence of an iMRI in an operating room can also impact surgeon behavior. Could surgeons be less aggressive initially because they know an iMRI will be obtained before the case ends? Until randomized controlled studies are performed, we may never know the degree to which iMRI increases extent of resection. Nor do we know if it increases rates of seizure freedom. Nevertheless, iMRI remains the current state-of-the-art in intraoperative neuronavigation, and, although it may be only an expensive steppingstone to a future better technology.

References

1. Weaver FM, Follett K, Stern M, et al. Bilateral deep brain stimulation vs best medical therapy for patients with advanced Parkinson disease: a randomized controlled trial. JAMA. 2009;301(1):63–73. https://doi.org/10.1001/jama.2008.929.
2. Laxton AW, Stone S, Lozano AM. The neurosurgical treatment of Alzheimer's disease: a review. SFN. 2014;92(5):269–81. https://doi.org/10.1159/000364914.
3. Spencer SS, Berg AT, Vickrey BG, et al. Initial outcomes in the multicenter study of epilepsy surgery. Neurology. 2003;61(12):1680–5. https://doi.org/10.1212/01.WNL.0000098937.35486.A3.
4. Doyle WK, Spencer DD. Anterior temporal resections. In: Epilepsy. A comprehensive textbook. Lippincott-Raven Publishers; 1997. p. 1807–17.
5. Helmstaedter C, Kurthen M, Lux S, Reuber M, Elger CE. Chronic epilepsy and cognition: a longitudinal study in temporal lobe epilepsy. Ann Neurol. 2003;54(4):425–32. https://doi.org/10.1002/ana.10692.

6. Olivier A. Extratemporal resections in the surgical treatment of epilepsy. In: Contemporary issues in neurological surgery: surgery for epilepsy. Blackwell Scientific Publications; 1991. p. 150–67.
7. Berkovic SF, McIntosh AM, Kalnins RM, et al. Preoperative MRI predicts outcome of temporal lobectomy: an actuarial analysis. Neurology. 1995;45(7):1358–63. https://doi.org/10.1212/wnl.45.7.1358.
8. Nimsky C, Ganslandt O, Cerny S, Hastreiter P, Greiner G, Fahlbusch R. Quantification of, visualization of, and compensation for brain shift using intraoperative magnetic resonance imaging. Neurosurgery. 2000;47(5):1070–9; discussion 1079–1080. https://doi.org/10.1097/00006123-200011000-00008.
9. Dorward NL, Alberti O, Velani B, et al. Postimaging brain distortion: magnitude, correlates, and impact on neuronavigation. J Neurosurg. 1998;88(4):656–62. https://doi.org/10.3171/jns.1998.88.4.0656.
10. Hill DL, Maurer CR, Maciunas RJ, Barwise JA, Fitzpatrick JM, Wang MY. Measurement of intraoperative brain surface deformation under a craniotomy. Neurosurgery. 1998;43(3):514–26; discussion 527–528. https://doi.org/10.1097/00006123-199809000-00066.
11. Nabavi A, Black PM, Gering DT, et al. Serial intraoperative magnetic resonance imaging of brain shift. Neurosurgery. 2001;48(4):787–97; discussion 797–798. https://doi.org/10.1097/00006123-200104000-00019.
12. Roberts DW, Hartov A, Kennedy FE, Miga MI, Paulsen KD. Intraoperative brain shift and deformation: a quantitative analysis of cortical displacement in 28 cases. Neurosurgery. 1998;43(4):749–58; discussion 758–760. https://doi.org/10.1097/00006123-199810000-00010.
13. Shahar T, Rozovski U, Marko NF, et al. Preoperative imaging to predict intraoperative changes in tumor-to-corticospinal tract distance: an analysis of 45 cases using high-field intraoperative magnetic resonance imaging. Neurosurgery. 2014;75(1):23–30. https://doi.org/10.1227/NEU.0000000000000338.
14. Elias WJ, Fu KM, Frysinger RC. Cortical and subcortical brain shift during stereotactic procedures. J Neurosurg. 2007;107(5):983–8. https://doi.org/10.3171/JNS-07/11/0983.
15. Mahboob S, McPhillips R, Qiu Z, et al. Intraoperative ultrasound-guided resection of gliomas: a meta-analysis and review of the literature. World Neurosurg. 2016;92:255–63. https://doi.org/10.1016/j.wneu.2016.05.007.
16. Black PM, Moriarty T, Alexander E, et al. Development and implementation of intraoperative magnetic resonance imaging and its neurosurgical applications. Neurosurgery. 1997;41(4):831–42; discussion 842–845. https://doi.org/10.1097/00006123-199710000-00013.
17. Steinmeier R, Fahlbusch R, Ganslandt O, et al. Intraoperative magnetic resonance imaging with the magnetom open scanner: concepts, neurosurgical indications, and procedures: a preliminary report. Neurosurgery. 1998;43(4):739–47; discussion 747–748. https://doi.org/10.1097/00006123-199810000-00005.
18. Hushek SG, Martin AJ, Steckner M, Bosak E, Debbins J, Kucharzyk W. MR systems for MRI-guided interventions. J Magn Reson Imaging. 2008;27(2):253–66. https://doi.org/10.1002/jmri.21269.
19. Nimsky C, Ganslandt O, Kober H, Buchfelder M, Fahlbusch R. Intraoperative magnetic resonance imaging combined with neuronavigation: a new concept. Neurosurgery. 2001;48(5):1082–9; discussion 1089–1091. https://doi.org/10.1097/00006123-200105000-00023.
20. Kaibara T, Saunders JK, Sutherland GR. Advances in mobile intraoperative magnetic resonance imaging. Neurosurgery. 2000;47(1):131–7; discussion 137–138. https://doi.org/10.1097/00006123-200007000-00028.
21. Hall WA, Liu H, Martin AJ, Pozza CH, Maxwell RE, Truwit CL. Safety, efficacy, and functionality of high-field strength interventional magnetic resonance imaging for neurosurgery. Neurosurgery. 2000;46(3):632–41; discussion 641–642. https://doi.org/10.1097/00006123-200003000-00022.

22. Hadani M, Spiegelman R, Feldman Z, Berkenstadt H, Ram Z. Novel, compact, intraoperative magnetic resonance imaging-guided system for conventional neurosurgical operating rooms. Neurosurgery. 2001;48(4):799.
23. Rogers CM, Jones PS, Weinberg JS. Intraoperative MRI for brain tumors. J Neuro-Oncol. 2021;151(3):479–90. https://doi.org/10.1007/s11060-020-03667-6.
24. Schmitz B, Nimsky C, Wendel G, et al. Anesthesia during high-field intraoperative magnetic resonance imaging experience with 80 consecutive cases. J Neurosurg Anesthesiol. 2003;15(3):255–62. https://doi.org/10.1097/00008506-200307000-00014.
25. Hauser WA, Hesdorffer DC. Epilepsy: frequency, causes, and consequences. Demos Medical Pub; 1990. Accessed April 23, 2023. https://books.google.com/books/about/Epilepsy.html?id=AbBrAAAAMAAJ
26. Berger MS, Ghatan S, Geyer JR, Keles GE, Ojemann GA. Seizure outcome in children with hemispheric tumors and associated intractable epilepsy: the role of tumor removal combined with seizure foci resection. Pediatr Neurosurg. 1991;17(4):185–91. https://doi.org/10.1159/000120593.
27. Berger MS, Ghatan S, Haglund MM, Dobbins J, Ojemann GA. Low-grade gliomas associated with intractable epilepsy: seizure outcome utilizing electrocorticography during tumor resection. J Neurosurg. 1993;79(1):62–9. https://doi.org/10.3171/jns.1993.79.1.0062.
28. Bautista RE, Cobbs MA, Spencer DD, Spencer SS. Prediction of surgical outcome by interictal epileptiform abnormalities during intracranial EEG monitoring in patients with extrahippocampal seizures. Epilepsia. 1999;40(7):880–90. https://doi.org/10.1111/j.1528-1157.1999.tb00794.x.
29. Rasmussen T. Tailoring of cortical excisions for frontal lobe epilepsy. Can J Neurol Sci. 1991;18(4 Suppl):606–10. https://doi.org/10.1017/s0317167100032819.
30. Buchfelder M, Ganslandt O, Fahlbusch R, Nimsky C. Intraoperative magnetic resonance imaging in epilepsy surgery. J Magn Reson Imaging. 2000;12(4):547–55. https://doi.org/10.1002/1522-2586(200010)12:4<547::AID-JMRI5>3.0.CO;2-H.
31. Roessler K, Hofmann A, Sommer B, et al. Resective surgery for medically refractory epilepsy using intraoperative MRI and functional neuronavigation: the Erlangen experience of 415 patients. Neurosurg Focus. 2016;40(3):E15. https://doi.org/10.3171/2015.12.FOCUS15554.
32. Nayel MH, Awad IA, Luders H. Extent of mesiobasal resection determines outcome after temporal lobectomy for intractable complex partial seizures. Neurosurgery. 1991;29(1):55–60; discussion 60–61. https://doi.org/10.1097/00006123-199107000-00009.
33. Wyler AR, Hermann BP, Somes G. Extent of medial temporal resection on outcome from anterior temporal lobectomy: a randomized prospective study. Neurosurgery. 1995;37(5):982–90; discussion 990–991. https://doi.org/10.1227/00006123-199511000-00019.
34. Siegel AM, Wieser HG, Wichmann W, Yasargil GM. Relationships between MR-imaged total amount of tissue removed, resection scores of specific mediobasal limbic subcompartments and clinical outcome following selective amygdalohippocampectomy. Epilepsy Res. 1990;6(1):56–65. https://doi.org/10.1016/0920-1211(90)90009-k.
35. Kanner AM, Kaydanova Y, deToledo-Morrell L, et al. Tailored anterior temporal lobectomy: relation between extent of resection of mesial structures and postsurgical seizure outcome. Arch Neurol. 1995;52(2):173–8. https://doi.org/10.1001/archneur.1995.00540260079020.
36. Awad IA, Rosenfeld J, Ahl J, Hahn JF, Lüders H. Intractable epilepsy and structural lesions of the brain: mapping, resection strategies, and seizure outcome. Epilepsia. 1991;32(2):179. https://doi.org/10.1111/j.1528-1157.1991.tb05242.x.
37. Wyler AR, Hermann BP, Richey ET. Results of reoperation for failed epilepsy surgery. J Neurosurg. 1989;71(6):815–9. https://doi.org/10.3171/jns.1989.71.6.0815.
38. Schramm J. Temporal lobe epilepsy surgery and the quest for optimal extent of resection: a review. Epilepsia. 2008;49(8):1296–307. https://doi.org/10.1111/j.1528-1167.2008.01604.x.
39. McKhann GM, Schoenfeld-McNeill J, Born DE, Haglund MM, Ojemann GA. Intraoperative hippocampal electrocorticography to predict the extent of hippocampal resection

in temporal lobe epilepsy surgery. J Neurosurg. 2000;93(1):44–52. https://doi.org/10.3171/jns.2000.93.1.0044.
40. Jones-Gotman M, Zatorre RJ, Olivier A, et al. Learning and retention of words and designs following excision from medial or lateral temporal-lobe structures. Neuropsychologia. 1997;35(7):963–73. https://doi.org/10.1016/s0028-3932(97)00024-9.
41. Awad IA, Katz A, Hahn JF, Kong AK, Ahl J, Lüders H. Extent of resection in temporal lobectomy for epilepsy. I. Interobserver analysis and correlation with seizure outcome. Epilepsia. 1989;30(6):756–62. https://doi.org/10.1111/j.1528-1157.1989.tb05335.x.
42. Jack CR, Sharbrough FW, Marsh WR. Use of MR imaging for quantitative evaluation of resection for temporal lobe epilepsy. Radiology. 1988;169(2):463–8. https://doi.org/10.1148/radiology.169.2.3174994.
43. Kelly JJ, Hader WJ, Myles ST, Sutherland GR. Epilepsy surgery with intraoperative MRI at 1.5 T. Neurosurg Clin. 2005;16(1):173–83. https://doi.org/10.1016/j.nec.2004.07.006.
44. Schwartz TH, Marks D, Pak J, et al. Standardization of amygdalohippocampectomy with intraoperative magnetic resonance imaging: preliminary experience. Epilepsia. 2002;43(4):430–6. https://doi.org/10.1046/j.1528-1157.2002.39101.x.
45. Kaibara T, Myles ST, Lee MA, Sutherland GR. Optimizing epilepsy surgery with intraoperative MR imaging. Epilepsia. 2002;43(4):425–9. https://doi.org/10.1046/j.1528-1157.2002.32401.x.
46. Levy R, Cox RG, Hader WJ, Myles T, Sutherland GR, Hamilton MG. Application of intraoperative high-field magnetic resonance imaging in pediatric neurosurgery: clinical article. J Neurosurg Pediatr. 2009;4(5):467–74. https://doi.org/10.3171/2009.4.PEDS08464.
47. Engel J. Update on surgical treatment of the epilepsies. Summary of the Second International Palm Desert Conference on the Surgical Treatment of the Epilepsies (1992). Neurology. 1993;43(8):1612–7. https://doi.org/10.1212/wnl.43.8.1612.
48. Kurwale NS, Chandra SP, Chouksey P, et al. Impact of intraoperative MRI on outcomes in epilepsy surgery: preliminary experience of two years. Br J Neurosurg. 2015;29(3):380. https://doi.org/10.3109/02688697.2014.1003034.
49. Walker DG, Talos F, Bromfield EB, Black PMcL. Intraoperative magnetic resonance for the surgical treatment of lesions producing seizures. J Clin Neurosci. 2002;9(5):515–20. https://doi.org/10.1054/jocn.2002.1134.
50. Sun GC, Chen XL, Zhao Y, et al. Intraoperative MRI with integrated functional neuronavigation-guided resection of supratentorial cavernous malformations in eloquent brain areas. J Clin Neurosci. 2011;18(10):1350–4. https://doi.org/10.1016/j.jocn.2011.01.025.
51. Sommer B, Grummich P, Coras R, et al. Integration of functional neuronavigation and intraoperative MRI in surgery for drug-resistant extratemporal epilepsy close to eloquent brain areas. Neurosurg Focus. 2013;34(4):E4. https://doi.org/10.3171/2013.2.FOCUS12397.
52. Cui ZQ, Ling ZP, Song HF, et al. Combining pyramidal tract mapping, microscopic-based neuronavigation, and intraoperative magnetic resonance imaging improves outcome of epilepsy foci resection in the sensorimotor cortex. Turk Neurosurg. 24(4) https://doi.org/10.5137/1019-5149.JTN.9517-13.0.
53. Roessler K, Sommer B, Merkel A, et al. A frameless stereotactic implantation technique for depth electrodes in refractory epilepsy using intraoperative magnetic resonance imaging. World Neurosurg. 2016;94:206–10. https://doi.org/10.1016/j.wneu.2016.06.114.
54. Sacino MF, Ho CY, Murnick J, Keating RF, Gaillard WD, Oluigbo CO. The role of intraoperative MRI in resective epilepsy surgery for peri-eloquent cortex cortical dysplasias and heterotopias in pediatric patients. Neurosurg Focus. 2016;40(3):E16. https://doi.org/10.3171/2016.1.FOCUS15538.
55. Sacino MF, Ho CY, Murnick J, et al. Intraoperative MRI–guided resection of focal cortical dysplasia in pediatric patients: technique and outcomes. J Neurosurg Pediatr. 2016;17(6):672–8. https://doi.org/10.3171/2015.10.PEDS15512.
56. Warsi NM, Lasry O, Farah A, et al. 3-T intraoperative MRI (iMRI) for pediatric epilepsy surgery. Childs Nerv Syst. 2016;32(12):2415–22. https://doi.org/10.1007/s00381-016-3263-3.

57. van Tonder L, Burn S, Iyer A, et al. Open resection of hypothalamic hamartomas for intractable epilepsy revisited, using intraoperative MRI. Childs Nerv Syst. 2018;34(9):1663–73. https://doi.org/10.1007/s00381-018-3786-x.
58. Roper SN. Surgical treatment of the extratemporal epilepsies. Epilepsia. 2009;50(Suppl 8):69–74. https://doi.org/10.1111/j.1528-1167.2009.02239.x.
59. Taylor DC, Falconer MA, Bruton CJ, Corsellis JA. Focal dysplasia of the cerebral cortex in epilepsy. J Neurol Neurosurg Psychiatry. 1971;34(4):369–87. https://doi.org/10.1136/jnnp.34.4.369.
60. Kuzniecky R, Garcia JH, Faught E, Morawetz RB. Cortical dysplasia in temporal lobe epilepsy: magnetic resonance imaging correlations. Ann Neurol. 1991;29(3):293–8. https://doi.org/10.1002/ana.410290311.
61. Roessler K, Sommer B, Grummich P, et al. Improved resection in lesional temporal lobe epilepsy surgery using neuronavigation and intraoperative MR imaging: favourable long term surgical and seizure outcome in 88 consecutive cases. Seizure. 2014;23(3):201–7. https://doi.org/10.1016/j.seizure.2013.11.013.
62. Van Wagenen WP. Surgical division of commissural pathways in the corpus callosum: relation to spread of an epileptic attack. Arch NeurPsych. 1940;44(4):740. https://doi.org/10.1001/archneurpsyc.1940.02280100042004.
63. Erickson TC. Spread of the epileptic discharge: an experimental study of the after-discharge induced by electrical stimulation of the cerebral cortex. Arch NeurPsych. 1940;43(3):429. https://doi.org/10.1001/archneurpsyc.1940.02280030003001.
64. Bogen JE. Cerebral commissurotomy in man: preliminary case report. Bull Los Angeles Neuro Soc. 1962;27:169–72.
65. Luessenhop AJ. Interhemispheric commissurotomy: (the split brain operation) as an alternative to hemispherectomy for control of in- tractable seizures. Am J Surg. 36:265–8.
66. Schwartz T. Corpus callosotomy: indications and technique. In: Schulder M, editor. Handbook of stereotactic and functional neurosurgery. CRC Press; 2003. https://doi.org/10.1201/9780203912416.
67. Collins S, Walker J, Barbaro N, Laxer K. Corpus callosotomy in the treatment of Lennox-Gastaut syndrome. Epilepsia. 1989;30:670–89.
68. Sorenson JM, Wheless JW, Baumgartner JE, et al. Corpus callosotomy for medically intractable seizures. Pediatr Neurosurg. 1997;27(5):260–7. https://doi.org/10.1159/000121264.
69. Spencer SS, Spencer DD, Williamson PD, Sass K, Novelly RA, Mattson RH. Corpus callosotomy for epilepsy. I Seizure effects. Neurology. 1988;38(1):19–24. https://doi.org/10.1212/wnl.38.1.19.
70. Roberts DW. The role of Callosal section in surgical treatment of epilepsies. Neurosurg Clin N Am. 1993;4(2):293–300. https://doi.org/10.1016/S1042-3680(18)30595-3.
71. Olivier A, Germano IM, Cukiert A, Peters T. Frameless stereotaxy for surgery of the epilepsies: preliminary experience. Technical note. J Neurosurg. 1994;81(4):629–33. https://doi.org/10.3171/jns.1994.81.4.0629.
72. Olivier A, Alonso-Vanegas M, Comeau R, Peters TM. Image-guided surgery of epilepsy. Neurosurg Clin N Am. 1996;7(2):229–43.
73. Hodaie M, Musharbash A, Otsubo H, et al. Image-guided, frameless stereotactic sectioning of the corpus callosum in children with intractable epilepsy. Pediatr Neurosurg. 2001;34(6):286–94. https://doi.org/10.1159/000056040.
74. Kossoff EHW, Shields WD. Nonpharmacologic care for patients with Lennox-Gastaut syndrome: ketogenic diets and vagus nerve stimulation. Epilepsia. 2014;55(s4):29–33. https://doi.org/10.1111/epi.12546.
75. Vining EP, Freeman JM, Pillas DJ, et al. Why would you remove half a brain? The outcome of 58 children after hemispherectomy-the Johns Hopkins experience: 1968 to 1996. Pediatrics. 1997;100(2 Pt 1):163–71. https://doi.org/10.1542/peds.100.2.163.
76. Villemure JG, Mascott CR. Peri-insular hemispherotomy: surgical principles and anatomy. Neurosurgery. 1995;37(5):975–81. https://doi.org/10.1227/00006123-199511000-00018.

77. Wennberg RA, Quesney LF, Villemure JG. Epileptiform and non-epileptiform paroxysmal activity from isolated cortex after functional hemispherectomy. Electroencephalogr Clin Neurophysiol. 1997;102(5):437–42. https://doi.org/10.1016/s0921-884x(97)96047-1.
78. Smith SJ, Andermann F, Villemure JG, Rasmussen TB, Quesney LF. Functional hemispherectomy: EEG findings, spiking from isolated brain postoperatively, and prediction of outcome. Neurology. 1991;41(11):1790–4. https://doi.org/10.1212/wnl.41.11.1790.
79. Munari C, Kahane P, Francione S, et al. Role of the hypothalamic hamartoma in the genesis of gelastic fits (a video-stereo-EEG study). Electroencephalogr Clin Neurophysiol. 1995;95(3):154–60. https://doi.org/10.1016/0013-4694(95)00063-5.
80. Tasch E, Cendes F, Li LM, et al. Hypothalamic hamartomas and gelastic epilepsy: a spectroscopic study. Neurology. 1998;51(4):1046–50. https://doi.org/10.1212/wnl.51.4.1046.
81. Rosenfeld JV, Harvey AS, Wrennall J, Zacharin M, Berkovic SF. Transcallosal resection of hypothalamic hamartomas, with control of seizures, in children with gelastic epilepsy. Neurosurgery. 2001;48(1):108–18. https://doi.org/10.1097/00006123-200101000-00019.
82. Régis J, Bartolomei F, de Toffol B, et al. Gamma knife surgery for epilepsy related to hypothalamic hamartomas. Neurosurgery. 2000;47(6):1343–51; discussion 1351–1352.
83. Curry DJ, Raskin J, Ali I, Wilfong AA. MR-guided laser ablation for the treatment of hypothalamic hamartomas. Epilepsy Res. 2018;142:131–4. https://doi.org/10.1016/j.eplepsyres.2018.03.013.
84. Gadgil N, Lam S, Pan IW, et al. Staged magnetic resonance-guided laser interstitial thermal therapy for hypothalamic hamartoma: analysis of ablation volumes and morphological considerations. Neurosurgery. 2020;86(6):808–16. https://doi.org/10.1093/neuros/nyz378.
85. Yao Y, Wang X, Hu W, et al. Magnetic resonance-guided laser interstitial thermal therapy for hypothalamic hamartoma: surgical approach and treatment outcomes. J Clin Med. 2022;11(21):6579. https://doi.org/10.3390/jcm11216579.
86. Smith JR, Flanigin HF, King DW, et al. An 8-year experience with depth electrodes in the evaluation of ablative seizure surgery candidates. Stereotact Funct Neurosurg. 1990;54–55:60–6. https://doi.org/10.1159/000100191.
87. Ross DA, Brunberg JA, Drury I, Henry TR. Intracerebral depth electrode monitoring in partial epilepsy: the morbidity and efficacy of placement using magnetic resonance image- guided stereotactic surgery. Neurosurgery. 1996;39(2):327–33; discussion 333–334. https://doi.org/10.1097/00006123-199608000-00018.
88. Brekelmans GJ, van Emde BW, Velis DN, Lopes da Silva FH, van Rijen PC, van Veelen CW. Comparison of combined versus subdural or intracerebral electrodes alone in presurgical focus localization. Epilepsia. 1998;39(12):1290–301. https://doi.org/10.1111/j.1528-1157.1998.tb01327.x.
89. Ajmone-Marsan C. Chronic intracranial recording and electrocorticography. In: Daly DD, Pedley TA, editors. Current practice of clinical electroencephalography.
90. Spencer SS, So NK, Engel JJ, Williamson PD, Levesque MF, Spencer DD. Depth electrodes. In: Surgical treatment of the epilepsies. Raven; 1993. p. 359–76.
91. Talairach J, David M, Tournoux P, Corredor H, Kvasina T. Atlas d Anatomie Stéréotaxique: Repérage Radiologique Indirect Des Noyaux Gris Centraux Des Régions Mésencéphalo-Sous-Optique et Hypothalamique de l Homme. Masson & Co; 1957.
92. Olivier A, Bertrand G. Stereotaxic device for percutaneous twist-drill insertion of depth electrodes and for brain biopsy. Technical note. J Neurosurg. 1982;56(2):307–8. https://doi.org/10.3171/jns.1982.56.2.0307.
93. Vakharia VN, Rodionov R, Miserocchi A, et al. Comparison of robotic and manual implantation of intracerebral electrodes: a single-centre, single-blinded, randomised controlled trial. Sci Rep. 2021;11(1):17127. https://doi.org/10.1038/s41598-021-96662-4.
94. Yeh HS, Taha JM, Tobler WD. Implantation of intracerebral depth electrodes for monitoring seizures using the Pelorus stereotactic system guided by magnetic resonance imaging. Technical note. J Neurosurg. 1993;78(1):138–41. https://doi.org/10.3171/jns.1993.78.1.0138.

95. Murphy MA, O'Brien TJ, Cook MJ. Insertion of depth electrodes with or without subdural grids using frameless stereotactic guidance systems – technique and outcome. Br J Neurosurg. 2002;16(2):119–25. https://doi.org/10.1080/02688690220131886.
96. Mehta AD, Labar D, Dean A, et al. Frameless stereotactic placement of depth electrodes in epilepsy surgery. J Neurosurg. 2005;102(6):1040–5. https://doi.org/10.3171/jns.2005.102.6.1040.
97. Hawasli AH, Bagade S, Shimony JS, Miller-Thomas M, Leuthardt EC. Magnetic resonance imaging-guided focused laser interstitial thermal therapy for intracranial lesions: single-institution series. Neurosurgery. 2013;73(6):1007. https://doi.org/10.1227/NEU.0000000000000144.
98. Jermakowicz WJ, Kanner AM, Sur S, et al. Laser thermal ablation for mesiotemporal epilepsy: analysis of ablation volumes and trajectories. Epilepsia. 2017;58(5):801–10. https://doi.org/10.1111/epi.13715.
99. Youngerman BE, Oh JY, Anbarasan D, et al. Laser ablation is effective for temporal lobe epilepsy with and without mesial temporal sclerosis if hippocampal seizure onsets are localized by stereoelectroencephalography. Epilepsia. 2018;59(3):595–606. https://doi.org/10.1111/epi.14004.
100. Kang JY, Wu C, Tracy J, et al. Laser interstitial thermal therapy for medically intractable mesial temporal lobe epilepsy. Epilepsia. 2016;57(2):325–34. https://doi.org/10.1111/epi.13284.
101. Willie JT, Laxpati NG, Drane DL, et al. Real-time magnetic resonance-guided stereotactic laser amygdalohippocampotomy for mesial temporal lobe epilepsy. Neurosurgery. 2014;74(6):569–85. https://doi.org/10.1227/NEU.0000000000000343.
102. Aum DJ, Reynolds RA, Mcevoy S, et al. Surgical outcomes of open and laser interstitial thermal therapy approaches for corpus callosotomy in pediatric epilepsy. Epilepsia. 2023;64(9):2274–85. https://doi.org/10.1111/epi.17679.
103. McCracken DJ, Willie JT, Fernald BA, et al. Magnetic resonance thermometry-guided stereotactic laser ablation of cavernous malformations in drug-resistant epilepsy: imaging and clinical results. Oper Neurosurg. 2016;12(1):39. https://doi.org/10.1227/NEU.0000000000001033.
104. Hooten KG, Werner K, Mikati MA, Muh CR. MRI-guided laser interstitial thermal therapy in an infant with tuberous sclerosis: technical case report. J Neurosurg Pediatr. 2018;23(1):92–7. https://doi.org/10.3171/2018.6.PEDS1828.
105. Tovar-Spinoza Z, Ziechmann R, Zyck S. Single and staged laser interstitial thermal therapy ablation for cortical tubers causing refractory epilepsy in pediatric patients. Neurosurg Focus. 2018;45(3):E9. https://doi.org/10.3171/2018.6.FOCUS18228.
106. Esquenazi Y, Kalamangalam GP, Slater JD, et al. Stereotactic laser ablation of epileptogenic periventricular nodular heterotopia. Epilepsy Res. 2014;108(3):547–54. https://doi.org/10.1016/j.eplepsyres.2014.01.009.
107. Makary M, Chiocca EA, Erminy N, et al. Clinical and economic outcomes of low-field intraoperative MRI-guided tumor resection neurosurgery. J Magn Reson Imaging. 2011;34(5):1022–30. https://doi.org/10.1002/jmri.22739.
108. Abraham P, Sarkar R, Brandel MG, et al. Cost-effectiveness of intraoperative MRI for treatment of high-grade gliomas. Radiology. 2019;291(3):689–97. https://doi.org/10.1148/radiol.2019182095.
109. Eljamel MS, Mahboob SO. The effectiveness and cost-effectiveness of intraoperative imaging in high-grade glioma resection; a comparative review of intraoperative ALA, fluorescein, ultrasound and MRI. Photodiagn Photodyn Ther. 2016;16:35–43. https://doi.org/10.1016/j.pdpdt.2016.07.012.
110. Sacino MF, Huang SS, Keating RF, Gaillard WD, Oluigbo CO. An initial cost-effectiveness analysis of intraoperative magnetic resonance imaging (iMRI) in pediatric epilepsy surgery. Childs Nerv Syst. 2018;34(3):495–502. https://doi.org/10.1007/s00381-017-3658-9.

Pediatric Cortical Dysplasia: PET Studies

Bharathi Dasan Jagadeesan and Csaba Juhász

Abstract Malformations of cortical development (MCDs) constitute an important cause of seizures in children and often form the underlying basis for many of the epilepsy syndromes found in childhood. Functional neuroimaging with positron emission tomography (PET) plays a critical role in the diagnosis and management of MCDs, including cortical dysplasias, in children with drug-resistant epilepsy. In this chapter, we begin with a focused discussion on the continuing relevance of well-established PET techniques such as 2-deoxy-2[^{18}F]fluoro-D-glucose (FDG) PET, particularly in the light of newer advances in structural imaging techniques. The limitations with these techniques are also discussed in some detail, followed by a discussion on recent advances such as quantitative imaging, hybrid PET-MRI scanners, PISCOM, and the utilization of artificial intelligence in the interpretation of these studies. Thereafter, we discuss the use and limitations of GABAA receptor imaging, and advances made by tryptophan PET imaging in cortical dysplasias and tuberous sclerosis. Finally, we conclude the chapter with a brief discussion on other promising new tracers and their likely roles in PET imaging of cortical dysplasias.

Keywords Cortical dysplasia · Epilepsy · FDG-PET · Tryptophan · GABAA receptors · Hybrid PET-MRI systems

1 Introduction

Cortical dysplasias are one of the commonest causes for pediatric epilepsy. Cortical dysplasias essentially result from malformations of cortical development (MCDs)

B. D. Jagadeesan (✉)
Departments of Radiology, Neurology and Neurosurgery, University of Minnesota, Minneapolis, MN, USA
e-mail: jagad002@umn.edu

C. Juhász
Departments of Pediatrics, Neurology, Neurosurgery, Wayne State University, PET Center and Translational Imaging Laboratory, Children's Hospital of Michigan, Detroit, MI, USA

P. A. Broderick (ed.), *Biomedical Imaging in Neurodegeneration*,
https://doi.org/10.1007/978-3-031-95878-6_6

Table 1 The initial international consensus classification of focal cortical dysplasias (FCDs) as reported by the International League Against Epilepsy

Type I (FCDI): abnormalities in radial or tangential cortical lamination, without evidence of any additional principal epileptogenic lesion in the brain, including
FCD with abnormal radial cortical lamination (FCDIa)
FCD with abnormal tangential cortical lamination (FCDIb).
FCD with abnormal radial and tangential cortical lamination (FCDIc).
Type II (FCDII): presence of disrupted cortical lamination and dysmorphic neurons and/or balloon cells, including.
FCD with dysmorphic neurons (FCDIIa).
FCD with dysmorphic neurons and balloon cells (FCDIIb).
Type III (FCDIII) has cortical lamination abnormalities together with another epileptogenic lesion, including.
Cortical lamination abnormalities in the temporal lobe associated with hippocampal sclerosis (FCDIIIa).
Cortical lamination abnormalities adjacent to a glial or glioneuronal tumor (FCDIIIb).
Cortical lamination abnormalities adjacent to a vascular malformation (FCDIIIc).
Cortical lamination abnormalities adjacent to any other lesion acquired during early life, for example, trauma, ischemic injury, encephalitis (FCDIIId).

[1]. On neuropathology, MCDs possess a spectrum of morphologic features associated with multiple putative etiologic features, including genetic and environmental influences. The classification scheme of MCDs was based on the affected primary step in neocortical development, such as neuronal organization, migration, or cortical organization [2]. The initial classifications of cortical dysplasias were based on a correlation of their histopathology with clinical, EEG, and neuroimaging features [3]. The first international consensus classification of cortical dysplasias by the International League Against Epilepsy (ILAE) included type I, II (both isolated forms), and type III cortical dysplasias (the latter associated with another principal lesion) and their various subtypes (Table 1) [4].

Subsequently, the classification scheme was refined repeatedly (see the review of Balestrini et al. [5]). The term "mild MCD" (mMCD) was introduced for lesions comprising an excess of heterotopic neurons in layer I (type I mMCD) or the white matter (type II mMCD). Most recently, the ad hoc task force of the ILAE Diagnostic Methods Commission introduced a classification update with an integrated, multilayered, genotype–phenotype approach that integrates findings from histopathology, genetic analysis of resected tissue, and presurgical MRI to characterize cortical dysplasias [6].

MCDs encompass lesions ranging from those, which are easily evident on structural neuroimaging, such as schizencephaly and bilateral diffuse polymicrogyria, to those, which may be detected only on histological sections such as microscopic focal cortical dysplasias and excessive single white matter neurons. While most type II cortical dysplasias can be detected by clinical MRI, the neuroimaging diagnosis of type I and III dysplasias remains challenging.

The role of functional neuroimaging in children with cortical dysplasias is threefold: (i) To enable detection of focal or multifocal neuropathology in children with epilepsy due to cortical dysplastic lesions and normal structural imaging or subtle MRI results. This will aid in detection of the location and true extent of the underlying pathology and assist further management including surgical intervention in these children. (ii) To enable further subclassification and phenotyping of epileptic syndromes with multifactorial etiology such as infantile spasms and Lennox–Gastaut syndrome, with the purpose of tailoring treatment according to focal or multifocal etiologies in individual patients. (iii) To study the functional correlates of lesions, whether or not they are detectable on MRI. This involves the documentation of functional abnormalities extending beyond structural lesions including perilesional and remote functional cortical and subcortical abnormalities and dual pathology. These remote abnormalities may be a part of an epileptic network [7] and denote emerging or established secondary epileptogenic foci. Thus, functional imaging can contribute to a deeper understanding of the mechanisms of epileptogenesis, seizure spread, secondary neuronal injury and also cognitive abnormalities found in patients with different cortical malformations.

Positron emission tomography (PET) is a functional neuroimaging technique that has revolutionized the management of some disorders associated with cortical dysplasias such as infantile spasms. The clinical application of PET imaging has also significantly enhanced our ability to detect cortical dysplasias in patients with epilepsy in general, while deepening our understanding of the pathophysiology of this group of developmental malformations of the brain. PET allows noninvasive, in vivo assessment and quantification of diverse biochemical processes in the brain (both globally and regionally) such as glucose consumption, oxygen consumption, neurotransmitter release or transport, postsynaptic receptor binding, inflammation, and amino acid metabolism. Furthermore, PET allows the assessment of such processes under various physiologic states, including at rest and following neuronal activation states. More importantly, PET imaging enables the detection of quantitative alterations in biochemical processes in various disease states. In the present review, we address the role of PET in cortical dysplasia and epilepsy starting with the most widely used tracer, 2-deoxy-2[^{18}F]fluoro-D-glucose (FDG).

2 FDG-PET in Cortical Dysplasia

In general, the abnormalities found on interictal FDG-PET studies of patients with cortical dysplasia consist of focal or multifocal areas of decreased FDG uptake ([8–10]) (Fig. 1). Decreased FDG uptake can reflect diminished neuronal activity resulting in decreased glucose utilization in regions of cortical dysplasia, which may not always be macroscopically evident. This might also be secondary to local decrease in synaptic density. However, ectopic areas of dysplastic cortical

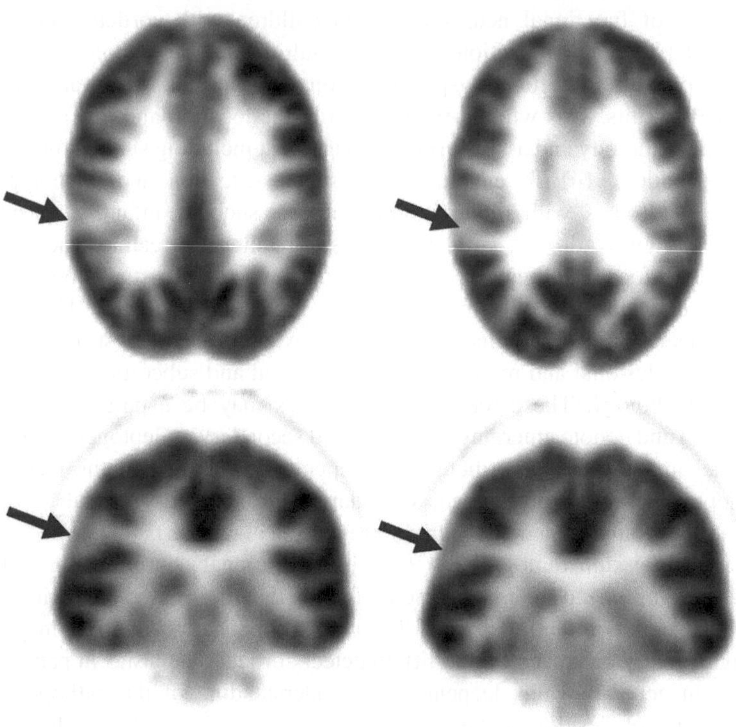

Fig. 1 FDG-PET images showing a small area focal cortical hypometabolism (arrows) in the right parietal cortex, in a patient with normal structural MRI, consistent with focal cortical dysplasia. Upper panel: axial PET images; lower panel: coronal PET images

tissue, which are found in the subcortical white matter, typically appear as areas of increased FDG uptake located beneath the overlying cortical activity [11, 12] (Fig. 2). Focal increases in FDG uptake may also be present in otherwise hypometabolic cortical areas if clinical or subclinical electroencephalographic seizures occur during the period of FDG uptake into the brain following tracer injection, as well as when there is frequent interictal spiking (Fig. 3) [12–14].

The pattern visualized on FDG-PET reflects glucose transport and tissue trapping during a prolonged period of time (approximately 30 min), thus representing a summation of cellular metabolic processes during the uptake period. Therefore, FDG-PET is not suitable to measure acute processes, which are better detected by blood flow imaging via PET (e.g., by using O-15 labeled water) or, more commonly in recent years, by functional MRI (fMRI). All patients with epilepsy should be subjected to scalp EEG recording during the period of FDG uptake, which lasts up to 60 minutes postinjection, and FDG-PET findings must always be interpreted in correlation with the EEG data thus obtained [15].

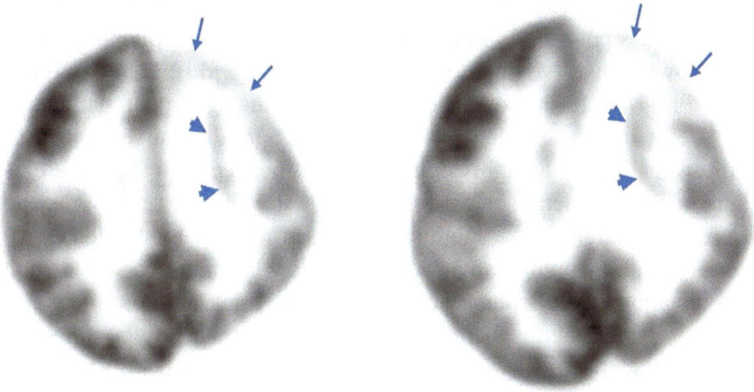

Fig. 2 Consecutive axial planes of an FDG-PET scan showing focal prominent uptake within the left frontal lobe white matter, consistent with band heterotopia (arrowheads), in a 10-year old patient with drug-resistant epilepsy originating from the left hemisphere. The overlying prefrontal cortex is severely hypometabolic (arrows)

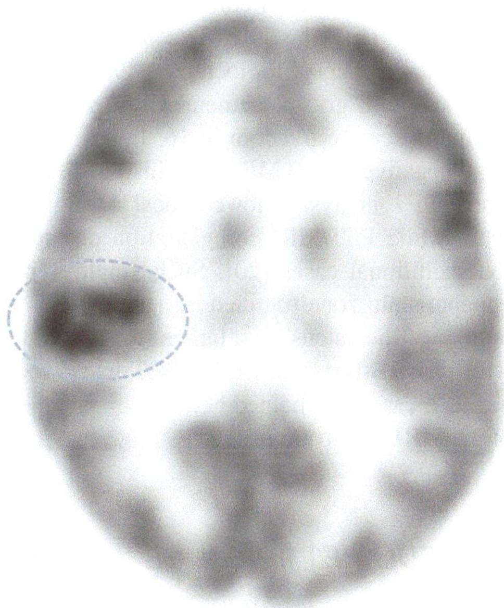

Fig. 3 A relative increase of cortical FDG uptake (circled area) in the right temporo-parietal region in a 7-year-old girl with intractable seizures and normal MRI. EEG recorded during the FDG uptake period showed no ictal activity but frequent interictal runs of right centro-temporo-parietal spikes, likely driving the focal hypermetabolism in that area

2.1 FDG-PET in "Nonlesional" Neocortical Epilepsy Due to Microscopic Cortical Dysplasias

Structural neuroimaging with MRI can be paradoxically normal in a significant number of patients with mild cortical developmental malformations [16, 17]. Patients with such normal MRI studies presenting with focal seizures are considered to have nonlesional focal epilepsy. However, it has been shown that successful seizure outcome is possible in many of these patients when epileptogenic cortex is resected following invasive monitoring with subdural grid EEG or depth electrodes [18]. FDG-PET has been extensively studied in this group of patients and has proved to be of immense value by (i) serving as an alternative to invasive EEG monitoring when it reveals highly circumscribed focal cortical abnormalities which are in agreement with ictal scalp EEG findings and seizure semiology, especially in very young children where subdural EEG grid placement should be avoided if possible; (ii) guiding preoperative subdural grid placement for EEG monitoring in the localization of epileptogenic cortex when PET abnormalities are clearly lateralized to a single hemisphere and localized to a relatively confined region of the cortical surface (typically encompassing one or two lobes); and (iii) helping to avoid the need for invasive monitoring when ictal scalp EEG monitoring does not reveal highly focal onset of seizures and PET suggests multifocal abnormalities involving both hemispheres and multiple lobes of the brain. These applications of FDG-PET are strongly suggested by the study of Hong et al. [19], who reported on 41 patients with nonlesional epilepsy using FDG-PET, scalp ictal EEG, and ictal single-photon emission computed tomography (SPECT). FDG-PET was found to have a sensitivity of 42.9% (31.3% for extratemporal epilepsy and 70% for temporal neocortical epilepsy) for localizing the epileptic focus when compared with findings from subdural grid placement. Good surgical outcome was found in 33 patients, including seizure freedom in 16. In a subsequent study of 23 consecutive patients with negative brain MRI and histologically verified type II cortical dysplasia, visual PET analysis showed focal or regional extratemporal hypometabolism in 78%, while PET/MRI coregistration detected a partially hypometabolic gyrus in four additional cases [20]. Such single hypometabolic gyri corresponded to the epileptogenic zone (defined by stereo-EEG) and showed cortical dysplasia at the bottom of the sulcus. In a larger study from the same group (including 62 patients with type II dysplasia), FDG-PET had a localization value in 84% of the MRI-negative cases and helped better delineate the boundaries dysplastic cortex in 65% of the MRI-positive cases [21].

2.2 *FDG-PET in Pediatric Temporal Lobe Epilepsy*

Neocortical dysplasias also constitute an important cause of childhood temporal lobe epilepsy. Porter et al. [22]reviewed the neuropathologic features of temporal lobectomy specimens from 33 pediatric patients with refractory temporal lobe epilepsy and found evidence for cortical dysplasia in 64% of the patients. FDG-PET can reliably localize neocortical dysplastic areas in patients with nonlesional temporal lobe epilepsy [23]. However, temporal lobe neocortical dysplasias may not only occur alone, but may coexist with hippocampal sclerosis in affected children [24]; MR imaging may reveal only the coexisting hippocampal atrophy in such cases. Patients with cortical dysplasia coexisting with hippocampal sclerosis were reported to have a different pattern of glucose hypometabolism than patients with hippocampal sclerosis alone; the former have their greatest decreases in FDG uptake in the lateral temporal lobe, whereas the latter show maximum decreases in the anterior temporal lobe [25]. Thus, FDG-PET may aid in the localization of epileptogenic dysplastic cortical regions in patients with temporal and extratemporal refractory partial epilepsies and normal MRI studies.

2.3 FDG-PET in Childhood Epilepsy Syndromes

Epileptic syndromes are defined on the basis of the clustering together of particular clinical findings with characteristic EEG changes in patients afflicted with these conditions. The clinical characteristics could include the type of seizures, cognitive abnormalities, and predominance of seizures in particular age groups. However, a direct correlation between clinical epilepsy syndromes and underlying neuropathology may not always be present. Syndromes such as epileptic spasms (previously known as infantile spasms) and Lennox–Gastaut syndrome have been noted to have a broad spectrum of underlying neuropathology associated with them [26, 27]. In these syndromes, the amenability of individual patients to surgical intervention and the need for better prognostication dictate the need for accurate localization of underlying neuropathology by imaging. Structural imaging studies seldom meet these requirements completely, and functional neuroimaging with FDG-PET has greatly filled this gap in our ability to understand and treat these patients.

2.3.1 West Syndrome (with Epileptic Spasms)

This syndrome is characterized by the triad of infantile tonic/myoclonic spasms, hypsarrhythmia on EEG, and developmental arrest with symptoms manifesting in early infancy. PET studies have dramatically altered the understanding and clinical management of infantile spasms, which used to be considered to be generalized seizures resulting from complex cortico-subcortical interactions. Using FDG-PET, it has been found that most infants diagnosed with "cryptogenic" spasms have, in fact, focal or multifocal cortical regions of decreased (or, occasionally increased) glucose utilization that are often consistent with areas of cortical dysplasia missed by MRI [28–30]. When a single region of abnormal glucose utilization is apparent on PET, corresponding to the EEG focus, surgical removal of the PET focus results not only in control of otherwise intractable seizures but also in complete or partial reversal of the associated developmental delay. When bilateral multifocal areas of hypometabolism are found, if all seizures arise from a single area, surgical resection may ameliorate the epilepsy, but there may be no significant improvement in cognitive development. However, most patients with bilateral multifocal lesions are not surgical candidates since no single focus of seizure origin can be found by EEG monitoring. When a generalized and symmetric hypometabolic pattern is found on

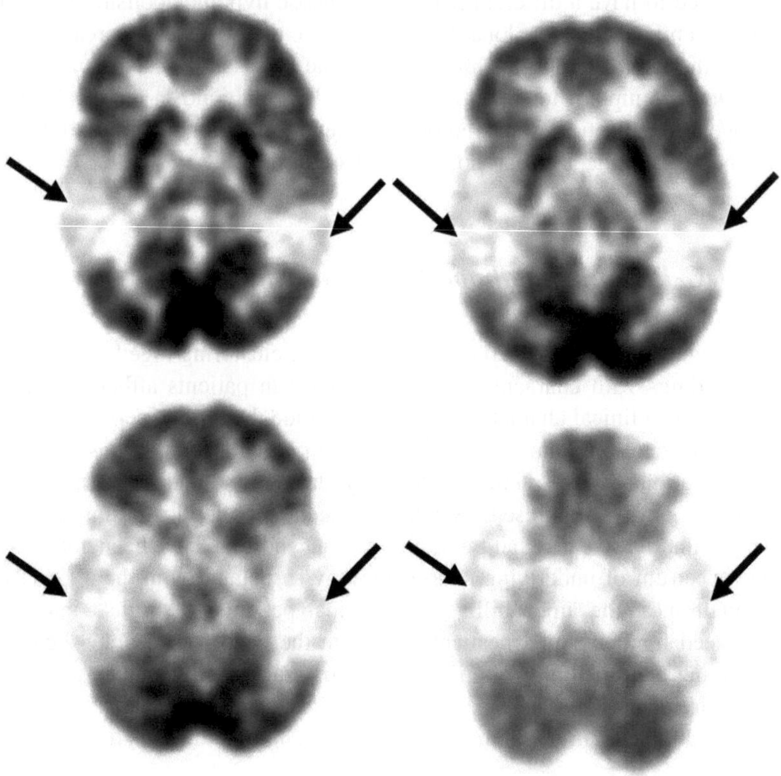

Fig. 4 FDG-PET planes at different brain levels in a child with infantile spasms and autistic features. The bilateral temporal cortex is severely hypometabolic (arrows)

FDG-PET, then neurometabolic or neurogenetic disorders are likely to constitute the underlying pathology, and lesional etiology is less likely (Fig. 4). Hence the pattern of glucose hypometabolism on FDG-PET not only enables evaluation for surgery but also enables a better understanding of the underlying pathophysiology in individual children with West syndrome.

Longitudinal PET studies in children with infantile spams demonstrated that focal glucose metabolic abnormalities, especially those in the occipital cortex, may be transient [30, 31]. Specifically, children with favorable clinical outcome showed transient focal cortical hypometabolism at the onset of the spasms, which reduced in extent or disappeared when the FDG-PET scan was repeated 1 year later [31]. These transient hypometabolic lesions were found to be equally common in patients with cryptogenic and symptomatic etiology. It is, therefore, likely that, in these children, regional hypometabolism represents malfunctioning cortex rather than structural lesions, and hence the surgical approach to children with infantile spasms and extensive cortical hypometabolism may need to be carefully addressed using intracranial electrodes. FDG-PET studies can also show hypermetabolism in the lenticular nuclei and in the brainstem area, suggesting that these structures may be associated with the basic mechanisms of epileptic spasms [32].

2.3.2 Lennox–Gastaut Syndrome (LGS)

Children with LGS have a triad consisting of multiple seizure types, developmental delay, and 1–2.5 Hz generalized "slow" spike and wave EEG pattern. Early PET studies suggested four metabolic subtypes in LGS: unilateral focal, unilateral diffuse, bilateral diffuse hypometabolism, and normal patterns [33, 34]. Focal cortical hypometabolism may be related to cortical dysplasia or other malformations of cortical development [35–37], and patients with the unilateral focal and unilateral diffuse patterns may be occasionally considered for cortical resection when ictal EEG findings are concordant with the area of hypometabolism. A recent FDG-PET study expanded on our understanding of LGS and supported the concept of LGS being a secondary network epilepsy, where the shared electroclinical manifestations reflect recruitment of a common epileptic brain network, despite diverse underlying etiologies [38]. In this study of 21 patients with LGS (mean age: 15 years) and 18 pseudo-controls, an objective group comparison (LGS vs. controls) revealed common hypometabolism in a network of regions including prefrontal and premotor cortex, anterior and posterior cingulate, inferior parietal lobule, and precuneus. The same brain regions were recruited during interictal bursts of generalized paroxysmal fast activity and tonic seizures on EEG-functional MRI studies [39], indicating that these regions are central to the electroclinical expression of LGS.

2.3.3 Tuberous Sclerosis

FDG-PET can enhance the detection of cortical abnormalities in tuberous sclerosis complex (TSC) by identifying cortical tubers, including small ones not visualized on conventional MRI, and detecting dysplastic cortex that may appear normal on MRI in the perituberal area. While epileptogenic regions were found to have increased volume of FDG-PET hypometabolism [40], visual assessment of FDG-PET images cannot distinguish between epileptogenic and nonepileptogenic tubers. Recently, a novel FDG-PET-based logistic regression model, including multiple PET-derived quantitative variables, was reported to have a good accuracy to predict the epileptogenic tubers and also surgical outcomes in a cohort of 43 patients with TSC and drug-resistant epilepsy [41]. In a previous study of children with TSC, who underwent electrocorticography before respective epilepsy surgery, glucose hypometabolism was associated with increased delta slowing and frequent spike activity independently and additively [42], thus providing an insight in the pathophysiology of TSC-driven glucose metabolic abnormalities. Beyond epileptic focus localization, FDG-PET also provides a comprehensive assessment of functional abnormalities in the brain and evaluates the integrity of the homotopic contralateral cortex, aiding in predicting potential postoperative cognitive deficits. FDG-PET has been also instrumental in understanding the neurobehavioral phenotypes of tuberous sclerosis complex (TSC), including autism, attention deficit hyperactivity disorder, and cognitive impairment. PET abnormalities reveal cortical and subcortical dysfunction, providing insights into the pathophysiology of autism in TSC.

2.3.4 Sturge–Weber Syndrome

Sturge–Weber syndrome (SWS) is a noninherited venous vascular disorder characterized by a leptomeningeal venous malformation often associated with enlarged deep veins, atrophy, and calcification in the affected hemisphere. Such abnormalities can also coincide with cortical developmental malformations, most commonly with cortical dysplasia and polymicrogyria [43, 44], although these cannot be readily detected by clinical MRI in most cases [45]. The majority of SWS-affected children develop focal seizures during the first 2 years of life, and some will suffer from drug-resistant epilepsy. Since SWS is commonly unilateral, resective epilepsy surgery can be highly effective to eliminate such seizures. FDG-PET plays a crucial role in determining the timing and need for surgical resection in such patients by providing valuable insights into the metabolic activity of the affected brain tissue. It can help assess the order of progression of cerebral glucose metabolism abnormalities during the early and late stages of SWS showing the presence of early cortical hypermetabolism, with subsequent transition to hypometabolism, which may serve as an imaging marker for subsequent severe epilepsy, which can be avoided with early surgical intervention [46, 47].Cortical hypometabolism in SWS often extends beyond the MRI-detected vascular and parenchymal structural abnormalities [48] and may help evaluate the integrity of the frontal lobe; this information can be pivotal when deciding the extent of resective surgery, i.e., hemispherectomy vs. posterior quadrant resection.

2.4 Functional Correlates of FDG-PET Abnormalities in Patients with Cortical Dysplasia

Comparison of MRI and FDG-PET findings in patients with cortical dysplasia in whom a clear structural abnormality is depicted on MRI imaging reveals that the abnormalities in FDG-PET often extend well beyond the spatial limits of the structural lesion. The extent of perilesional glucose metabolism found in these patients with intractable lesional epilepsy has been shown to have a correlation with the lifetime number of seizures: the greater the lifetime number of seizures, the more extensive the area of FDG hypometabolism [49]. Repeated seizures can result in cortical glucose hypometabolism remote from the EEG focus, a finding supported by studies showing recovery of metabolism in remote brain regions after successful elimination of the primary focus [50, 51]. In longitudinal PET studies of patients who did not undergo epilepsy surgery, robust interval changes in the extent of cortical hypometabolism are often observed. In particular, young patients with persistent frequent seizures showed a progressive expansion of hypometabolic cortex, involving additional ipsi- and even contralateral brain regions [52, 53]. On the other hand, about one-fourth of those with a major interval decrease of seizure frequency showed less extensive hypometabolism on the follow-up scan years later, supporting the concept that some of the observed metabolic abnormalities were purely functional and aggravated by seizures [53]. Importantly, the cortical extent of glucose hypometabolism at baseline correlated with developmental levels assessed at the time of both baseline and follow-up scans, suggesting that optimized medical treatment to control seizures may contribute to better neurocognitive outcome if no surgery can be offered.

These findings altogether suggest that interictal cortical hypometabolism is a dynamic functional change, and a significant part of it is likely to be the consequence rather than cause of recurring seizures. More recently, it has been suggested that changes in FBP1 (fructose-1,6-bisphosphatase 1), which is an enzyme that plays a key role in regulating glucose metabolism, could underly the changes in FDG uptake noted in interictal PET in patients with mesial temporal lobe sclerosis [54]. This enzyme is upregulated in the hippocampus of patients with refractory epilepsy, and its high expression has been associated with low FDG uptake and poor surgical outcomes. Further understanding of this phenomenon could also enhance our knowledge on the mechanisms of secondary epileptogenesis and cognitive decline in patients with epilepsy. This in turn could lead to the development of newer treatment paradigms.

The often extensive and nonspecific nature of FDG-PET abnormalities in many patients with extratemporal cortical dysplasia has led to the search and testing for other, more specific radiotracers, with the potential ability of differentiating those abnormal brain areas, which are the primary cause for epilepsy from those areas that show changes in function as a consequence of epilepsy. In addition, the application of hybrid PET/MRI imaging represents an additional clinical tool to refine and enhance the value of PET imaging in human epilepsy.

2.5 *Advances in the Utilization of FDG-PET in Conjunction with MRI*

2.5.1 Hybrid PET MRI Systems

Hybrid PET/MRI systems have several advantages for epilepsy diagnosis and treatment. PET/MRI combines the high spatial resolution of MRI with the functional information provided by PET, allowing for more accurate localization of the epileptogenic zone and better understanding of the disease. Studies have shown that hybrid PET/MRI has high sensitivity and specificity for epileptic focus localization, especially in patients with MRI-negative temporal lobe epilepsy [55]. The technique has also been found to significantly increase specificity and positive predictive value in pediatric patients with epilepsy [56].

Hybrid PET/MRI has been used for stereo-EEG implantation guidance, and concordance between FDG-PET/MRI and EEG findings was significantly predictive of a successful surgery outcome [57]. The technique has also been found to change surgical decision-making compared with FDG-PET coregistered with MRI by increasing epileptogenic lesion detection, especially focal cortical dysplasia [58].

A definite advantage of hybrid PET/MRI is that its use can avoid multiple scanning sessions (including multiple sedations in children) and improve the workflow in presurgical evaluation by combining PET, MRI, EEG-fMRI, and electrical source imaging based on data recorded in a single session using an MR-compatible EEG system and a hybrid PET/MR scanner. However, simultaneous acquisition may not necessarily improve colocalization considering the static nature of the brain [59].

2.5.2 Quantitative MRI and PET

Quantitative MRI can be used in combination with FDG-PET to improve the detection of subtle epileptogenic lesions. Voxel-based morphometry is a quantitative MRI technique that can analyze the gray matter concentration and detect abnormalities in the gray-white matter junction. This technique can improve the detection of subtle lesions that may not be visible on structural MRI [60, 61]. Combining quantitative MRI with FDG-PET can further improve the detection of epileptogenic lesions. A recent study reported a practical and effective multimodal imaging approach to detect subtle epileptic abnormalities by combined, quantitative MRI and FDG-PET analyses in patients with negative preoperative MRI. MRI scans underwent morphometric analysis to detect subtle focal cortical dysplasia causing blurring of the gray/white matter junction. Complete resection of MRI/PET-positive regions was associated with seizure-free outcome, and this effect was more robust in the extratemporal subgroup as compared to those with temporal lobe epilepsy [62].

2.5.3 PISCOM

PISCOM (PET and ictal SPECT coregistration with MRI) is a novel procedure for epilepsy imaging that combines ictal SPECT and interictal PET imaging coregistered with structural MRI studies to localize the seizure onset zone with high precision [63]. In contrast to the traditional SISCOM (Subtraction Ictal SPECT Co-registered to MRI) procedure, PISCOM offers several distinct advantages, including (i) improved localization: a study, which compared the two techniques, showed that PISCOM demonstrated successful identification of the seizure onset zone in 83% of patients, compared to 74% with SISCOM, indicating a higher success rate; (ii) reduced indeterminate activity: PISCOM showed fewer misleading results due to significantly less indeterminate activity arising from propagation, background, or artifacts compared to SISCOM, indicating improved specificity; and (iii) retention of spatial accuracy: PISCOM has shown precise spatial coincidence with SISCOM at the sublobar level, and the extension of the seizure onset voxel delimited by PISCOM and SISCOM was found to be equivalent [64].

2.5.4 Artificial Intelligence-Enhanced PET Interpretation in Epilepsy

The use of deep learning (DL) and other machine learning (ML) methods in the enhancement of various neuroimaging modalities that are used in the study of patients with epilepsy including PET has been recently reviewed by multiple authors [65–67]. The specific DL methods used in the studies encompassed various approaches, including feed-forward neural networks (FFNNs), convolutional neural networks (CNNs), and transfer learning models. These DL methods were applied to process neuroimaging data from modalities such as structural MRI, functional MRI, diffusion-weighted MRI, and positron emission tomography in the context of epilepsy research. They observed that with regard to PET, a study by Zhang et al. used a DL model to localize epileptogenic zones in temporal lobe epilepsy (TLE) patients based on FDG-PET images. The DL model achieved a high performance score with an area under the curve (AUC) of 0.93. The average Dice coefficient for the DL model (0.51) was higher than that of statistical parametric mapping (SPM) (0.24) and that of two specialists (0.31–0.44). These results suggest that the DL model outperformed the traditional SPM method and human experts in localizing epileptogenic zones in TLE patients [68]. A study by Kerr et al. used molecular neuroimaging with PET to diagnose right and left temporal lobe epilepsy. The DL model achieved a diagnostic accuracy of 88% for right TLE and 83% for left TLE, which was superior to visual analysis of interictal PET (85% and 79%) and structural MRI (77% and 70%) [69]. A study by Beheshti et al. (2020) achieved a prediction accuracy of 96% for lateralization using an SVM classifier and top-ranked glucose metabolic brain voxels from whole-brain PET images of 56 MRI-negative mTLE patients [70].

AI can also be used to integrate information from structural MRI and FDG-PET studies. Tan et al. aimed to diagnose focal cortical dysplasias (FCD) in patients with MRI-negative temporal lobe epilepsy (TLE) by combining positron emission tomography (PET) features with structural MRI (sMRI) data. They calculated morphology and intensity-based features from the cortical surfaces of 28 histopathologically confirmed FCD patients and used a cohort of 23 patients with MRI-negative TLE as the control group. By training a support vector machine (SVM) using leave-one-out cross-validation (LOOCV), the model achieved a lesion detection rate of 93% when using a combination of sMRI and PET features, outperforming a model using only sMRI features. This study demonstrates the potential of integrating PET data with sMRI for improved diagnosis of FCD in epilepsy patients [71].

Another AI-derived technique that has been studied for its utility in enhancing the role of FDG-PET in epilepsy is SIPCOM. SIPCOM stands for Subtraction Interictal PET Co-registered to MRI. It is a technique used to compare predicted and clinical PET images in epilepsy patients. The SIPCOM images are generated by subtracting the predicted PET image generated by deep learning algorithms from the actual clinical PET image. This technique allows for the identification of areas of abnormal metabolism that may be indicative of the epileptogenic zone. To generate the SIPCOM images, a generative adversarial network (GAN) is trained using healthy controls to predict the PET image from the corresponding T1w MRI [72].

The application of deep learning (DL) and machine learning (ML) in epilepsy research is accompanied by several limitations. Firstly, the reliance on large datasets for training DL models can be challenging due to the limited availability of comprehensive and diverse neuroimaging data in epilepsy studies. Additionally, the interpretability of DL models remains a concern, as the complex architectures of these models can make it difficult to understand the underlying decision-making process. Furthermore, the generalizability of DL and ML models across different patient populations and imaging modalities is an ongoing challenge, as these models may not perform consistently across diverse datasets. Lastly, the integration of DL and ML algorithms into clinical practice requires rigorous validation and standardization to ensure their reliability and safety for use in patient care. These limitations highlight the need for further research and development to address the challenges associated with the implementation of DL and ML in epilepsy studies and clinical settings.

2.6 *Unique Considerations with Utilization of FDG-PET in Children with Epilepsy*

FDG uptake in the brain can vary with age, and there are templates available for pediatric patients to account for these age-related differences. Studies have demonstrated that regional glucose uptake is most symmetric at around 1 year of age, and asymmetries emerge as the brain matures in a region-specific manner. Several frontal regions show an increase in left>right metabolic asymmetry, while selected regions in the temporal, parietal, and occipital cortex show a right>left asymmetry by teenage years [73]. To address these age-related differences, templates for pediatric patients have been developed to define normal, age-specific regional cortical asymmetries in children aged 1–18 years. These templates take into account the changes in glucose uptake that occur as the brain matures, allowing for more accurate interpretation of FDG-PET images in pediatric patients [73].

3 GABA Receptor Imaging with Flumazenil Pet in Children with Cortical Dysplasia and Epilepsy

Gamma amino butyric acid (GABA) is the major inhibitory neurotransmitter in the human brain, and GABAergic mechanisms play a key role in regulating central nervous system excitability and susceptibility to seizures [74]. The action of GABA is mediated in part by the $GABA_A$ receptor complex. Flumazenil is a benzodiazepine antagonist that binds to the alpha subunit of the $GABA_A$ receptor. Hence, PET imaging of $GABA_A$ receptor binding can be performed using the tracer [^{11}C]flumazenil (FMZ). Patients undergoing FMZ-PET should not take drugs (such as benzodiazepines) that directly interact with the $GABA_A$ receptors. Among antiepileptic medications, chronic vigabatrin treatment has been shown to be associated with decreased regional GABAA receptor binding in young children with partial seizures or infantile spasms [75]; however, similar effects have not been reported in adults, whose $GABA_A$ binding capacity is generally lower than that of children under normal circumstances [76]. Indeed, age-related changes in FMZ binding have been reported in humans and, in normal adult volunteers, baseline FMZ binding values are 25–50% lower (depending on the brain region) than those in children around 2 years of age (measured in the nonepileptic hemisphere of children with epilepsy) [77].

The utility of FMZ-PET in cortical dysplasias can again be studied in terms of its utility in (i) detection of primary (and secondary) epileptic foci in patients with partial epilepsies and normal structural imaging, (ii) studying epilepsy syndromes with a view to improve subclassification and prognostication, and (iii) studying the clinical and electrophysiologic correlations of FMZ-PET abnormalities regardless of the visualization (or nonvisualization) of structural brain malformations on MRI.

3.1 FMZ-PET in the Presurgical Detection of Epileptic Foci

Initial observations suggested that FMZ-PET may be more sensitive and specific than FDG-PET in the detection of epileptic foci in temporal lobe epilepsy; these observations were followed by similar findings on the utility of FMZ-PET in frontal lobe epilepsy [78] (Fig. 5). This initiated the use of FMZ-PET in patients with cortical dysplasias and intractable epilepsy. In one of the largest FMZ-PET studies in epileptic patients, Ryvlin et al. [79] compared the utility of structural MRI, FDG-PET, and FMZ-PET in 100 patients with intractable partial seizures, including 34 with normal MRI, and found that FMZ-PET outperformed FDG-PET in the detection of cortical epileptic foci in extratemporal lobe neocortical epilepsies with normal structural imaging, including those with focal cortical dysplasia. In a subsequent study, Muzik et al. [80]compared the performance of FMZ- and FDG-PET in the detection of neocortical epileptic foci in children with extratemporal lobe epilepsy using the gold standard of intracranial EEG monitoring.

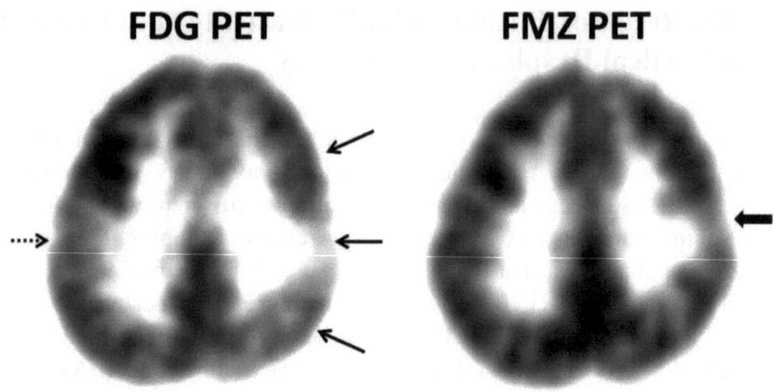

Fig. 5 FDG-PET and FMZ-PET images from a child with intractable frontal lobe epilepsy. FDG-PET showed bilateral hypometabolism, with extensive left cortical abnormalities, including the left fronto-parietal cortex (thin arrows) and the right posterior frontal cortex (dashed arrow). FMZ-PET showed a focal area of decreased binding in the left posterior frontal cortex (arrow). Subsequent intracranial EEG showed seizure onset from this latter region

They found that FMZ-PET was significantly more sensitive than FDG-PET for the detection of cortical regions of seizure onset and frequent spiking. Therefore, the application of PET, in particular FMZ-PET, in guiding subdural electrode placement in refractory extratemporal lobe epilepsy can enhance coverage of the epileptogenic zone. The exact nature of abnormal FMZ binding likely depends on the nature of the underlying cortical malformation, since either decreased or increased regional FMZ binding has been shown in epileptogenic dysplastic cortical regions [81]. Interestingly, a study in patients with cortical dysplasias undergoing FMZ-PET imaging reported increased periventricular FMZ binding deep in the white matter. [82] These findings may reflect disordered neuronal migration, leading to the presence of heterotopic neurons containing $GABA_A$ receptors. A subsequent study also suggested that presence of periventricular white matter FMZ binding on PET was associated with poor surgical outcome in patients with mesial temporal lobe epilepsy [83].

3.2 *FMZ-PET and Dual Pathology*

Dual pathology refers to the coexistence of hippocampal sclerosis and an epileptogenic neocortical lesion [84], and this condition is present in 15–35% of patients with intractable epilepsy, including those with a history of infantile spasms [85–87]. Unsuspected dual pathology is considered one of the major factors responsible for poor surgical outcome in such patients. Visual as well as quantitative analysis of FMZ-PET sensitively detected decreased benzodiazepine receptor binding in the hippocampus as well as the neocortex (often consistent with subtle cortical dysplasia) in patients with dual pathology [88]. Thus, in addition to its role in detecting epileptogenic foci in children with normal MRI and cortical dysplasia, FMZ-PET can also play a useful role in the detection of additional neocortical lesions in patients with temporal lobe epilepsy due to hippocampal sclerosis.

3.3 FMZ-PET in Childhood Epilepsy Syndromes

In children with infantile spasms associated with cortical dysplasia, FMZ-PET can indicate regions of decreased (or increased) FMZ binding consistent with subtle cortical abnormalities not detected by MRI [89]. A potential advantage of using FMZ-PET in small children is the high regional FMZ binding that allows better visualization of some brain regions such as the medial temporal lobe. The exact role of FMZ-PET as compared to FDG-PET in children with West syndrome, however, requires further study to determine whether FMZ-PET offers additional localizing information.

3.4 Clinical and Electrophysiological Correlates of FMZ-PET Abnormalities

In patients with extratemporal epilepsy, FMZ-PET abnormalities are found to be less extensive than FDG-PET abnormalities and also more specific in the delineation of cortical areas of seizure onset and frequent interictal spiking as defined by intracranial EEG [80, 90]. The extent of nonresected areas with decreased flumazenil binding in the lobe of seizure onset is a good predictor of seizure outcome following cortical resection, i.e., poor surgical outcome is associated with nonresection of cortex showing decreased flumazenil binding in the lobe of seizure onset [91]. Thus, FMZ-PET is an excellent complementary imaging method in patients with normal MRI and extensive glucose hypometabolism on FDG-PET.

In children with epilepsy and a readily apparent lesion on structural imaging, i.e., lesional neocortical epilepsy, decreased FMZ binding present in the lesion frequently extends into the adjacent cortex [92]. Furthermore, these perilesional FMZ-PET defects often show an eccentric and not concentric distribution. Subdural EEG recordings following grid placement showed that the ictal onset almost always involved the perilesional FMZ abnormality (perilesional epileptogenicity) and that electrodes on the MRI-defined structural lesion are often electrographically silent [92, 93]. Resection of the perilesional FMZ abnormalities with spiking resulted in excellent seizure-free outcome.

In addition to perilesional abnormalities, remote FMZ-PET abnormalities can occur outside the lobe of seizure onset. These remote FMZ-PET abnormalities are associated with an early age at seizure onset and often appear in ipsilateral, synaptically connected regions from the lesion or primary focus area. These remote cortical regions are commonly involved in early seizure propagation or interictal spiking [94]. Despite these promising results, FMZ-PET has not gained widespread clinical utilization and its use decreased over the years due to several limitations.

3.5 Limitations of Clinical Applications of ^{11}C-FMZ-PET

One of the main limitations of FMZ-PET is its limited availability and accessibility compared to other imaging modalities, partly due to the short half-life of C-11 (20 minutes), which may restrict its widespread use in clinical practice. Additionally, the need for specialized equipment and expertise in performing and interpreting FMZ-PET scans may pose challenges in certain healthcare settings. Another drawback of FMZ-PET is its reliance on benzodiazepine receptors, which may not fully capture the complexity of epileptogenic processes in the brain. The specificity of FMZ-PET for detecting epileptogenic foci may be influenced by factors such as age-related changes in FMZ binding, as baseline FMZ binding values are lower in adults than in children [77]. This age-related variability in FMZ binding could potentially impact the interpretation of FMZ-PET results, particularly in pediatric patients.

Furthermore, the interpretation of FMZ-PET scans requires careful consideration of individual patient characteristics and clinical context. The nature of the change in FMZ uptake in epileptogenic dysplastic cortical regions, whether decreased or increased regional FMZ binding, may introduce challenges in the accurate localization of epileptogenic foci and the planning of surgical interventions.

To overcome some of the above limitations, several ^{18}F-labeled flumazenil analog PET tracers have been developed, including 5-(2′-[^{18}F]fluoroethyl) flumazenil, 3-(2′-[^{18}F]fluoroflumazenil, and [^{18}F]flumazenil (^{18}F-FMZ) [95]. These tracers offer advantages over ^{11}C-flumazenil, such as longer half-life (110 minutes for F-18), better image resolution due to shorter positron range, and easier availability for clinical use [96]. However, with the exception of ^{18}F-FMZ, they also have some disadvantages, such as lower receptor affinity, higher nonspecific binding, and faster metabolism and kinetics in the brain. In particular, ^{18}F-FMZ has been established as the tracer of choice for patients with refractory epilepsy in some neurological centers in Europe [97], but study results have been also reported from other parts of the world with this tracer. For example, a study of 31 patients with temporal lobe epilepsy from Australia reported that visual assessment of static standardized uptake value images on ^{18}F-FMZ-PET had high specificity (94%) and moderate sensitivity (68%) for the localization of the epileptogenic zone, with a more restricted abnormality than FDG-PET performed in the same patients [98]. In another study of seven patients with refractory temporal lobe epilepsy, lateralization of the seizure onset zone by PET matched in 86% for ^{18}F-FMZ (and 71% for FDG) with stereo-EEG data [99]. These promising initial results warrant larger prospective studies to fully evaluate the clinical utility of F-18-labeled flumazenil analog tracers in the presurgical evaluations of patients with focal epilepsy.

4 Imaging Tryptophan Metabolism and Serotonin Receptors in Cortical Dysplasia

Imaging tryptophan and its metabolite serotonin (i.e., 5-hydroxytryptamine, 5-HT) in epileptic patients has developed along two different directions: (i) the imaging of tryptophan uptake and metabolism in epileptogenic lesions and (ii) the imaging of serotonin receptor binding.

4.1 Imaging Tryptophan Uptake and Metabolism

The PET tracer alpha[^{11}C]-methyl-L-tryptophan (AMT) has emerged as an important tool in the study of certain types of epileptogenic lesions and foci. In particular, AMT-PET has emerged as one of the best currently available methods for the noninvasive localization of epileptogenic tuber(s) in a subset of children with tuberous sclerosis and intractable seizures [100–103]. Unlike most other clinically tested PET tracers, which show predominantly decreased uptake in epileptogenic cortical areas interictally, AMT-PET scanning reveals locally increased uptake of AMT in and/or around epileptogenic tubers, while showing normal or decreased uptake in nonepileptogenic tubers (Fig. 6a–c). In a study of 17 children with TSC who underwent resective epilepsy surgery after AMT-PET scanning, tubers with at least 10% increase of AMT uptake (as compared to uptake measured in normal cortex) were all epileptogenic [102]. Using a cutoff threshold of 1.02 for tuber/cortex AMT uptake ratio provided an accuracy of 83% for detecting tubers that needed to be resected to achieve a seizure-free outcome. Studies of resected epileptic tissues have suggested that the likely mechanism for this increased AMT uptake is the accumulation of tryptophan metabolites of the inflammatory kynurenine pathway [104, 105]. Local increases in tryptophan metabolism through this pathway can occur due to immune activation of the rate-limiting enzyme indoleamine-2,3-dioxygenase [106] and result in the production of the excitotoxic and convulsant NMDA agonist quinolinic acid. Hence, the increased AMT uptake found in epileptogenic cortical lesions likely reflects low-level neuroinflammation.

Detection of epileptic foci by increased AMT uptake extends beyond TSC and has been reported in a variety of cortical malformation, most commonly in cortical dysplasia (Fig. 6d–f). Fedi et al. [107] found increased AMT uptake corresponding to the epileptogenic area in 7 of 18 patients with neocortical epilepsy (including 4 with MRI-evident cortical dysplasia). Subsequently, we studied 27 children with extratemporal lobe epilepsy, including 7 with cortical dysplasia using AMT-PET,

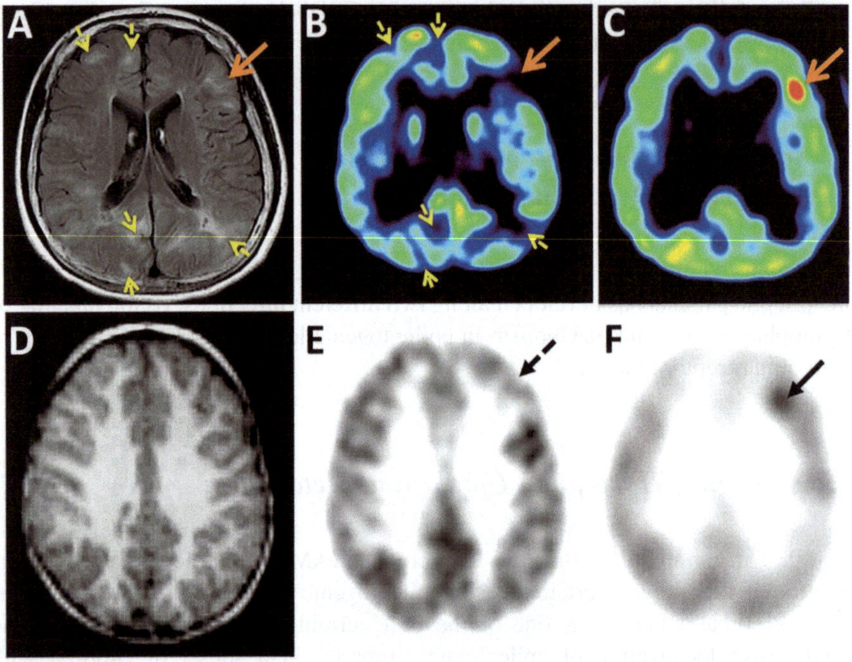

Fig. 6 Detection of an epileptogenic tuber (**a–c**) and focal cortical dysplasia (**d–f**) by high AMT uptake on PET imaging. (**a**) Axial FLAIR MRI showed multiple lesions in both hemispheres (arrows indicate six of the lesions). (**b**) All of these lesions were severely hypometabolic on FDG-PET. (**c**) AMT-PET showed increased tracer uptake in the left frontal hypometabolic tuber (orange arrows), suggesting epileptogenicity in this lesion. In a young child with suspected frontal lobe seizures but no clear lesion detected by MRI, although mild blurring of the gray/white matter junction was noticed in the left frontal lobe (**d**), FDG-PET (**e**) showed mild hypometabolism in the left frontal cortex (arrow); (**f**) in the same area, AMT-PET showed increased interictal uptake (arrow) suggesting epileptogenicity in a small dysplastic lesion that remained undetected by MRI. (Reproduced with permission from: Juhász C, John F. Utility of MRI, PET, and ictal SPECT in presurgical evaluation of non-lesional pediatric epilepsy. Seizure. 2020;77:15–28)

FDG-PET, structural MRI, and prolonged ictal and interictal scalp EEG monitoring, including subdural grid monitoring and postoperative histopathological examination of the resected tissue [108]. Focal cortical increase of AMT uptake was objectively identified in 15 patients (55%). Occurrence of increased AMT uptake did not correlate with interictal spike frequency. Of the three patients with normal MRI and histologically proven cortical dysplasia, AMT-PET identified the lesion in two, and FDG-PET in two. In those patients with structurally evident cortical dysplasia, FDG abnormalities were found to be widespread and nonspecific in all the cases but one, whereas AMT-PET localized specifically to the epileptogenic dysplastic region. Overall, while AMT-PET may be less sensitive than FDG-PET in children with extratemporal neocortical epilepsies, it is considerably more specific for the epileptogenic region. Additionally, the finding of focally increased AMT uptake in cortical regions, in the absence of abnormalities in structural MRI or FDG-PET, likely indicates underlying cortical malformations. The high specificity of increased AMT uptake on PET imaging was confirmed in a larger subsequent study of epileptic children, both with and without an MRI-defined lesion [109]. The specificity of AMT-PET for detecting the seizure onset lobe was equally high in lesional (97%) and nonlesional groups (100%), whereas sensitivity was higher in the lesional than the nonlesional group (47% versus 29%). AMT-PET localized and visualized epileptogenic regions in 25% of patients with nonlocalizing MRI. Given its highly specific nature, this focal uptake may guide preoperative grid placement for EEG monitoring.

A subsequent detailed histopathology analysis of various epileptogenic cortical malformations from surgery specimens obtained from epileptic children who underwent resection following AMT-PET showed increased AMT uptake was common in type IIB dysplasia and polymicrogyria with heterotopia, while no increase was seen in some other malformations (mild MCD and types IA-IIA dysplasia or subependymal heterotopias) [110]. Interestingly, type II dysplasias (in particular type IIB) are associated with robust upregulation of inflammatory pathways and markers that may play a prominent role in the epileptogenesis of these lesions [111, 112]. Thus, AMT-PET may be useful to identify neuroinflammatory epileptic lesions potentially amenable for anti-inflammatory treatment.

AMT-PET can also be useful to localize nonresected epileptic cortex in patients with a previously failed cortical resection [113], including those with focal cortical dysplasia. This is particularly important since other tracers such as FDG or FMZ are rarely useful in such a clinical setting.

Finally, novel, F-18-labeled tryptophan analog tracers, which serve as a more clinically feasible alternative to C-11-AMT, are now beginning to emerge, including 5-[^{18}F]fluoro-α-methyl-tryptophan and others, which have shown promise in the imaging of the indoleamine 2,3-dioxygenase pathway in preclinical studies [114]. One of these F-18-labeled tryptophan analogs, 1-(2-[^{18}F]fluoroethyl)-tryptophan, has recently reached clinical trials in imaging brain tumors [115, 116] and can be tested for human epilepsy imaging in the near future.

4.2 Serotonin Receptor Imaging in Cortical Dysplasia

PET imaging studies using the tracer [^{18}F]FCWAY, a selective 5-HT1_A receptor antagonist, have found cortical and subcortical (including raphe nuclei) changes of 5-HT1_A receptors in patients with developmental malformations of the neocortex. The cortical abnormalities in tracer uptake have been found in areas corresponding to the anatomically malformed area as well as other anatomically normal areas [117]. The abnormalities in tracer binding in subcortical regions are especially interesting, since they are consistent with a role of the serotonergic system in the epileptic network to include subcortical structures. The ^{11}C analog of [^{18}F]FCWAY, [carbonyl-^{11}C]WAY-100635 (^{11}C-WAY), is also used for studying serotonin receptor binding in the brain [118]. Nevertheless, most human PET studies with 5-HT_{1A}-targeting tracers focused on temporal lobe epilepsy, partly because of their ability to assess regional serotonergic changes and epilepsy-associated depression [119].

5 Other Tracers for Imaging Cortical Dysplasia and Epilepsy

5.1 Glutamate Receptors

Glutamate is the main excitatory neurotransmitter in the brain, and it has a number of receptors that fall into two broad categories: ionotropic receptors, which are ligand-gated ion channels, and metabotropic receptors, which are coupled to intracellular messenger systems. Efforts to develop specific PET tracers targeting various subtypes of glutamate receptors go back to the 1970s, but potential clinical applications lagged, partly because of the complexity of these systems (see review of Kim et al. [120]). In recent years, however, several novel glutamate receptor radiotracers have been tested and showed promise in human epilepsy imaging. In the first human study of 11 patients with focal epilepsy, using the selective NMDA receptor channel tracer ^{18}F-GE179, individual analyses revealed a single cluster of focal binding increase in four patients (Fig. 7); two with MRI lesions, and two with normal MRI [121]. In a group analysis, comparing the epilepsy group with controls, global increases of NMDA tracer binding were also found in the same study, consistent with increased NMDA channel activation. In another, exploratory clinical study of patients with epilepsy, increased uptake of [^{11}C]K-2, a radiotracer for AMPA receptors, was found in the epileptogenic focus of patients with mesial temporal lobe epilepsy [122]; in vivo PET values were closely correlated with the local AMPA receptor distribution assessed in surgical specimens from the same individuals. In addition, since metabotropic glutamate receptor type 5 (mGluR5) abnormalities have been described in brain tissues resected from epilepsy patients with cortical dysplasia, PET studies with radiotracers targeting these receptors are particularly relevant for imaging epileptic foci associated with cortical dysplasia. The first, proof-of-concept human PET study was performed in 10 patients with an MRI diagnosis of cortical dysplasia using the radioligand [^{11}C]ABP688 [123]. In this group, reduced radiotracer binding was found in the dysplastic lesion in 7/10 patients, including reduced binding in 4/5 operated patients, where surgical specimens revealed type II cortical dysplasia.

5.2 Synaptic Vesicle Glycoprotein 2A (SV2A) PET Tracers

SV2A is widely expressed in brain synapses, and the interest in SV2A for epilepsy research and imaging stems from several facts: SV2A is the molecular target of

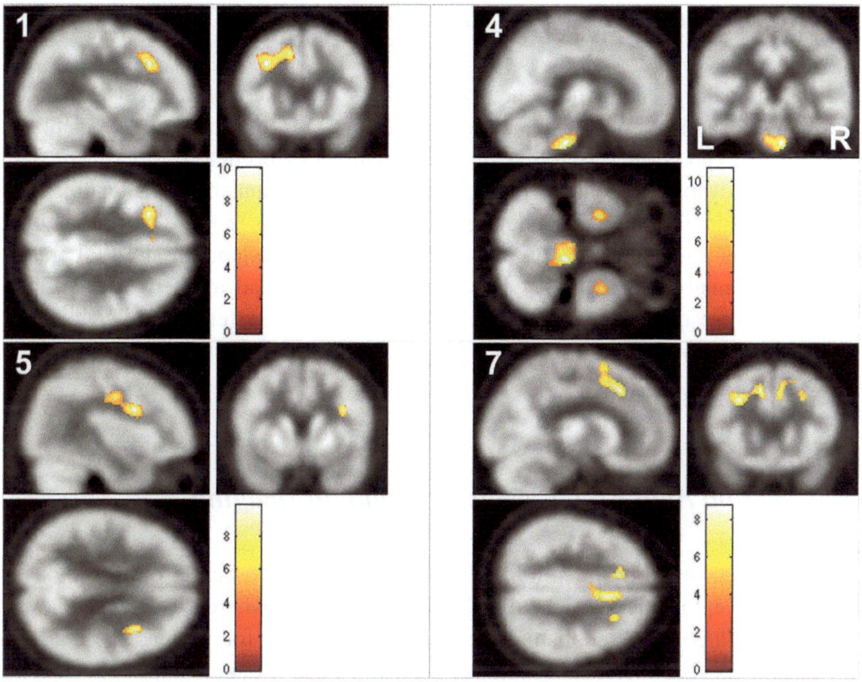

Fig. 7 Focal increases in [^{18}F]GE-179 binding in four patients with focal epilepsy, detected by a voxel-by-voxel comparison (1 vs. 10 controls, $p < 0.001$ voxelwise uncorrected). Significant clusters ($p < 0.05$) are overlaid onto a custom [^{18}F]GE-179 template. L, left; R, right. (Reproduced with permission from McGinnity et al. [121])

levetiracetam, a common antiseizure drug; also, SV2A knockout mice suffer from seizures from postnatal day 7; and the brain expression of SV2A is decreased in epilepsy patients [124]. Tracers that target synaptic vesicle glycoprotein 2A (SV2A), a protein involved in synaptic vesicle exocytosis, have been investigated for their potential in the evaluation of epileptic foci. These tracers offer a unique perspective on presynaptic function and synaptic transmission in epileptic brain regions, complementing the information provided by traditional tracers such as FDG. For example, the SV2A PET tracer [^{11}C]UCB-J has been used to assess the status of presynaptic function in mesial temporal epileptic foci, which showed decreased tracer binding that exceeded the severity of decreased FDG uptake [125]. In a recent study of 17 children with MRI-negative type II cortical dysplasia, PET imaging using ^{18}F-SynVesT-1, an F-18-labeled PET tracer for SV 2A, showed promise to complement FDG-PET for localizing the epileptogenic region in this very challenging patient group [126].

5.3 *Cannabinoid Receptor PET Tracers*

Cannabinoids are endogenous neuromodulators that exert anticonvulsant effects through the activation of cannabinoid receptors. Since prominent cannabinoid receptor 1 expression was found in malformations of cortical development, particularly in dysplastic neurons [127], and blockade of cannabinoid signaling may have a therapeutic effect on epileptogenic cortical dysplasia [128], there has been an increasing interest in imaging these receptors in the epileptic brain. PET tracers targeting cannabinoid receptors have been developed and applied in preclinical studies to investigate the role of the endocannabinoid system in epileptogenesis [129]. The promising results can encourage the application of such tracers in clinical epilepsies.

5.4 *Amino Acid PET Beyond Tryptophan Imaging*

Increased uptake of [^{11}C]methionine has been shown in a few cases of histologically verified epileptogenic cortical dysplasias [130, 131], where FDG-PET showed decreased glucose metabolism. The exact mechanism for increased uptake of methionine is yet undetermined although it may be due to ongoing epileptic activity, disruption of the blood–brain barrier in regions showing increased uptake, and/or upregulated amino acid transporters. Increased uptake may also be due to an immature pattern of amino acid transport [132] in developmentally abnormal dysplastic tissue. Additionally, radiolabeled methionine is accumulated by other tissues with high metabolic activity, such as brain tumors [133]. One study suggested that it is possible to differentiate between cortical dysplasia and low grade glioneuronal tumors using [^{11}C]methionine PET due to a greater extent of amino acid utilization in low-grade tumors [134].

References

1. Blumcke I, Spreafico R, Haaker G. Histopathological findings in brain tissue obtained during epilepsy surgery. N Engl J Med. 2017;377:1648–56.
2. Barkovich AJ, Kuzniecky RI, Dobyns WB, et al. A classification scheme for malformations of cortical development. Neuropediatrics. 1996;27:59–63.
3. Palmini A, Najm I, Avanzini G. Terminology and classification of the cortical dysplasias. Neurology. 2004;62:S2–8.
4. Blümcke I, Thom M, Aronica E. The clinicopathologic spectrum of focal cortical dysplasias: a consensus classification proposed by an AD hoc Task force of the ILAE diagnostic methods Commission. Epilepsia. 2011;52:158–74.
5. Balestrini S, Barba C, Thom M, et al. Focal cortical dysplasia: a practical guide for neurologists. Pract Neurol. 2023;23:293–302.
6. Najm I, Lal D, Vanegas MA. The ILAE consensus classification of focal cortical dysplasia: an update proposed by an AD hoc Task force of the ILAE diagnostic methods Commission. Epilepsia. 2022;63:1899–919.

7. Zhao B, Zhang C, Wang X, et al. Orbitofrontal epilepsy: distinct neuronal networks underlying electroclinical subtypes and surgical outcomes. J Neurosurg. 2020;135:1–11.
8. Swartz BE, Khonsari A, Brown C, et al. Improved sensitivity of 18FDG-positron emission tomography scans in frontal and 'frontal plus' epilepsy. Epilepsia. 1995;36:388–95.
9. Casse R, Rowe CC, Newton M, et al. Positron emission tomography and epilepsy. Mol Imaging Biol. 2002;4:388–51.
10. Macdonald-Laurs E, Warren AEL, Francis P, et al. The clinical, imaging, pathological and genetic landscape of bottom-of-sulcus dysplasia. Brain. 2023;147:1264.
11. Volder AG, Gadisseux JF, Michel CJ, et al. Brain glucose utilisation in band heterotopia: synaptic activity of "double cortex". Pediatr Neurol. 1994;11:290–4.
12. Poduri A, Golja A, Takeoka M, et al. Focal cortical malformations can show asymmetrically higher uptake on interictal fluorine-18 fluorodeoxyglucose positron emission tomography (PET). J Child Neurol. 2007;22:232–7.
13. Namer IJ, Valenti-Hirsch MP, Scholly J, et al. Hypermetabolism during resting-state FDG-PET suggesting intrinsic epileptogenicity in focal cortical dysplasia. Clin Nucl Med. 2014;39:993–5.
14. Chugani HT. Hypermetabolism on pediatric PET scans of brain glucose metabolism: what does it signify? J Nucl Med. 2021;62:1301–6.
15. Hur YJ, Lee JS, Lee JD, et al. Quantitative analysis of simultaneous EEG features during PET studies for childhood partial epilepsy. Yonsei Med J. 2013;54:572–7.
16. Oertzen VJ, Urbach H, Junbluth S. Standard magnetic resonance imaging is inadequate for patients with refractory focal epilepsy. J Neurol Neurosurg Psychiatry. 2002;73:643–7.
17. Russo GL, Tassi L, Cossu M. Focal cortical resection in malformations of cortical development. Epileptic Disord. 2003;5:115–23.
18. Siegel AM, Jobst BC, Thadani VM, et al. Medically intractable, localization-related Epilepsy with Normal MRI: Presurgical evaluation and surgical outcome in 43 patients. Epilepsia. 2001;42:883–8.
19. Hong KS, Lee SK, Kim JY, et al. Pre-surgical evaluation and surgical outcome of 41 patients with non-lesional neocortical epilepsy. Seizure. 2002;11:184–92.
20. Chassoux F, Rodrigo S, Semah F, et al. FDG-PET improves surgical outcome in negative MRI Taylor-type focal cortical dysplasias. Neurology. 2010;75:2168–75.
21. Chassoux F, Landré E, Mellerio C, et al. Type II focal cortical dysplasia: electroclinical phenotype and surgical outcome related to imaging. Epilepsia. 2012;53:349–58.
22. Porter BE, Judkins AR, Clancy RR, et al. Dysplasia: a common finding in intractable pediatric temporal lobe epilepsy. Neurology. 2003;61:365–8.
23. Kim YK, Lee DS, Lee S. Differential features of metabolic abnormalities between medial and lateral temporal lobe epilepsy: quantitative analysis of (18)F-FDG PET using SPM. J Nucl Med. 2003;44:1006–12.
24. Bocti C, Robitaille Y, Diadori P, et al. The pathological basis of temporal lobe epilepsy in childhood. Neurology. 2003;60:191–5.
25. Diehl B, Lapresto E, Najm I. Neocortical temporal FDG-PET hypometabolism correlates with temporal lobe atrophy in hippocampal sclerosis associated with microscopic cortical dysplasia. Epilepsia. 2003;44:559–64.
26. Jellinger K. Neuropathological aspects of infantile spasms. Bain Dev. 1987;9:349–57.
27. Meencke HJ. Morphological aspects of etiology and the course of infantile spasms (West syndrome). Neuropediatrics. 1985;16:59–66.
28. Chugani HT, Shewmon DA, et al. Infantile spasms: I. PET identifies focal cortical dysgenesis in cryptogenic cases for surgical treatment. Ann Neurol. 1990;27:406–41.
29. Chugani HT, Shewmon DA, Shields WD, et al. Surgery for intractable infantile spasms: neuroimaging perspectives. Epilepsia. 1993;34:764–71.
30. Sakaguchi Y, Kidokoro H, Ogawa C. Longitudinal findings of MRI and PET in west syndrome with subtle focal cortical dysplasia. AJNR Am J Neuroradiol. 2018;39:1932–7.
31. Metsahonkala L, Gaily E. Focal and global cortical hypometabolism in patients with newly diagnosed infantile spasms. Neurology. 2002;58:1646–51.

32. Chugani HT, Shewmon DA, Sankar R, et al. Infantile spasms: II. Lenticular nuclei and brain stem activation on positron emission tomography. Ann Neurol. 1992;31:212–9.
33. Chugani HT, Mazziotta JC Jr, et al. The Lennox-Gastaut syndrome: metabolic subtypes determined by 2-deoxy-2[18F]fluoro-D-glucose positron emission tomography. Ann Neurol. 1987;21:4–13.
34. Theodore WH, Rose D, Patronas N, et al. Cerebral glucose metabolism in the Lennox-Gastaut syndrome. Ann Neurol. 1987;21:14–21.
35. Kobayashi K, Ohtsuka Y, Ohno S. Clinical spectrum of epileptic spasms associated with cortical malformation. Neuropediatrics. 2001;32:236–44.
36. You SJ, Lee JK, Ko TS. Epilepsy surgery in a patient with Lennox-Gastaut syndrome and cortical dysplasia. Brain Dev. 2007;29:167–70.
37. Lee YJ, Kang HC, Lee JS, et al. Resective pediatric epilepsy surgery in Lennox-Gastaut syndrome. Pediatrics. 2010;125:e58–66.
38. Balfroid T, Warren AEL, Dalic LJ, et al. Frontoparietal 18F-FDG-PET hypo-metabolism in Lennox-Gastaut syndrome: further evidence highlighting the key network. Epilepsy Res. 2023;192:107131.
39. Warren AEL, Harvey AS, Vogrin SJ, et al. The epileptic network of Lennox-Gastaut syndrome: cortically driven and reproducible across age. Neurology. 2019;93:e215–26.
40. Chandra PS, Salamon N, Huang J, et al. FDG-PET/MRI coregistration and diffusion-tensor imaging distinguish epileptogenic tubers and cortex in patients with tuberous sclerosis complex: a preliminary report. Epilepsia. 2006;47:1543–9.
41. Wang Z, Li Y, He Z. Predictive model for epileptogenic tubers from all tubers in patients with tuberous sclerosis complex based on 18F-FDG PET: an 8-year single-centre study. BMC Med. 2023;21:500.
42. Nishida M, Asano E, Juhász C, et al. Cortical glucose metabolism correlates negatively with delta-slowing and spike-frequency in epilepsy associated with tuberous sclerosis. Hum Brain Mapp. 2008;29:1255–64.
43. Murakami N, Morioka T, Suzuki SO, et al. Focal cortical dysplasia type IIa underlying epileptogenesis in patients with epilepsy associated with Sturge-Weber syndrome. Epilepsia. 2012;53:e184–8.
44. Wang DD, Blümcke I, Coras R, et al. Sturge-weber syndrome is associated with cortical dysplasia ILAE type IIIc and excessive hypertrophic pyramidal neurons in brain resections for intractable epilepsy. Brain Pathol. 2015;25:248–55.
45. Pinto AL, Chen L, Friedman R, et al. Sturge-weber syndrome: brain magnetic resonance imaging and neuropathology findings. Pediatr Neurol. 2016;58:25–30.
46. Chugani HT, Mazziotta JC, Phelps ME. Sturge-Weber syndrome: a study of cerebral glucose utilization with positron emission tomography. J Pediatr. 1989;114:244–53.
47. Alkonyi B, Chugani HT, Juhász C. Transient focal cortical increase of interictal glucose metabolism in Sturge-Weber syndrome: implications for epileptogenesis. Epilepsia. 2011;52:1265–72.
48. Juhász C, Haacke EM, Hu J, et al. Multimodality imaging of cortical and white matter abnormalities in Sturge-Weber syndrome. AJNR Am J Neuroradiol. 2007;28:900–6.
49. Juhász C, Chugani DC, Muzik O, et al. Electroclinical correlates of flumazenil and fluorodeoxyglucose PET abnormalities in lesional epilepsy. Neurology. 2000;55:825–34.
50. Hajek M, Wieser HG, Khan N, et al. Preoperative and postoperative glucose consumption in mesiobasal and lateral temporal lobe epilepsy. Neurology. 1994;44:2125–32.
51. Spanaki MV, Kopylev L, DeCarli C, et al. Postoperative changes in cerebral metabolism in temporal lobe epilepsy. Arch Neurol. 2000;57:1447–52.
52. Benedek K, Juhász C, Chugani DC, et al. Longitudinal changes in cortical glucose hypometabolism in children with intractable epilepsy. J Child Neurol. 2006;21:26–31.
53. Govil-Dalela T, Kumar A, Behen ME, et al. Evolution of lobar abnormalities of cerebral glucose metabolism in 41 children with drug-resistant epilepsy. Epilepsia. 2018;59:1307–15.
54. Rao W, Chen B, Xiao L, et al. Expression of fructose-1,6-bisphosphatase 1 is associated with [18F]FDG uptake and prognosis in patients with mesial temporal lobe epilepsy. Eur Radiol. 2023;33:3396–406.

55. Flaus A, Mellerio C, Rodrigo S, et al. 18F-FDG PET/MR in focal epilepsy: a new step for improving the detection of epileptogenic lesions. Epilepsy Res. 2021;178:106819.
56. Pedersen C, Aboian M, Messina SA, et al. PET/MRI applications in pediatric epilepsy. World J Nucl Med. 2023;22(2):78–86.
57. Zhang M, Liu W, Huang P, et al. Utility of hybrid PET/MRI multiparametric imaging in navigating SEEG placement in refractory epilepsy. Seizure. 2020;81:295–303.
58. Tóth M, Barsi P, Tóth Z, et al. The role of hybrid FDG-PET/MRI on decision-making in presurgical evaluation of drug-resistant epilepsy. BMC Neurol. 2021;21:363.
59. Oldan JD, Shin HW, Khandani AH, et al. Subsequent experience in hybrid PET-MRI for evaluation of refractory focal onset epilepsy. Seizure. 2018;61:128–34.
60. Tahry RE, Santos SF, Vrielynck P, et al. Additional clinical value of voxel-based morphometric MRI post-processing for MRI-negative epilepsies: a prospective study. Epileptic Disord. 2020;22:156–64.
61. Sun K, Ren Z, Yang D, et al. Voxel-based morphometric MRI post-processing and PET/MRI co-registration reveal subtle abnormalities in cingulate epilepsy. Epilepsy Res. 2021;171:106568.
62. Czarnetzki C, Spinelli L, Huppertz H-J, et al. Yield of non-invasive imaging in MRI-negative focal epilepsy. J Neurol. 2024;271:995–1003.
63. Perissinotti A, Niñerola-Baizán A, Rubí S, et al. PISCOM: a new procedure for epilepsy combining ictal SPECT and interictal PET. Eur J Nucl Med Mol Imaging. 2018;45:2358–67.
64. Aparicio J, Niñerola-Baizán A, Perissinotti A, et al. Presurgical evaluation of drug-resistant paediatric focal epilepsy with PISCOM compared to SISCOM and FDG-PET. Seizure. 2022;97:43–9.
65. Sukprakun C, Tepmongkol S. Nuclear imaging for localization and surgical outcome prediction in epilepsy: a review of latest discoveries and future perspectives. Front Neurol. 2022;13:1083775.
66. Knight A, Gschwind T, Galer P, et al. Artificial intelligence in epilepsy phenotyping. Epilepsia. 2024;
67. Kerr WT, McFarlane KN. Machine learning and artificial intelligence applications to epilepsy: a review for the practicing epileptologist. Curr Neurol Neurosci Rep. 2023;23:869–79.
68. Zhang Q, Liao Y, Wang X, et al. A deep learning framework for 18F-FDG PET imaging diagnosis in pediatric patients with temporal lobe epilepsy. Eur J Nucl Med Mol Imaging. 2021;48:2476–85.
69. Kerr WT, Nguyen ST, Cho AY, et al. Computer-aided diagnosis and localization of lateralized temporal lobe epilepsy using interictal FDG-PET. Front Neurol. 2013;4:31.
70. Beheshti I, Sone D, Maikusa N, et al. Accurate lateralization and classification of MRI-negative 18F-FDG-PET-positive temporal lobe epilepsy using double inversion recovery and machine-learning. Comput Biol Med. 2021;137:104805.
71. Tan Y-L, Kim H, Lee S, et al. Quantitative surface analysis of combined MRI and PET enhances detection of focal cortical dysplasias. NeuroImage. 2018;166:10–8.
72. Flaus A, Jung J, Ostrowky-Coste K, et al. Deep-learning predicted PET can be subtracted from the true clinical fluorodeoxyglucose PET co-registered to MRI to identify the epileptogenic zone in focal epilepsy. Epilepsia Open. 2023;8:1440–51.
73. Pilli VK, Jeong J, Konka P, et al. Objective PET study of glucose metabolism asymmetries in children with epilepsy: implications for normal brain development. Hum Brain Mapp. 2019;40:53–64.
74. Sivilotti L, Nistri A. GABA receptor mechanisms in the central nervous system. Prog Neurobiol. 1991;36:35–92.
75. Juhász C, Muzik O, Chugani DC, et al. Prolonged vigabatrin treatment modifies developmental changes of GABAA-receptor binding in young children with epilepsy. Epilepsia. 2001;42:1320–6.
76. Verhoeff NPLG, Petroff OAC, Hyder F, et al. Effects of Vigabatrin on the GABAergic system as determined by [123I]Iomazenil SPECT and GABA MRS. Epilepsia. 1999;40:1433–8.

77. Chugani DC, Muzik O, Juhász C, et al. Postnatal maturation of human GABAA receptors measured with positron emission tomography. Ann Neurol. 2001;49:618–26.
78. Savic I, Thorell JO, Roland P. [11C]Flumazenil positron emission tomography visualizes frontal epileptogenic regions. Epilepsia. 1995;36:1225–32.
79. Ryvlin P, Bouvard S, Bars DL. Clinical utility of flumazenil-PET versus [18F]fluorodeoxyglucose-PET and MRI in refractory partial epilepsy. A prospective study in 100 patients. Brain. 1998;121:2067–81.
80. Muzik O, da Silva EA, Juhasz C, et al. Intracranial EEG versus flumazenil and glucose PET in children with extratemporal lobe epilepsy. Neurology. 2000;54:171–9.
81. Richardson MP, Koepp MJ, Duncan JS, et al. Benzodiazepine receptors in focal epilepsy with cortical dysgenesis: an 11C-flumazenil PET study. Ann Neurol. 1996;40:188–98.
82. Hammers A, Koepp MJ, Richardson MP, et al. Grey and white matter flumazenil binding in neocortical epilepsy with normal MRI. A PET study of 44 patients. Brain. 2003;126:1300–18.
83. Njiwa JY, Gray KR, Costes N, et al. Advanced [(18)F]FDG and [(11)C]flumazenil PET analysis for individual outcome prediction after temporal lobe epilepsy surgery for hippocampal sclerosis. NeuroImage Clin. 2014;7:122–31.
84. Lévesque MF, Nakasato N, Vinters HV, et al. Surgical treatment of limbic epilepsy associated with extrahippocampal lesions: the problem of dual pathology. J Neurosurg. 1991;75:364–70.
85. Raymond AA, Fish DR, Stevens JM, et al. Association of hippocampal sclerosis with cortical dysgenesis in patients with epilepsy. Neurology. 1994;44:1841.
86. Lawn N, Londono A, Sawrie S, Morawetz R, Martin R, Gilliam F, Faught E, Kuzniecky R. Occipitoparietal epilepsy, hippocampal atrophy, and congenital developmental abnormalities. Epilepsia. 2000;41:1546–53.
87. Vernet O, Farmer J-P, Montes JL, et al. Dysgenetic mesial temporal sclerosis: an unrecognized entity. Childs Nerv Syst. 2000;16:719–23.
88. Juhász C, Nagy F, Muzik O, et al. [11C]Flumazenil PET in patients with epilepsy with dual pathology. Epilepsia. 1999;40:566–74.
89. Juhász C, Chugani HT, Muzik O, et al. Neuroradiological assessment of brain structure and function and its implication in the pathogenesis of west syndrome. Brain Dev. 2001;23:488–95.
90. Arnold S, Berthele A, Drzezga A, et al. Reduction of benzodiazepine receptor binding is related to the seizure onset zone in extratemporal focal cortical dysplasia. Epilepsia. 2000;41:818–24.
91. Juhász C, Chugani DC, Muzik O, et al. Relationship of flumazenil and glucose PET abnormalities to neocortical epilepsy surgery outcome. Neurology. 2001;56:1650–8.
92. Juhász C, Chugani DC, Muzik O, et al. Electroclinical correlates of flumazenil and fluorodeoxyglucose PET abnormalities in lesional epilepsy. Neurology. 2000;55:825–35.
93. Asano E, Chugani DC, Juhász C, et al. Epileptogenic zones in tuberous sclerosis complex: subdural EEG versus MRI and FDG PET. Epilepsia. 2000;41
94. Juhász C, Asano E, Shah A. Focal decreases of cortical GABAA receptor binding remote from the primary seizure focus: what do they indicate? Epilepsia. 2009;50:240–50.
95. Haskali MB, Roselt PD, O'Brien TJ. Effective preparation of [18F]Flumazenil using copper-mediated late-stage radiofluorination of a stannyl precursor. Molecules. 2022;27
96. Gendron T, Destro G, Straathof NJW. Multi-patient dose synthesis of [18F]Flumazenil via a copper-mediated 18F-fluorination. EJNMMI Radiopharm Chem. 2022;7:5.
97. Hodolic M, Topakian R, Pichler R. (18)F-fluorodeoxyglucose and (18)F-flumazenil positron emission tomography in patients with refractory epilepsy. Radiol Oncol. 2016;50:247–53.
98. Vivash L, Gregoire MC, Lau EW, et al. 18F-flumazenil: a γ-aminobutyric acid A-specific PET radiotracer for the localization of drug-resistant temporal lobe epilepsy. J Nucl Med. 2013;54:1270–7.
99. Avendaño-Estrada A, Velasco F, Velasco AL. Quantitative analysis of [18F]FFMZ and [18F]FDG PET studies in the localization of seizure onset zone in drug-resistant temporal lobe epilepsy. Stereotact Funct Neurosurg. 2019;97:232–40.

100. Asano E, Chugani DC, Muzik O. Multimodality imaging for improved detection of epileptogenic foci in tuberous sclerosis complex. Neurology. 2000;54:1976–84.
101. Fedi M, Reutens DC, Andermann F. Alpha-[11C]-Methyl-L-tryptophan PET identifies the epileptogenic tuber and correlates with inter-ictal spike frequency. Epilepsy Res. 2003;52:203–13.
102. Kagawa K, Chugani DC, Asano E, et al. Epilepsy surgery outcome in children with tuberous sclerosis complex evaluated with alpha-[11C]methyl-L-tryptophan positron emission tomography (PET). J Child Neurol. 2005;20:429–38.
103. Chugani HT, Luat AF, Kumar A, et al. α-[11C]-Methyl-L-tryptophan–PET in 191 patients with tuberous sclerosis complex. Neurology. 2013;81:674–80.
104. Chugani DC, Muzik O. Alpha[C-11]methyl-L-tryptophan PET maps brain serotonin synthesis and kynurenine pathway metabolism. J Cereb Blood Flow Metab. 2000;20:2–9.
105. Chugani DC. α-methyl-L-tryptophan: mechanisms for tracer localization of epileptogenic brain regions. Biomark Med. 2011;5:567–75.
106. Juhász C, Buth A, Chugani DC, et al. Successful surgical treatment of an inflammatory lesion associated with new-onset refractory status epilepticus. Neurosurg Focus. 2013;34:E5.
107. Fedi M, Reutens D, Okazawa H. Localizing value of alpha-methyl-L-tryptophan PET in intractable epilepsy of neocortical origin. Neurology. 2001;57:1629–36.
108. Juhász C, Dc C, Muzik O. Alpha-methyl-L-tryptophan PET detects epileptogenic cortex in children with intractable epilepsy. Neurology. 2003;60:960–8.
109. Wakamoto H, Chugani DC, Juhász C, et al. Alpha-methyl-l-tryptophan positron emission tomography in epilepsy with cortical developmental malformations. Pediatr Neurol. 2008;39:181–8.
110. Chugani HT, Kumar A, Kupsky W, et al. Clinical and histopathologic correlates of 11C-alpha-methyl-L-tryptophan (AMT) PET abnormalities in children with intractable epilepsy. Epilepsia. 2011;52:1692–8.
111. Iyer A, Zurolo E, Spliet WG, et al. Evaluation of the innate and adaptive immunity in type I and type II focal cortical dysplasias. Epilepsia. 2010;51:1763–73.
112. Zimmer TS, Broekaart DWM, Luinenburg M, et al. Balloon cells promote immune system activation in focal cortical dysplasia type 2b. Neuropathol Appl Neurobiol. 2021;47:826–39.
113. Juhász C, Chugani DC, Padhye U. Evaluation with alpha-[11C]Methyl-l-tryptophan positron emission tomography for reoperation after failed epilepsy surgery. Epilepsia. 2004;45:124–30.
114. John F, Muzik O, Mittal S, et al. Fluorine-18-labeled PET radiotracers for imaging tryptophan uptake and metabolism: a systematic review. Mol Imaging Biol. 2020;22:805–19.
115. Jiang H, Guo Y, Cai H, et al. Automated radiosynthesis of 1-(2-[18F]fluoroethyl)-L-tryptophan ([18F]FETrp) for positron emission tomography (PET) imaging of cancer in humans. J Labelled Comp Radiopharm. 2023;66:180–8.
116. Muzik O, Shields AF, Barger GR, et al. The first human application of an F-18-labeled tryptophan analog for PET imaging of cancer. Mol Imaging Biol. 2024;26:29–35119.
117. Dufournel D, Merlet I, Costes N, et al. In vivo PET study of 5-HT1A receptors in malformations of cortical development. Epilepsia. 2002;43
118. Martinez D, Hwang D, Mawlawi O. Differential occupancy of somatodendritic and postsynaptic 5HT(1A) receptors by pindolol: a dose-occupancy study with [11C]WAY 100635 and positron emission tomography in humans. Neuropsychopharmacology. 2001;24:209–29.
119. Lothe A, Didelot A, Hammers A, et al. Comorbidity between temporal lobe epilepsy and depression: a [18F]MPPF PET study. Brain. 2008;131:2765–82.
120. Kim JH, Marton J, Ametamey SM, et al. A review of molecular imaging of glutamate receptors. Molecules. 2020;25
121. McGinnity CJ, Koepp MJ, Hammers A, et al. NMDA receptor binding in focal epilepsies. J Neurol Neurosurg Psychiatry. 2015;86:1150–7.
122. Miyazaki T, Nakajima W, Hatano M. Visualization of AMPA receptors in living human brain with positron emission tomography. Nat Med. 2020;26:281–8.

123. DuBois JM, Rousset OG, Guiot MC, et al. Metabotropic Glutamate Receptor Type 5 (mGluR5). In: Cortical abnormalities in focal cortical dysplasia identified in vivo with [11C]ABP688 Positron-Emission Tomography (PET) imaging. Cereb Cortex. p. 4170–9.
124. Serrano ME, Kim E, Petrinovic MM, et al. Imaging synaptic density: the next holy grail of neuroscience? Front Neurosci. 2022;16
125. Finnema SJ, Nabulsi NB, Eid T, et al. Imaging synaptic density in the living human brain. Sci Transl Med. 2016;8
126. Xiao L, Yang J, Zhu H, et al. [18F]SynVesT-1 and [18F]FDG quantitative PET imaging in the presurgical evaluation of MRI-negative children with focal cortical dysplasia type II. Eur J Nucl Med Mol Imaging. 2024;51:1651.
127. Zurolo E, Iyer AM, Spliet WG, et al. CB1 and CB2 cannabinoid receptor expression during development and in epileptogenic developmental pathologies. Neuroscience. 2010;170:28–41.
128. García-Rincón D, Díaz-Alonso J, Paraíso-Luna J, et al. Contribution of altered endocannabinoid system to overactive mTORC1 signaling in focal cortical dysplasia. Front Pharmacol. 2019;9:1508.
129. Cleeren E, Casteels C, Goffin K, et al. Positron emission tomography imaging of cerebral glucose metabolism and type 1 cannabinoid receptor availability during temporal lobe epileptogenesis in the amygdala kindling model in rhesus monkeys. Epilepsia. 2018;59:959–70.
130. Sasaki M, Kuwabara Y, Yoshida T, et al. Carbon-11-methionine PET in focal cortical dysplasia: a comparison with fluorine-18-FDG PET and technetium-99m-ECD SPECT. J Nucl Med. 1998;39:974–7.
131. Madakasira PV, Simkins R, Narayanan T, et al. Cortical dysplasia localized by [11C]methionine positron emission tomography: case report. AJNR Am J Neuroradiol. 2002;23:844–6.
132. O'Tuama LA, Phillips PC, Smith QR, et al. L-methionine uptake by human cerebral cortex: maturation from infancy to old age. J Nucl Med. 1991;32:16–22.
133. Ito K, Matsuda H, Kubota K. Imaging spectrum and pitfalls of (11)C-methionine positron emission tomography in a series of patients with intracranial lesions. Korean J Radiol. 2016;17:424–34.
134. Phi JH, Paeng JC, Lee HS, et al. Evaluation of focal cortical dysplasia and mixed neuronal and glial tumors in pediatric epilepsy patients using 18F-FDG and 11C-methionine PET. J Nucl Med. 2010;51:728–34.

Neuromolecular Imaging Tau in Amyloidosis with Quantum Polymer Sensors in Parkinson's and in the Epilepsies

Patricia A. Broderick

Abstract Innovative LIVE biomarkers are essentially needed for the diseases of the neurodegenerative dementias, Parkinson's and the epilepsies since these diseases may correlate with the archetypal neurodegenerative disease of dementia, Alzheimer's. We have become acutely aware that, these biomarkers for diseases of dementias seem to be ubiquitous in brain sites without distinction ... tau peptides, beta amyloids, and Lewy bodies are found in all three diseases. (Mayo Clinic-Alzheimer's disease. Symptoms and Causes. Mayo Clinic. 2020. Amyloidosis describes the twisted pathological turns that take place during naturally occurring physiological events in the brain, body and even the skin of patients who present with neurodegeneration. Therefore, the present chapter discusses a next generation "uptick" nanotechnology that demonstrates a unique, real time imaging inventive art enabling a different kind of look at the neuroanatomical sites such as the brain, actually, to see inside neuronal circuitry at the nanoscale. Neuromolecular imaging of the tau peptide LIVE in the brain of the neurodegenerative dementias, Parkinson's and the epilepsies is presented using a quantum sensing nanotechnology to discern the temporal and spatially sensitive voltaic image of the polymeric tau metal complex, which is video-tracked by the BRODERICK PROBE® nanoprobe online and in real time. Tau is imaged LIVE in dorsal striatum of the living Parkinson's animal as well as within the anterior temporal neocortex of the epilepsy patient, intraoperatively, with NYU Institutional Review Board (IRB) approval. The Broderick nanoprobe capacitates redox matching of the metallopolymer of the tau peptide, LIVE, inferably at one of the oxidation states of the transition metal polymeric compounds of cobalt to further polymerize with the phosphorylated Broderick nanoprobe. Studies of the striking consequences among intensities of phosphorylated tau are quantified. An operationally stable voltaic image of the

P. A. Broderick (✉)
Department of Molecular, Cellular and Biomedical Science, The City University of New York Medical School, New York, NY, USA

CUNY Grad Ctr.: Psychology: Cognitive and Behavioral Neuroscience, New York, NY, USA

CUNY Neuroscience Collaborative Graduate Program: Biology, New York, NY, USA
e-mail: broderick@med.cuny.edu

P. A. Broderick (ed.), *Biomedical Imaging in Neurodegeneration*,
https://doi.org/10.1007/978-3-031-95878-6_7

soluble healthy tau Metallopolymer at an electrochemically positive waveform is reported. Imaging tau on line along with tau quantification, made possible in the living, functional animal and human being, is relevant to both individual and societal longevity, and this relevance is inexorably critical to and on behalf of humanity.

Keywords Biomarkers · Tau · Aggregation · Biogenic amines · Ascorbic acid · Alzheimer's · Parkinson's · Epilepsy · Brain · Dorsal striatum · Striatum · Basal ganglia · Anterior temporal neocortex · Cortex · Temporal lobe · Animals · Patients · Sensors · Imaging · Neuroimaging · Surgery · Algorithms · Tauopathy · Diagnosis · Therapy · Detection · Electrochemistry · Polarity · Voltaic · Voltage · Oxidation · Reduction · Redox · Amyloids · Lewy Bodies · Neurochemistry · Seizures · Nanotechnology · Analytical chemistry · Solubility · Phosphorylation · Polymers · Polymerization · Potentiostat · Electroactive · Oxidation potentials

1 Introduction

1.1 Amyloidosis

Amyloidosis is a rare disease characterized by a buildup of abnormal amyloid deposits in the body. Amyloid deposits can build up in the heart, brain, kidneys, spleen and other parts of the body. A person may have amyloidosis in one organ or several. Amyloidosis tends to make subjects look and feel older and not fit into the modern society where outer beauty takes precedence over inner beauty. Amyloidosis is best described as the process of folding and unfolding of *protein* molecules. It's the protein molecules that do most of the work in the human cell. After all, the word protein comes from the Greek prota, meaning "of primary importance." Proteins are created as chains of amino acids, and these chains usually fold spontaneously into what is called their "native form" in milliseconds or a few seconds. The protein's function depends *sensitively on its shape*. For example, enzymes and the molecules they alter are often described as fitting together like a lock and key. By the same token, misfolded proteins, such as tau peptides, beta amyloids and Lewy bodies, are behind some of the most dreaded neurodegenerative diseases, such as Alzheimer's, Parkinson's, the epilepsies [1] and mad cow disease.

1.2 The Process of Folding and Unfolding Proteins (Fig. 1)

1.3 The Broderick Polymer

About this next generation "uptick" nanotechnology, it is further specified as neuromolecular imaging, abbreviated NMI. The BRODERICK PROBE® biomedical library of sensors are the principal part of this LIVE and online novel

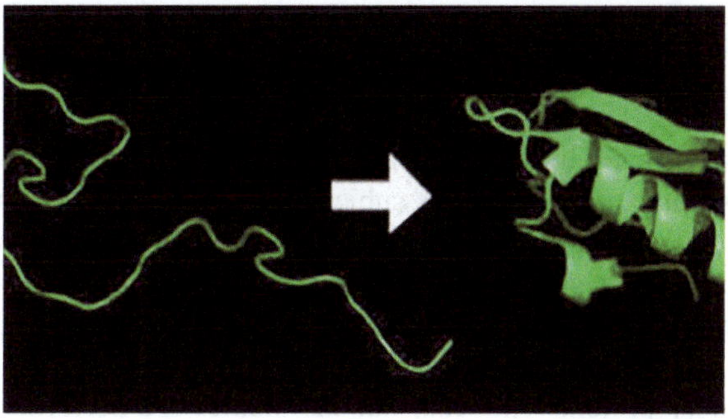

Fig. 1 An unfolded protein is a string made up of amino acids. When the protein folds, it forms structures like the alpha helix and the beta sheet (the flat ribbon) in the image to the right. These secondary structures then double back on themselves to form the final structure. Wikimedia Commons. (Adapted under the terms of the Creative Commons Attribution License. Washington University in St. Louis, MO, USA)

imaging nanotechnology. The original formulae of these sensors comprise carbon/lipid/phosphatidyl polymers, and this dramatic advance has transformed spectrometry into a spectral analysis in the form of LIVE electrochemistry using carbon allotropes in lipid matrices [3, 5, 6] to video-track time-sensitive movement of neuromolecules while at the same time, the sensors image movement nuclei and further while the subject is actually walking and/or in altered states of consciousness and further to another Broderick sensing device, a semiconductor carbon allotrope, fullerene [24] and further to a photoluminescent fiber optic protein probe [8, 10]. Each and every sensor is inventive art formalized under the umbrella of the BRODERICK PROBE® trademark.

1.4 The Present Theragnostics

When we look at the diagnosis of Parkinson's and Alzheimer's disease, the conventional empirical and theoretical art virtually remains at a loss for in-depth theragnostics. Positron emission tomography (PET) images tau and Aβ plaques to assess underlying patterns of tau and Aβ plaques but radioactive tracers are needed to facilitate [45]. When we discern Parkinson's disease, Lewy Body Dementia (LBD) and Alzheimer's disease including other neurodegenerative disorders of asynchrony similar to epileptogenic seizures, we are met with a conundrum as how to diagnose and/or treat because criteria may confuse. In fact, we thought we had this therapy down to a dance step, as the Lewy body seemed to be a comfortable distinction for Parkinson's versus the others. As Lewy bodies, so true of Parkinsonian patients and animals, comprise alpha(α)synuclein, we were able to say in the past that Parkinsonian subjects present with Lewy bodies, Alzheimer's subjects present with Aβ plaques plus tau, and epilepsy subjects exhibit tau.

1.5 Enter—the Prionopathies and the Synucleinopathies

What has happened? These previously untreatable diseases present with horrendous complexities such that Parkinsonian patients now present with Lewy bodies plus Aβ plaques and tau, and Alzheimer's subjects present with Aβ plaques plus Lewy bodies plus tau. To compound the felony, the trio of deadly diseases come from within, infections from within, a type of protein unfolding derived from prions called prionopathies and derived from alpha synuclein called synucleinopathies. To this point, *postmortem* studies have assayed Lewy bodies, alpha synuclein, the amyloid precursor proteins and Tau, but the studies are performed *postmortem*, which means after death, and the limitations of "life after death" may simply add to already misdiagnoses.

1.6 What Is Tau?

Tau is a group of six highly soluble protein isoforms produced by alternative splicing from the gene *MAPT* (microtubule-associated protein tau). Microtubules are tubular polymers, comprising tubulin that self-assemble on a scaffold-like design to make up the infrastructure of a cell; these tubular polymers act as a kind of railway to transport organelles throughout the cytoplasm. These intracellular structures have a stable region and a region that remains dynamic, both important to playing their role in a cell. It is noteworthy that the conventional role of tau resided in the maintenance of microtubule *stability* in axons, an established thesis that is highly questioned in current times [28, 38, 47, 48]. Tau peptides and proteins are abundant in the neurons of the central nervous system (CNS). It is generally known that tau is less common elsewhere but is also expressed at very low levels in CNS astrocytes and oligodendrocytes. As defined herein, *tau is a healthy protein! It is healthy until it becomes unhealthy due to dysregulation.*

1.7 The Broderick Nanoprobe: The Art of Phosphorylation

Thus, tau, when unhealthy, is the biomarker for tau pathologies or tauopathies. Tauopathies are neurodegenerative disorders characterized by the deposition of *abnormal* tau in the brain. Pathologies and dementias of the nervous system, such as, for example, Alzheimer's disease, Parkinson's disease and tau-associated epilepsy are associated with tau that has become hyperphosphorylated insoluble aggregates, neurofibrillary *tangles*. The path from phosphorylation to hyperphosphorylation to aggregation is puzzling; such is the reason why the BRODERICK PROBE® polymer is important. The nanoprobe acts to image through mechanisms of phosphorylation, enabling precise images showing differential degrees of phosphorylation [11, 12]. Therefore, we may study degrees of phosphorylation to the point at which aggregation occurs, the proverbial "hitting the nail on the head." *The BRODERICK PROBE® enables the online view, that is, actually the ability to see the point at which there is little or no return from tau phosphorylation as it proceeds to aggregation.* Importantly, the data that we present herein is the distinction between the healthy and unhealthy tau. Whereas many elegant studies focus on unhealthy tau, we report the biomarker image of a tau metallopolymer, that is, the healthy tau that provides neuroprotection from aggregation in the tauopathies of Parkinson's disease and the epilepsies. Novel sensing nanotechnologies are used that are based on analytical chemistry in the living brain.

1.8 The Rationale

Therefore, the purpose of this chapter is to introduce the online and real-time voltaic image of the tau peptide by the BRODERICK PROBE® biomedical sensors in the dorsal striatum of the living Parkinsonian animal model as well as in the living epilepsy patient intraoperatively. Identifying tau online brings real-time diagnostic, therapeutic and prognostic answers to a world that is daily increasing in cases of dementia not at all material to its derivation. The quest is imperative, as any decision not to pursue this line of "research questions," may be our proverbial "Waterloo." Indeed, this unwise alternative may well be the demise of a cognitive human race, at least as it is recognized presently as a kingdom domain, inclusive of separate and distinct species of plants and animals [34] albeit there are now six kingdoms classified [2]. Humans are in the animal kingdom.

2 Materials and Methods

2.1 Neuromolecular Imaging (NMI)

This is a novel biotechnology that uses conventional microvoltammetry as its basis. Neuromolecular imaging (NMI) employs a semiderivative voltammetric circuit to apply a potential difference to a sensor. The term voltammetry denotes a specific electrical circuit, and there are several different types of circuits by which a potential difference can be applied to a sensor. Voltammetry does not define neurotransmitter selective sensors nor does voltammetry denote the detection or imaging of neurotransmitters per se. NMI, on the other hand, includes detection and imaging of neuromolecules. Thus, a significant difference between conventional microvoltammetry and NMI is the development of new sensors to image neurotransmitters within seconds over long periods of time in vivo, in vitro and in situ.

The BRODERICK PROBE® sensors relate to several formulations of unique, patented and trademarked miniature carbon sensors comprising a series of compounds that include, among others, classes of molecules in the biochemical categories of lipid, glycolipid, lipoprotein and saturated and unsaturated fatty acid. These inventions are able to detect electrochemical signals for a vast number of neurotransmitters, neuromodulators and metabolites, including neuropeptides. NMI primarily but not exclusively detects release mechanisms for neurotransmitters, as shown by γ-butyrolactone studies [4]. NMI has shown, for the first time, a distinct difference, in tissue resected from epilepsy patients, between hippocampal white versus gray matter [13, 22].

2.2 About the Salience of the Nanoprobe

The BRODERICK PROBE® series of biosensors works by electrochemical detection and comprises carbon and/or carbon allotropes in one example and fiber optic proteins in another of the many examples. In the present study, the original carbon sensors, the carbon/lipid/phosphatidyl polymers, were used. This biomedical device is unique from other techniques and methodologies in that this series of biosensors can image signals in vivo, in vitro *or* in-situ within sub-seconds in real time, and the sensors are so tiny that they can be used anywhere and for any length of time, and controlled studies have found these operationally stable biomedical biosensors to *be void of causing* gliosis (scar tissue) or producing infection (bacterial growth) [9, 16, 18, 23].

Further details and examples of Broderick probes may be found [3, 5–13, 15–19, 22–24, 30] including circuit diagrams and methods of making Broderick probes. Broderick probes may be in electrical contact with an auxiliary electrode and/or a reference electrode. It will be apparent to those of ordinary skill in the art, particularly in view of the cited patent documents, that "Broderick probe" is a term that relates to a number of electrodes that vary by composition and the type of circuit in which it is employed and that these variations give rise to differences in detection properties.

2.3 About the Brand at the Nanoscale

This inventive art relates to the recent leaps in areas not only in spectroscopy but also in microscopy, and this inventive art gives scientists a nanotechnology that takes advantage of LIVE phenomena that occur naturally in the brain and spinal cord where matter is organized at the nanoscale. The science of advanced phosphoronanomaterials allows these phenomena to be studied based on quantum mechanics and allows focus on an expanded surface area. Indeed, the majority of biological processes do occur at the nanoscale, bringing us to brand new constructs and designs and the like in medicine, imaging, computing, printing, chemical catalysis, electrocatalysis and materials synthesis. We cannot think of "nanotechnology" as just working at ever smaller dimensions. This would be a mistake.

Rather, imagine working at the nanoscale that engages scientists in a brave new world, talking with each other to utilize unique platforms to cure disease. Imagine uniting diagnostic and therapeutic platforms such as physical, biochemical, chemical, mechanical, electrical, piezoelectric, photoelectric, stimulating and optical properties that naturally occur at nanoscale. The BRODERICK PROBE® works at the nanoscale, enabling tunability of carbon allotropes [3, 5, 6, 19] and opsin laser photodiodes [8, 10] that see inside the brain without opening or at least minimally touching the brain as well as semiconductors such as fullerene [15] to fulfill the hope of a better quality of life beyond degeneration. In this clinical report, the original carbon allotropic polymer expands the realm of the uses of the BRODERICK PROBE® to advance the quest for diagnoses and treatment of tau-related physiologies and pathologies in the healthy and the unhealthy living subject.

2.4 About the Brand: Advantages over the Prior and the Present Art

Advantages over other methodologies are clear. The literature cites that the dialysis (microdialysis) method has disadvantages relative to sampling time resolution. For one example, dialysis offers somewhat limited time resolution. The use of highly sensitive analytical techniques (e.g., neuromolecular imaging with the Broderick probe) images in temporal resolutions of seconds and less, spatial resolutions of nanospheres and current resolution of nanoamperes. Dialysis samples may not provide real time information regarding changes in the neurochemical environment. Also, dialysis is an invasive technique and therefore causes some damage to the tissue under study, and critically, the dialysis process creates an area around the membrane in which all solutes capable of crossing the membrane are depleted [25].

Post-mortem studies are daily used to distinguish between Alzheimer's disease and the age-old expression of "dementia" as these diseases relate to the pathological presence of Aβ plaques and neurofibrillary tangles versus Lewy bodies, respectively. Tau and Aβ plaques are seen as a double prion disease *postmortem* by Dr. Prusiner, who discovered the prion [42, 43]. Alzheimer's patients are often seen *postmortem* to exhibit alpha synucleated Lewy bodies in addition to Aβ plaques, thus resulting in further confusion, compounded with the innate hindrance of studying any disease *postmortem*. Physiological and pathological processes may not be translatable to life *per definitionem*. Tau is a physiological and pathological process.

2.5 The BRODERICK PROBE® Provides a Remedy

The BRODERICK PROBE® provides a remedy since tau is imaged LIVE in the healthy, positively charged, cationic redox-matching state, the soluble physiological state of the tau metallopolymer. NMI of tau's unhealthy anionic redox matching dynamic movement may be the unhealthy pathophysiological state since this state is the insoluble aggregate of neurofibrillary tau tangles. The process is perhaps synaptic, likely away from the microtubule, and comprises, in at least one embodiment as, but not limited to, one of a complex comprised of metal (Broderick patent claims [11]) as well as the process described in [12]. The metal complex forms of tau are rare in the contemporary field of tau theragnostics, and the field is appraised [29].

In this chapter, neuroprotection is seen since the soluble metallopolymer of tau images as a positively charged ionic compound. This finding is electrochemically based. Solubility, electric charge, phosphorylation and the size of the ionic compound factor into neurodegenerative processes. Indeed, neuroprotection from neurodegeneration may also occur via phosphorylation of the tau peptide or repeats of the tau domain, repeats that are positively charged. One example of this phenomenon is phosphorylation that detaches tau protein from microtubules (*Ser262, Ser214*) disenabling aggregation into paired helical neurofibrillary tangles [46]. The intensity of the phosphorylation is dependent on electric charge and therefore, the intensity of the phosphorylation is related to solubility and insolubility further distinguishing healthy tau from unhealthy tau. Phosphorous signaling is inherently and uniquely performed by the nanoprobe ([16], Chapter 6). Coulomb's Law correlates electric charge and solubility in an inverse proportionality.

2.6 The BRODERICK PROBE® Provides a Remedy to the Tauopathies of Parkinson's and Epilepsy

Indeed, imaging the neurotransmitter facets of the *tauopathies of Parkinson's and epilepsy* is performed with nanotechnology and nanobiotechnology, which are studied extensively in the Broderick laboratory. The laboratory is well known for delineating the neuroprotective neurotransmitter elements of the disease as well as its degenerative elements in the living animal model and in the epilepsy patient intraoperatively during brain surgery in humans. Nanosensing and subtyping epilepsy patients are part and parcel of our major work in the field of human epilepsy as well ([16], Chapter 12). Often beset with depression and consequent to treatment for psychosis-induced epilepsy, depression and epilepsy are no strangers to the brain as they often exist together in the epilepsy person. Indeed, the first studies of brain stimulation paradigms originated in psychotic epilepsy patients. Reports that psychotic/epilepsy patients were studied stereotaxically in the temporal lobe during surgery enlighten the nuances within comorbidity of psychotic epilepsy. The studies are described in a penetrating treatise on the dovetail events between sexuality and brain reward in epilepsy patients who present with psychosis ([16], Chapter 11).

2.7 The Precision of the Nanoprobe

The precision of this nanoprobe is enhanced further beyond the previously used techniques because the molecular transmission imaging is derived from the actual basal ganglia neurons responsible for motor function. The BRODERICK PROBE® sensed striatal neurons in order to study the functional activity of efferent dorsal striatal neuronal circuitry from the substantia nigra pars compacta (SNPC). Parkinson's model subjects were lesioned at the SNPC using 6-hydroxydopamine in order to inhibit neurons in the dorsal striatal area originating from the SNPC from producing dopamine; surgical technicians at the North Carolina facility, the International Charles River Laboratories, performed the expert lesion technology. Details of our experimental designs for these tauopathies are provided prolifically [17, 18, 23].

2.8 The Path to Seeing the Tau LIVE in the Parkinson's Subject

The neuroprobe/nanosensor signals from the brain are remembered by the Broderick nanoprobe polymer, the carbon polymer, the carbon/lipid/phosphatidyl polymer. The following drawings are provided by way of example only and without limitation, wherein like reference numerals (when used) indicate corresponding elements throughout the several views. The reference numerals are described in detail within the patents referenced [3, 5, 6, 8, 10–12, 15, 19, 22, 24].

2.8.1 The Operation of the BRODERICK PROBE® Nanoprobe Polymer (Fig. 2)

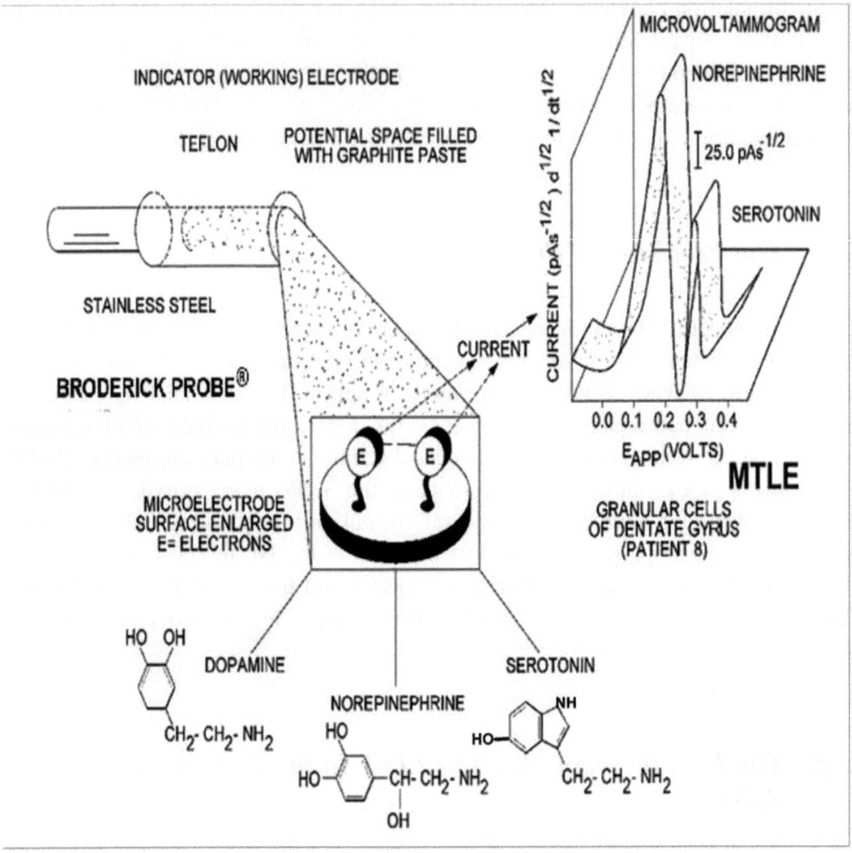

Fig. 2 Shows the operation of a Broderick nanoprobe polymer in imaging neurotransmitters and a scan resulting therefrom. Note that the operation of the original carbon allotropic polymer operates at the quantum nanoscale in both neurodegenerative diseases of PD and epilepsy. In fact, we show a scan recorded LIVE from our mesial temporal lobe patient herein at the upper right. The author repeats the dictum that the majority of biological processes occur at the nanoscale, bringing us to brand new constructs and designs and the like in medicine, imaging, computing, printing, chemical catalysis, electrocatalysis and materials synthesis. We cannot think of "nanotechnology" as just working at ever smaller dimensions. We shall focus on the expanded surface in the next diagram

2.8.2 QUANTUM Sensing with the BRODERICK Nanoprobe (Fig. 3)

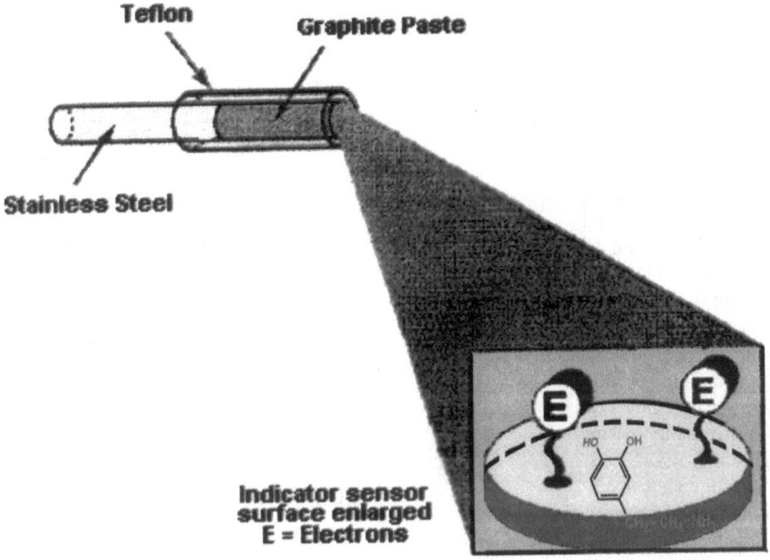

Fig. 3 The enlarged or, one may say, the enlarged surface on the nanoprobe enables molecular recognition sites for neuromolecules and molecules to be imaged at specific oxidation or reduction potentials Specificity is met within a voltage range. A voltaic (voltage) biomarker is empirically determined for each neuromolecule or molecule by quantum sensing with the Broderick Nanoprobe

2.8.3 Norepinephrine Is Separated from Dopamine and Serotonin

In electrochemistry, "E/V" stands for "electrode potential in volts," representing the electrical potential difference between a working electrode and a reference electrode, essentially measuring the tendency of a chemical reaction to occur at that electrode, with the unit being volts (V). Key points about E/V: "E" represents the electrode potential, and "V" denotes the unit of measurement, volts (Fig. 4).

Therefore, the catecholamines, DA and NE, are clearly differentiated from each other and, further, from electroactive signaling for the indoleamine, 5-HT [6]. It is concluded that there is no interference from chemicals closely allied molecules with half-wave/oxidation potentials for biogenic amines, as demonstrated by thousands of in vitro, in situ and in vivo studies accomplished in the Broderick Laboratory. The electrical circuit, the driver, uses semidifferential/semiderivative treatment of microvoltammetric NMI data in conjunction with a graphite stearate nanobiosensor in this example. The data are relevant to the interpretation of electrochemical signals in neuroanatomical substrates, richly innervated by dopaminergic, noradrenergic and serotonergic neuronal circuitry comprising multifaceted neuroactive intermolecular anion/cation exchange processes. Current response is concentration dependent according to the Cottrell Equation. The data are raw data, as seen in real time with the inventive art of polymeric carbon/lipid/phosphatidyl matrices as one of the novel series of *BRODERICK* PROBE® sensors. This nanobiotechnology is

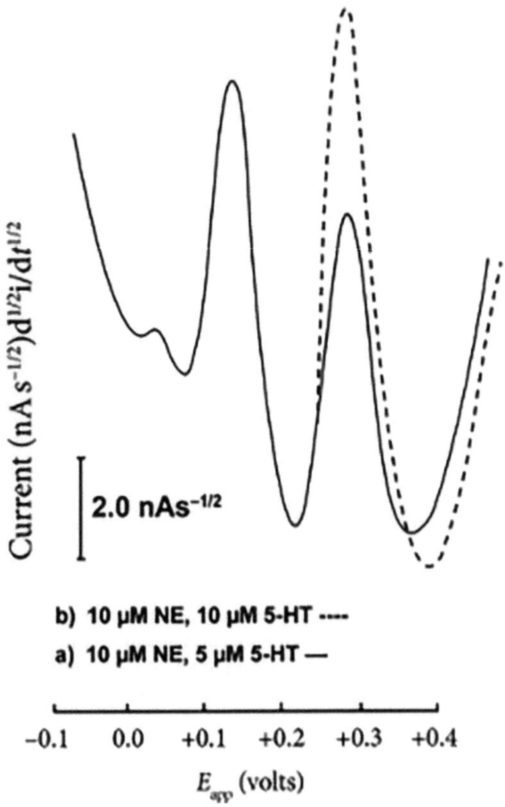

Fig. 4 Voltaic (voltage) biomarkers for the indoleamine neurotransmitter serotonin (5-HT) as well as the voltage biomarker for the catecholamine, norepinephrine is shown herein. Operational stability and selectivity are part and parcel of the nanotechnology that is empowered by the operation of our nanoprobe. The data shows that the neurotransmitters, dopamine (DA) and norepinephrine (NE) are separated by a range of 50 millivolts. Increasing concentrations of NE and 5-HT in saline phosphate buffer in the electrochemical cell (manufactured on site and calibrated in accordance with commercial apparatus) show correlated sharp increases in the designated signals for NE and 5-HT according to experimentally derived oxidation potentials specific for each neurotransmitter/neurochemical

known for its operational stability in vivo, in vitro and in situ *and is also* known for its excellent temporal and spatial resolution [7–12, 19, 24].

2.8.4 Dopamine (Catecholamine) Is Separated from Ascorbic Acid and (Indoleamine), Serotonin (Fig. 5)

Part **a** shows a representative in vitro recording, derived from the BRODERICK PROBE® N-stearoylcerebroside sensor. The sensor was pretreated with the PEA/BSA mixture. Three well-defined cathodic peaks are shown. In order, from the lowest oxidation potential recorded to the highest, the NMI electrochemical signal for ascorbate was first; the DA signal was second and the 5-HT signal was third. Recording specifications for this image were: sensitivity [2 nA/V]; time constant [1 sec]; and scan rate [10 mv/sec]. Charging current (-0.2 ± 0.0 V) is not shown.

The recording shows results from the pretreated BRODERICK PROBE® N-stearoylcerebroside. The recording was derived from the oxidation of ascorbate, DA and 5-HT at the following concentrations: ascorbate = 11.2 uM, DA = 150 nM and 5-HT = 150 nM. Therefore, ascorbate concentration was approximately 75 times

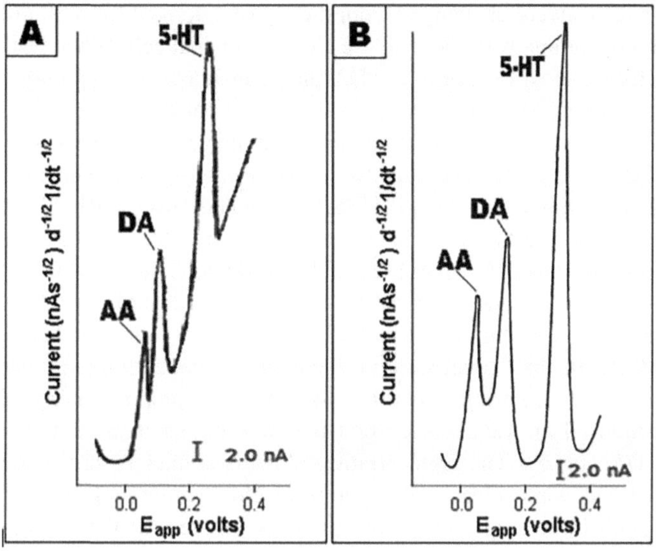

Fig. 5 In vitro electrochemical images (NMI) of ascorbate, DA and 5-HT in saline/phosphate buffer derived from the *BRODERICK PROBE® N-stearoylcerebroside* sensor pretreated with PEA/BSA (**a**) and from the *BRODERICK PROBE® laurate sensor* (**b**). The signature for each neurochemical, defined as the oxidation potential in mv, is shown on the *x* axis. Current is shown on the *y* axis; concentration of each neurochemical is proportional to current according to the Cottrell equation. Several and repetitive studies were performed. Three distinct and separate cathodic oxidation peaks were produced by the pretreated BRODERICK PROBE® N-stearoylcerebroside and the laurate sensors. The N-stearoylcerebroside and the laurate sensors exhibited operational stability, sensitivity and selectivity. (Adapted from Broderick, PA, *J. Neural Transmission*, 2008, under the attribute of the Creative Commons License)

higher than that of either monoamine. Currents produced by each neurochemical at the concentrations noted were ascorbate = 6 nA, DA = 10 nA and 5-HT = 18 nA. Thus, all three neurochemicals were detected at the same electrochemical parameters of sensitivity = 2.0 nA/V, 1 sec time constant and 10 mv/sec. Thus, greater concentrations of ascorbate were used for all three neurochemicals to be seen within the detection ranges of monoamines noted. The data also show that detection of 5-HT produced higher current values than did DA at equivalent concentrations.

In this same study using the PEA/BSA pretreated N-stearoylcerebroside sensor, ascorbate was first detected at 2.2 uM when each monoamine concentration was 50 nM. Recording specifications for these images were sensitivity [2 nA/V]; time Constant [1 sec]; and scan rate [10 mv/sec] (recordings not shown).

In an additional study, using the PEA/BSA pretreated N-stearoylcerebroside sensor, ascorbate was detected at 0.07 uM when the concentration of each monoamine was 20 nM. Moreover, continuous and ascorbate images occurred with increasing amplitude, as increasing concentration of ascorbate were added into the electrochemical cell, up to the final concentrations of ascorbate of 0.07 uM. Dopamine and 5-HT images were unchanged from the signals, derived from 20 nM, when no additional monoamine was added to the electrochemical cell. When DA was added to the electrochemical cell, only the DA image increased in amplitude. Ascorbate images were observed at the same electrochemical parameters as was used for the monoamines. Recording specifications for these images were: sensitivity [2 nA/V]; time constant [1 sec]; and scan rate [10 mv/sec] (recordings not shown).

In further studies using the PEA/BSA pretreated BRODERICK PROBE® N-stearoylcerebroside sensor, ascorbate was pipetted into the electrochemical cell at concentrations of 0.07 uM, 1.5 uM, 2.2 uM, 3.7 uM and a final 3.7 uM. Monoamine concentrations were sequentially 10 nM, 10 nM and 20 nM. Ascorbate was first observed at a concentration of 1.5 uM when the concentration of each monoamine was 40 nM. At the final concentration, ascorbate concentration (11.2 uM) was 275 times greater in concentration than DA (40 nM); this proportionality of ascorbate and DA produced an ascorbate current that was higher than, and still separated from, the DA current. The concentration of monoamines recorded remained the same at 40 nM while clear and separate electrochemical signals for ascorbate increased with increasing concentrations of ascorbate. All three neurochemicals were detected at the same electrochemical parameters. Recording specifications for these images were sensitivity [2 nA/V]; time constant [1 sec]; and scan rate [10 mv/sec] (recordings not shown).

Figure 5, In b, a representative in vitro recording, derived from the BRODERICK PROBE® laurate sensor pretreated with the PEA/BSA mixture is reported herein. Three well-defined cathodic peaks were delineated. In order of lowest to highest oxidation potential, the NMI electrochemical signal for ascorbate was the first to be recorded, the DA signal was second and the 5-HT signal was third. Recording specifications, specifically for this imaging, were: sensitivity [2 nA/V]; time constant [1 sec]; and scan rate [10 mv/sec]. Charging current (-0.2 ± 0.0 V) is not shown.

In Part b, note that the recording shown above is derived from the pretreated BRODERICK PROBE® laurate sensor, and the recording was further derived from the sequential oxidation of ascorbate at concentrations of 0.07 uM, 1.5 uM, 2.2 uM, 3.7 uM and a final 3.7 uM and also from the concomitant and sequential oxidation of DA and 5-HT each at concentrations of 10 nM, 10 nM and 20 nM. Therefore, the recording shows images for ascorbate, DA and 5-HT at the following concentrations: ascorbate = 11.2 uM, DA = 40 nM and 5-HT = 40 nM. It is important to point out though that the distinct image for ascorbate began to appear at a concentration of 0.07 uM when the concentration of the monoamines was 40 nM. Corresponding currents produced by each neurochemical at the final concentrations were, ascorbate = 9 nA, DA = 12 nA and 5-HT = 24 nA. All three neurochemicals were detected at the same electrochemical parameters of 2.0 nA/V, 1 sec time constant and 10 mv/sec. Higher concentrations of ascorbate were needed for all three neurochemicals to be seen simultaneously within the detection ranges of monoamines noted.

Moreover, detection of ascorbate was tested in a series of concentrations from 0.07 uM to 11.2 uM when each monoamine concentration was 40 nM when using the pretreated laurate sensor. In this study, ascorbate first appeared at 1.5 uM when each monoamine concentration was 40 nM; clear and separate electrochemical signals for ascorbate, DA and 5-HT were observed at the same electrochemical parameters as were used for the monoamines. Recording specifications for these images were sensitivity [2 nA/V]; time constant [1 sec]; and scan rate [10 mv/sec] (recordings not shown).

In addition, since the BRODERICK PROBE® laurate sensor was selected for in vivo studies of oxidative stress, studies were also performed on the laurate sensor, in vitro, when this formulation of sensor was not pretreated with the PEA/BSA mixture. The results showed that three well-defined cathodic peaks were delineated. In order of lowest oxidation potential to the highest, the NMI electrochemical signal for ascorbate was the first to be recorded, the DA signal was recorded second and the 5-HT signal was recorded third. Recording specifications, specifically for this imaging, were: sensitivity [0.2 nA/V]; time constant [1 sec]; scan rate [10 mv/sec] (data not shown).

The results from this study, in which the BRODERICK PROBE® laurate sensor was not pretreated with PEA/BSA, showed that when ascorbate was added to the electrochemical cell to a concentration of 5 uM and each monoamine was aliquoted into the electrochemical at 100 nM, the approximate currents produced were: ascorbate = 7.0 nA, DA = 10.0 nA and 5-HT = 40.0 nA. Comparable current ranges for recording specifications at electrochemical sensitivities of 2.0 nA/V, such as detection ranges for sensors pretreated with PEA/BSA, are: ascorbate = 0.7 nA, DA = 1.0 nA and 5-HT = 4.0 nA.

2.8.5 The Protein Optical Nanoprobe (Fig. 6)

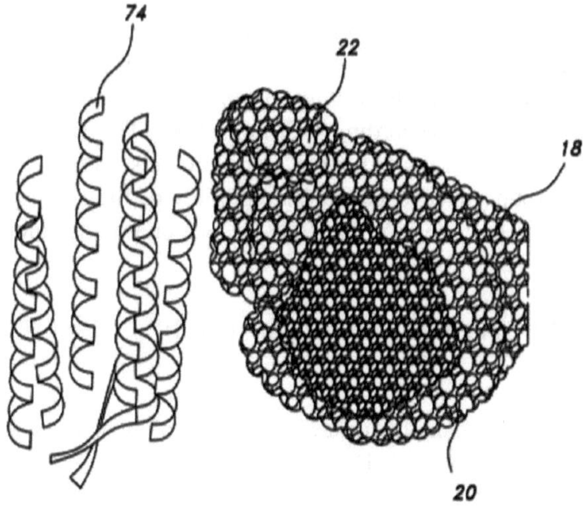

Fig. 6 The Photoluminescent Fiber Optic Protein Probe. A schematic of the opsin, rhodopsin, is shown to the left (74). The Broderick Semiconductor Fullerene Probe is shown to the right in the center of the schematic as encased in the Broderick probe's original carbon/lipid/phosphatidyl polymer. The biopolymer is *numbered 18,20,22*. Each of these sensing devices are trademarked, the BRODERICK PROBE®

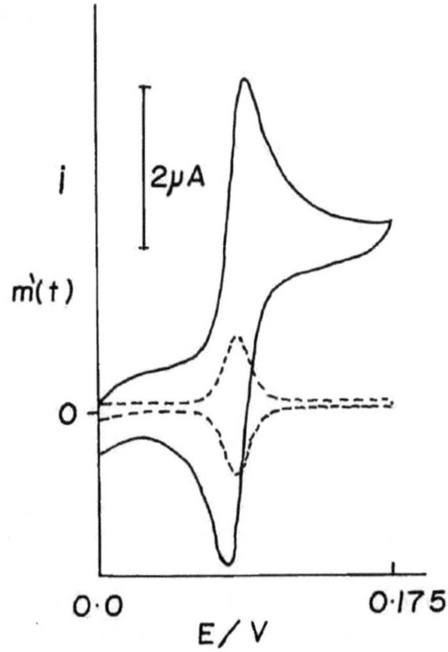

Fig. 7 Experimental *cyclic voltammogram* (full line) for a solution of ImM K4Fe(N)6 in 0.1 M KC1. Sweep rate = 5 mVs'1 at a C electrode vs SCE. The units for the *dashed semiderivative plot* are in nAs ~ 1/2. [O'Gorman, John, *Novel Electrochemical Methods*, A thesis presented to Dublin City University in submission for the MSc degree. October 10th, 1998 [40]. (Adapted under the terms of the Creative Commons Attribution License])

2.8.6 Cyclic Voltammetry Compared with Semiderivative Voltammetry (Fig. 7)

Notes on the Figure: A word about the meaning of semiderivative voltammetry. This circuit is a unique circuit of resistors and capacitors that is based on fractional calculus of the first half derivative of the linear sweep. The units are written as $nAs^{-1/2}$. This is the circuit that the author uses by Oldham [41]. However, the author has further modified the circuit. This circuit is described [3, 5, 6].

The slope of the voltage/current function in the semiderivative circuit is related to resistance according to Ohm's Law. The Broderick nanotechnology uses, for example, a semiderivative circuit in this case of imaging the polymerized and phosphorylated cobalt metal complex of the tau peptide, hereafter called the tau metallopolymer, for the study of dementia in the tauopathies.

2.8.7 The Cottrell Equation (Fig. 8)

Notes

Fractional calculus describes the function of D as the first half derivative of the linear sweep that is D_0.

$$i_t = \frac{nFAC_0 D_0^{1/2}}{3.14^{1/2} t^{1/2}}$$

where i = current at time t

n = number of electrons, eq/mol

F = Faraday's constant, 96,486 C/eq

A = electrode area, cm^2

C = concentration of O, mol/cm^3

D = diffusion coefficient of O, cm^2/s

Fig. 8 Where $D_o^{-1/2}$is the slope of the waveform. This is important wherein one chooses the type of circuit tor sensing. The Broderick technology uses a semiderivative circuit that contains a series of customized ladder network of resistors and capacitors that decreases resistance on the slope resulting in sharper curves. Then the decision on the Voltaic Biomarker is transparent. Broderick agrees with O'Gorman that the semidifferential/semiderivative circuit is more accurate than that of cyclic voltammetry. Semidifferential differs from the semiderivative circuit in a feedback loop. These circuits or the names of these circuits are generally used interchangeably

The slope is according to the first half derivative of the function of D at D0.

Current vs. voltage is interpreted by the slope of the line in the formula y = mx + b. Slope is revised by a system resistors and capacitors as in between D_0 and D_1.

The slope of the voltage vs. current graph is equivalent to the inverse of the resistance. The slope of the current vs. potential difference graph is equal to the resistance. Current, voltage and resistance are related through Ohm's Law. Voltage is equal to the resistance of a circuit multiplied by its current.

2.9 *Neuromolecular Imaging Live*

2.9.1 The Broderick Nanoprobe Is Placed LIVE in the Parkinson's Animal Model: Sensing Tauopathy in the Parkinson's Subject (Fig. 9)

- *Experimental Design for the Study of Tauopathy in the Parkinson's Subject*

Neurochemical profiles in PD and non-PD were imaged using the same experimental paradigm and design This advanced biotechnology enabled the imaging of dopamine (DA), serotonin (5-HT), homovanillic acid (HVA) a metabolite of DA, L-tryptophan (L-TP) a precursor to 5-HT and peptides, dynorphin A 1–17 (Dyn A) and somatostatin (somatostatin releasing inhibitory factor) (SRIF). Each neurotransmitter or neurochemical was imaged at a signature electroactive oxidation/half-wave voltaic potential in dorsal striatum of the PD as compared with the non-PD animal. Both endogenous and bromocriptine-treated neurochemical

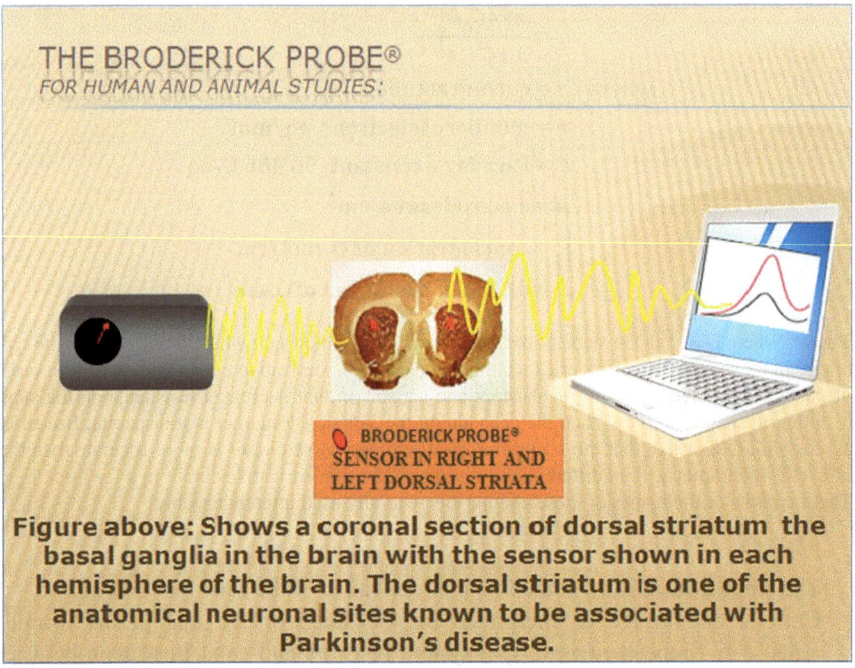

Fig. 9 This figure shows how the Broderick nanoprobe works in concert with neuromolecular imaging (NMI). Simply, in one example, apply a voltage to the registered trademark BRODERICK PROBE® nanobiosensor; voltage is converted into current via the nanoprobe electrically connected to a potentiostat through a commutator. A software application displays the NMI signals on a computer screen

profiles in PD and non-PD were imaged using the same experimental paradigm and detection choices of electrical circuit parameters. The signal labeled "peptide" is inferably the Voltaic Image of the tau Metallopolymer.

2.9.2 The Broderick Nanoprobe Is Placed LIVE in the Epilepsy Patient: Sensing Tauopathy in the Epilepsy Patient (Fig. 10)

- *Experimental Design for the Study of Tauopathy in the Epilepsy Patient*

Using semiderivative and linear voltammetric circuits, NMI and BRODERICK PROBE® biosensors, the selective detection of specific neurotransmitters and neurochemicals in discrete parts of the intact brain are imaged within a temporal resolution of seconds in a 42-year-old female patient with temporal lobe epilepsy and generalized tonic-clonic seizures. This patient underwent partial cortical resection with intraoperative in vivo studies using NMI and BRODERICK PROBE®

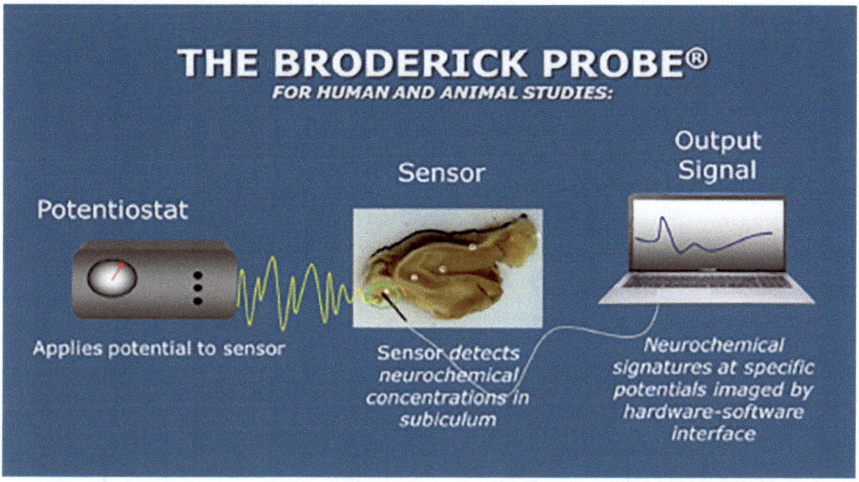

Fig. 10 Shows the *modus operandi* for neuromolecular imaging (NMI) and the BRODERICK PROBE®. A potential is applied to the sensor by a potentiostat, and a current is formed. The digitized current is read instantaneously on a computer. Tau in the epilepsy patient is first reported in Broderick PA (2021). Tau peptide signals are seen in Parkinson's subjects by Broderick Probe® sensors. (*MOJ Clin Med Case Rep 11(2)* 46–50 [12] with permission from Medcrave; and copyrighted by the author, Professor Broderick)

biosensors. These studies began with electroencephalographic (EEG) monitoring wherein the site of cortical resection was defined with subdural grid epilepsy electrodes, placed on the surface of the patient's brain with epilepsy strip electrodes placed deeper in the temporal lobe (i.e., the hippocampus).

Then, Broderick polymeric biosensors, BRODERICK PROBE® with a diameter 5 times less than epilepsy depth electrodes for invasive EEG were placed by direct visualization in the exposed cortical region with and without epileptic spike activity in regions destined for resection. About 6 to 10 recordings with γ-irradiated (11.6–12.7 kilograys), laminar biocompatible carbon-based BRODERICK PROBE® laurate biosensors were taken at a cortical depth of microns to less than 2 mm for a 20–30-minute time period. Glial Fibrillary Acidic Protein (GFAP) stains of tissue were postsurgically conducted in order to assess gliosis or other neuronal damage after electrode placement. Based on the pathology report from the NYU surgical team, "the BRODERICK PROBE® biosensors did not cause gliosis, sclerosis, brain/neuronal damage, or histiocyte infiltration and may well provide evidence that the nanoprobe is a safe biomedical tool [14, 20–22, 32].

3 Results

3.1 LIVE Parkinson's Tauopathy in Real Time

3.1.1 Temporally and Spatially Sensitive Voltaic Image of the Tau Metallopolymer

Voltaic Image of the Phosporylated Tau Metallopolymer in PD (Left) and not in not Non-PD (Right) (Fig. 11)

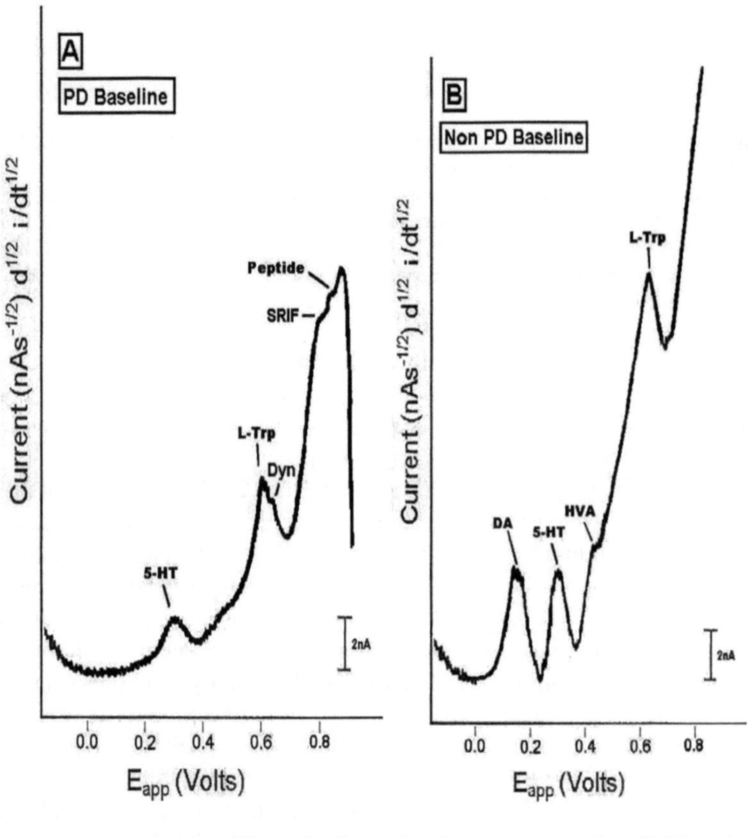

Fig. 11 Reports the strikingly different (**a**) Tauopathy electroactive signals in Parkinson's vs. (**b**) Natural brain electroactive signals without Parkinson's tau. The *tau peptide is marked "peptide."* The tau peptide is the metal complex of polymerized and phosphorylated tau. The Broderick polymer images the *voltaic tau peptide, the* tau metallopolymer marker of the phosphorylated tau, inferably one metal complex of cobalt, + 0.83 V is demonstrated. The voltage to current equation spans the −0.3 to + 1 V range using SHE, the standard hydrogen electrode vs Ag/Cl. (This figure and Figs. 12, 13, 14, and 15 are adapted with permission from Broderick and Kolodny [17], Under the terms of the Creative Commons Attribution License)

Notes The signal marked "peptide" is the tau-associated peptide/protein, one preferably of a cobalt metal complex. This is the first report of tau signaling LIVE in electroactive images in a LIVE subject presenting with and exhibiting the cardinal symptomatology of the brain disease called Parkinson's. In one embodiment, the electrode or sensor, nanoprobe, detects the activities of the Metallopolymer tau proteins/peptides (i.e., "tau") [11, 12].

The dopamine (DA) neurotransmitter is absent in lesioned group (PD). This is an expected outcome, of course, since dopaminergic terminals in dorsal striatum were lesioned by the Charles River expert technicians upon request, a priori. In fact, this laboratory has reported that the neurotransmitter serotonin is absent or depleted at times, especially when there is corresponding motor incoordination, as is the case with neurodegenerative disorders, specifically, the disease of Parkinson's in this example. The direct cause-and-effect discovery between the release of a serotonin molecule with a specific classified functional movement as it happens in real time is critical to the meaning of the data presented since *Temporal Synchrony*, shown between serotonin and movement, may well define the healthy brain [7]. The importance of such real time correlations between brain and behavior has reached the textbook [18].

Voltaic Image of the Phosphorylated Tau Metallopolymer After Administration of the Direct Dopamine Receptor, Bromocriptine at the Low Dose (Fig. 12)

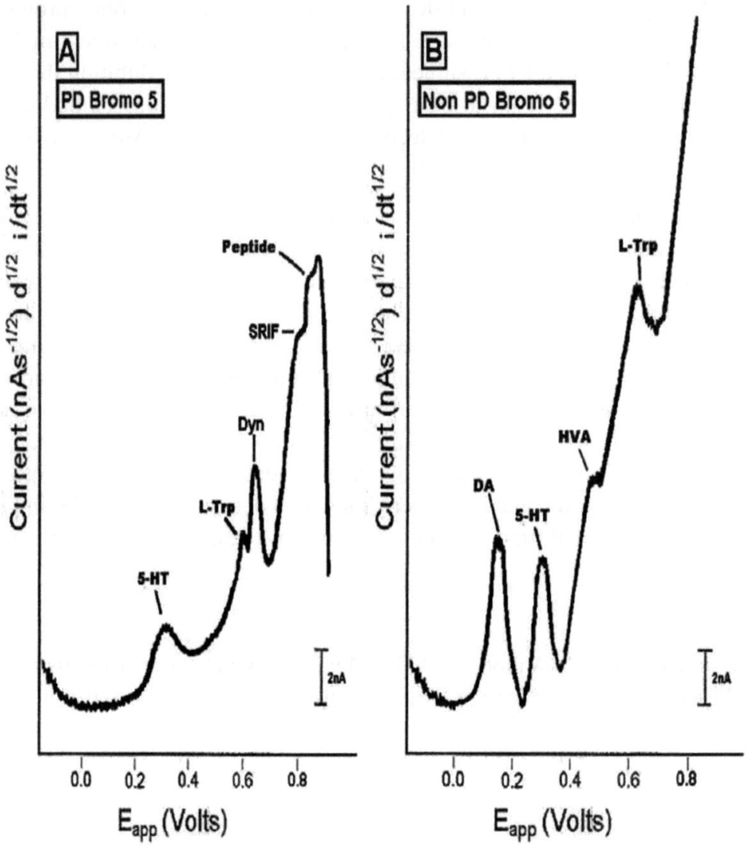

Fig. 12 A dose of 5 mg/kg ip was administered to the animal in the first study. In the second study, a dose of 5 mg/kg followed by a dose of 10 mg/kg, ip was administered to the animal. For animal studies, these doses are considered the low and high dose. The initial clinical dose for bromocriptine can start at 1.25 mg and then continue daily at levels of 2.5–7.5 mg daily. Nonetheless, there are reports of daily doses of bromocriptine for PD patients between 7.0 and 30 mg daily. Clinically, the low dose for PD patients has been reported to be about 30 mg daily [35], whereas the high dose has been reported to be about 52 mg daily [39]. The scan reports the middle of the low dose of bromocriptine. *The peptide is a metal complex of the polymeric complex of tau* and the peptide is imaged at the + 0.83 peak waveform thus enabling an innovative voltaic biomarker for Parkinson's disorders, reflecting tau polymerization with the Broderick nanoprobe polymer. The polymeric tau metal complex increased steadily above baseline in the PD group. A priori, the author will point to the presence of peptides that appear at the higher end of the oxidation potentials. The non-PD animal does not reflect the peptide group. The imaging of peptides in PD provides a significant and dramatic biomarker for PD. It is important to note that the scan from −0.2 mV to 1 mV provides a clear picture in one glance of what is happening in the Parkinson's brain as compared with the functioning natural brain of the subject that, indeed, does not present with PD. The natural state of the non-PD animal exhibits the temporally synchronous physiology of the biogenic amine schema

Voltaic Image of the Phosphorylated Tau Metallopolymer After Administration of the Direct Dopamine Receptor, Bromocriptine at the High Dose (Fig. 13)

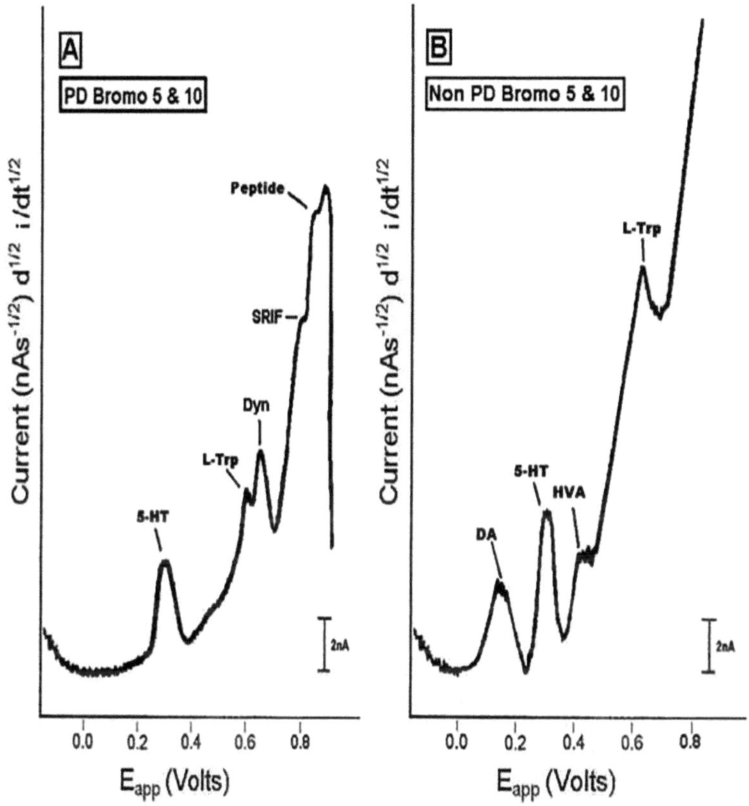

Fig. 13 Shows results from the administration of bromocriptine (5 mg/kg ip followed by 10 mg/kg, ip), the high dose, in dorsal striatal brain of PD versus non-PD. The results from the administration of bromocriptine, at the high dose, showed that: in PD, 5-HT increased about 75% above baseline, L-TP decreased about 10% from baseline, Dyn A increased about 50% above baseline and the peptide, *voltaic image of the polymeric tau metal complex,* at + 0.83 V increased significantly (about 50%) above baseline and then decreased slightly after this point only to be enhanced throughout. The point at which the tau metallopolymer dips, occurs at the hour point after administration of bromocriptine. The metallopolymer of tau seemed not to show the biphasic response that the pharmaceutical, bromocriptine, is known to express

Voltaic Image of the Phosphorylated Tau Metallopolymer after Administration of the Direct Dopamine Receptor, Bromocriptine, at the Low Dose: Line Graphs and Percentages Enable Quantification of the Tau Metallopolymer in the Tauopathies (Fig. 14)

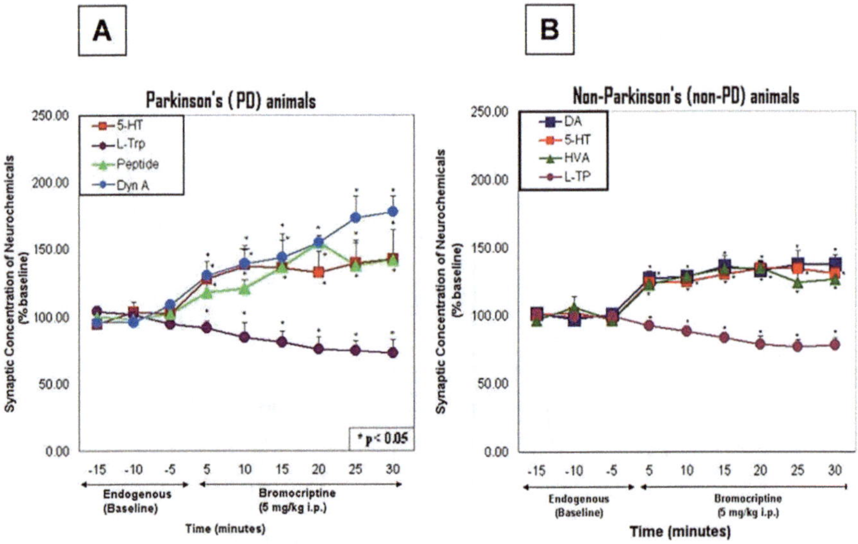

Fig. 14 Reports line graphs showing the time course of bromocriptine effects in PD versus non-PD animals at the low dose. A picture is worth a thousand words. The data show that tauopathies generate peptides while the natural schema for the neurotransmitter milieu of a subject without the disease of Parkinson's, may have little or no need for peptide production at least when Parkinson's disease is not present. In this graph, L-Tryptophan is downregulated in both PD and non-PD and greater effects on this precursor to serotonin are shown in PD rather than that which is shown in the non-PD group. The graphs begin with the endogenous baseline of natural neurotransmitter release, which shows an operational and temporal dynamic, synchronous rhythm. The time course of neurotransmitter release is imaged from the dorsal striatal brain of the living subject, online and LIVE after the administration of the pharmaceutical, bromocriptine. Thus, the beginning oxidation potentials, which enable the biogenic amines to oxidize, do not redox match in the Parkinson's subject. Interestingly, the results showed that bromocriptine, at 5 mg/kg ip, low dose, had significantly greater effects on 5-HT and L-TP in PD animals than in non-PD animals (ANOVA; $p < 0.0001$), even though endogenously, 5-HT and L-TP were higher in concentration in non-PD than in PD

Voltaic Image of the Phosphorylated Tau Metallopolymer After Administration of the Direct Dopamine Receptor, Bromocriptine, at the High Dose: Line Graphs and Percentages Enable Quantification of the Tau Metallopolymer (Fig. 15)

Notes This finding points out that the neurochemical image of the tau metallopolymer can be imaged in a dose-dependent manner. This data recognizes, for the first time, that the intensity of the phosphorylation of the tau peptide may be assessed quantitatively. A relationship between increased tau and enhancement of dopamine has been reported [33]. The ionic compound, discovered herein that images as a tau metallopolymer, signals at a positive oxidation potential, significantly indicating that the ionic compound is in a neuroprotective state for the tauopathies of Parkinson's and the epilepsies. Non-PD does not image any of the biomarkers assessed in these studies. The results of the studies for the other two peptides, SRIF and dynorphin, are discussed in detail in the previous interventionary studies by Broderick and Kolodny that are noted above.

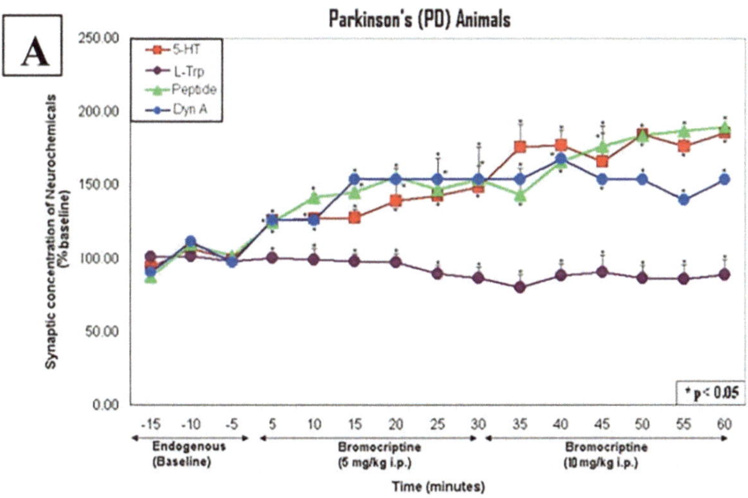

Fig. 15 (**a**) This figure reports and presents line graphs of the temporal course of neurochemical events for the PD animal (left) at the 5 & 10 mg/kg dose. The data focus on the middle of the *high* dose. (**b**) This figure reports and presents line graphs of the temporal course of neurochemical events for the non-PD animal at the 5 and 10 mg/kg dose. The data focuses on the middle of the *high* dose. Bromocriptine, at the high dose, also showed significantly greater effects on 5-HT and L-TP in PD animals versus non-PD animals (ANOVA; $p < 0.0001$). It is suggested that 5-HT-ergic function may be compensating for the lack of endogenous DA and HVA in dorsal striatal brain of PD in addition to compensating for the decreased DA and HVA biphasic effect of the high dose of bromocriptine in dorsal striatal brain of non-PD animals. It is also highly suggested that the polymeric metal complex of the tau peptide is enhanced as a result of dopaminergic depletion. These results corroborate with those of others [27, 37]. A significant difference between our work and others is the demonstration of the LIVE tau peptide, which redox matches as a metallopolymer, polymerized with the BRODERICK PROBE® polymer

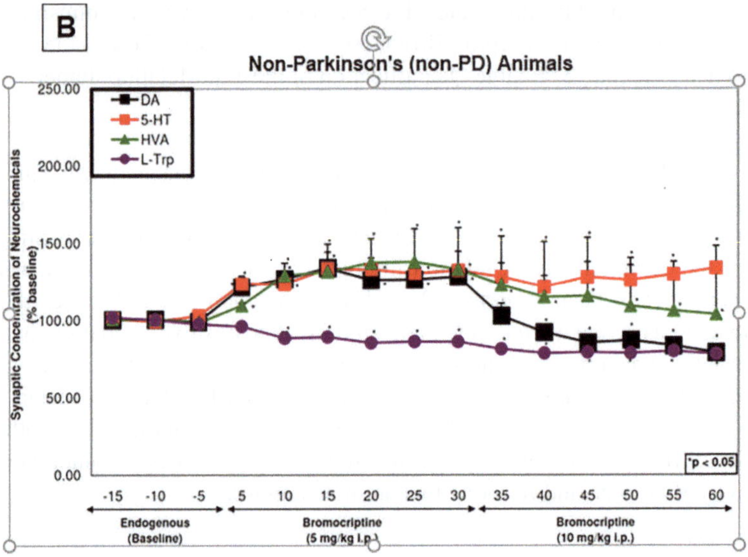

Fig. 15 (continued)

The amino acid, L-Tryptophan shows interesting results *as* the amino acid *is not enhanced* either in the Parkinson's group or in the non-PD group. The precursor to the neurotransmitter, serotonin, L- Tryptophan is reported herein to downregulate this precursor to serotonin. It seems that the downregulation is not a dramatic one in either group. The data show that neurodegeneration generates peptides and the data shows that the non-PD group does not produce peptides at least according to the peptides studied in this group.

The literature supports our findings. Tau increases are associated with dopamine depletion while tau ablation may not alter motor capacity [37]. The responses of tau and dopamine in patients are reported recently when using four-repeat tauopathies, further reporting that tau and dopamine alterations behave in tandem [27]. Increases in tau phosphorylation behave in tandem with enhanced dopamine [33]. The data regarding tau and dopaminergic motor effects is reported early on by utilizing aged tau knockout mice [37]. Minor tau phosphorylation changes were reported after the administration of the antipsychotic medication, pimavanserin [46]. Fyn-Tau modifies pentylene-induced seizures in early onset epileptogenesis [44]. Finally, comorbidity and underlying mechanisms of tauopathy and epilepsy are reviewed [31].

3.2 *Sensing the Parkinson's Subject Is Correlated with Sensing the Epilepsy Patient*

3.2.1 The Laurate BRODERICK PROBE®

Manufacture of BRODERICK PROBE® Laurate Sensors

Laurate sensors were manufactured on site. Lauric acid is a saturated fatty acid (formula: $CH_3(CH_2)_{10}COOH$). Laurate sensors, for these studies, comprised <1 mg aliquot of stock, a mixture of ultra-pure carbon (1.5 g), oil (Nujol) containing vitamin E (1.24 g) and lauric acid (100 mg). Details of the construction of the laurate sensor have been published [3, 5, 6].

NMI Properties of BRODERICK PROBE® Laurate Sensors

Oxidation Potentials for these sensors, in vitro, are, in the order of detection from lowest to highest potential, DA [0.13 V]; 5-HT [0.29 V]. Oxidation potentials for these sensors, in vivo, are, in the order of detection from lowest to highest oxidation potential, DA [0.15 V], 5-HT [0.3 V].

Laurate Scan (Fig. 16)

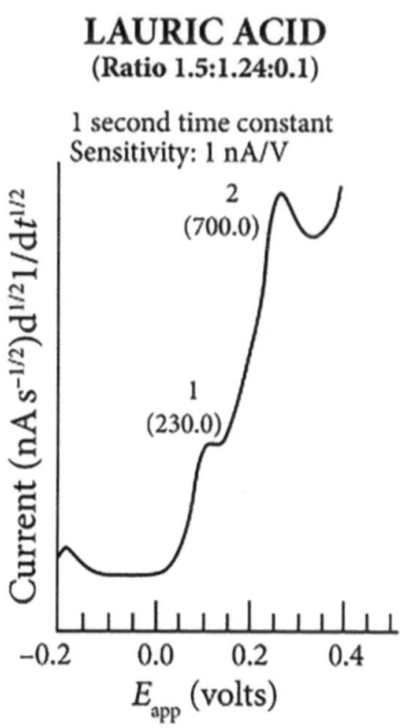

Fig. 16 Shows the operational oxidation potentials of the Laurate BRODERICK PROBE® biosensor for DA and 5-HT respectively. The numbers in parenthesis equal peak area

3.2.2 The Exposed Temporal Lobe of the Epilepsy Brain During Intraoperative Surgery (Fig. 17)

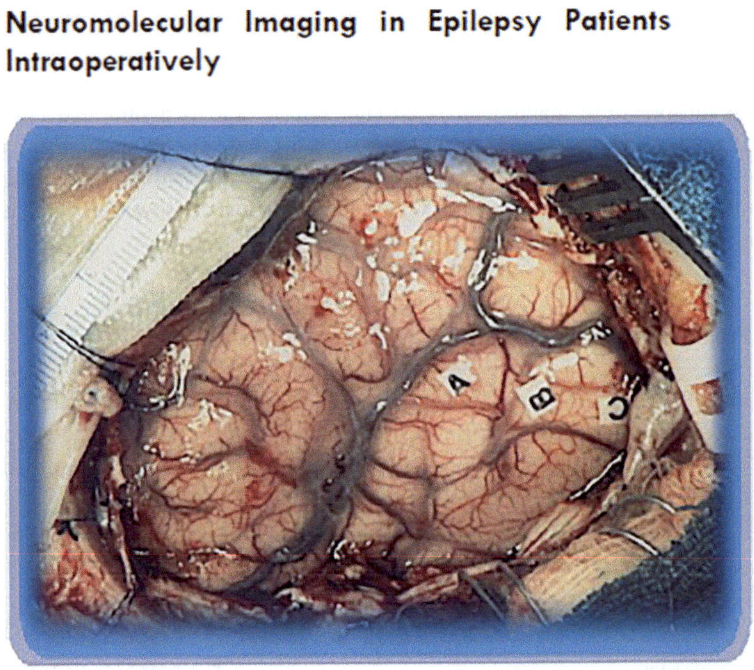

Fig. 17 Shows a photograph of the LIVE exposed temporal anterior cortical brain of the epilepsy patient during surgery. The BRODERICK PROBE® passes lightly, 2 um, over the temporal neocortex

3.2.3 Imaging Data Derived from the Anterior Cortex of the Epilepsy Patient: Surgery

LIVE In Vivo Data Online During Intraoperative Surgery in the Epilepsy Patient

The BRODERICK PROBE® is translated from the laboratory to the operating room (Fig. 18)

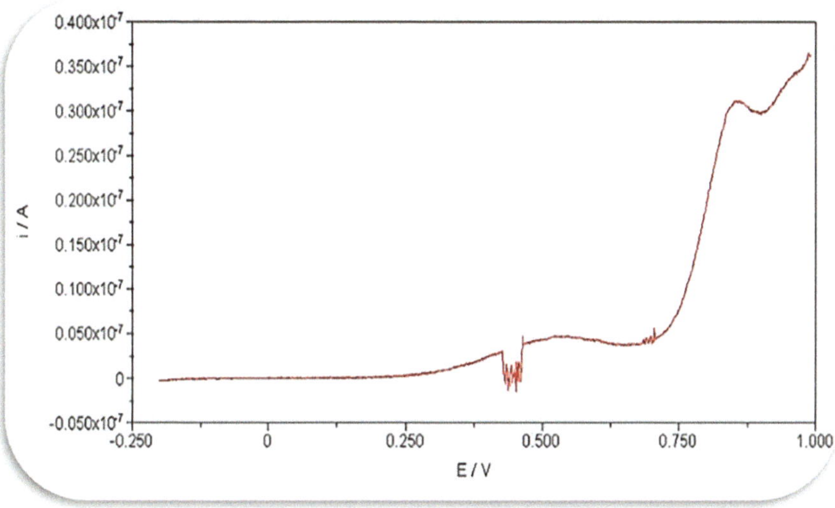

Fig. 18 Intraoperative NMI in vivo; the image is derived from the human epilepsy patient using the BRODERICK PROBE®. Linear circuit: On the x-axis: oxidation potential in millivolts (mV). On the y-axis: current in μamps. Starting at oxidation potential, 0.450 V, HVA, L-TP, DYN, & SRIF and the tau peptide are imaged. L-TP and peptides are present in higher concentrations than the monoamines or indoleamines. Biogenic amines are below detectable limits. The peptide is preferably the tau metallopolymer in the living epilepsy patient during surgery at NYU Tisch Hospital with IRB approval

Calibration Curves (Fig. 19)

4 Synopsis

The electroactive signal processing from the Parkinson's animal model and the epilepsy patient shows the tau peptide Metallopolymer is imaged, LIVE, in living animals and humans, as a metallopolymer at the voltaic biomarker on or about +0.83 volts as the ionic compound is polymerized with the phosphorylated BRODERICK PROBE® polymer. The quantum sensing of tau enables nanotechnology to be performed at the nanoscale. A large surface is enabled that identifies the tau peptide at an ionic recognition site preferably as a metal complex. Studies of the phosphorylated and polymerized tau peptide after bromocriptine, a direct receptor agonist of dopamine, show that tau steadily increases in real time while dopamine is in a depleted state upon neurodegeneration. This work provides quantitative data for tau by calculating percentage increases in tau before and after bromocriptine in Parkinson's subjects. The intensity of the phosphorylated state may be, indeed, assessed with the Broderick inventive art. Such is an attribute about tau that has long been desired since the degree of the intensity of tau phosphorylation is proportional to the degree of dopamine enhancement [33] *post* dopamine depletion and tau enhancement.

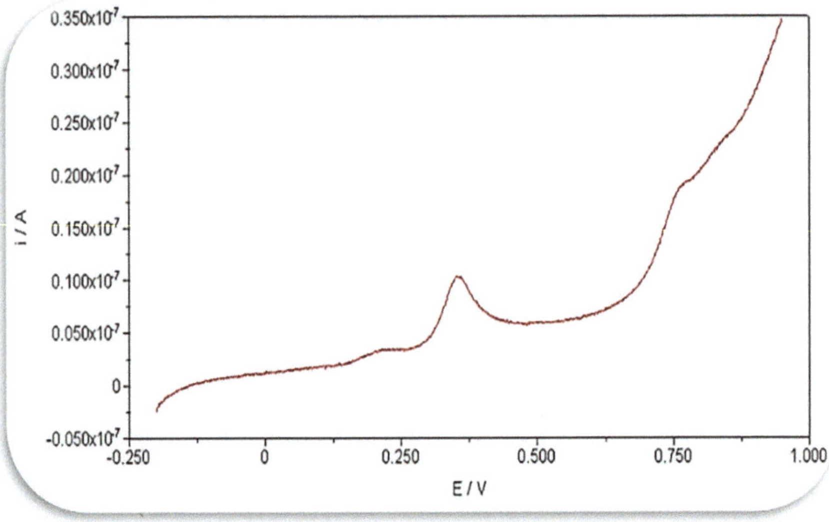

Fig. 19 NMI and BRODERICK PROBE®: In vitro *calibration studies for neurotransmitters and neurochemicals imaged in epilepsy patient tissue* in situ *and epilepsy patients* in vivo. Linear circuit: On the x-axis: Oxidation potential in millivolts (mV). On the y-axis: Current in μamps times 10. Starting from right to left, biogenic amines, metabolites and precursors and peptides are pipetted into phosphate buffer in the electrochemical call and neurotransmitters are subsequently studied in a series of scans. The higher concentrations of SRIF and tau were below detectable limit. The data confirms the patient data as L-TP and peptides are present in higher concentration than the monoamines or indoleamines in the brain of the epilepsy patient

Further, and not to be underestimated, neuroimaging tau intraoperatively in epilepsy patients with the BRODERICK PROBE® during surgery LIVE, is another first. Quantitative data for tau in the epilepsies is reported herein. This technological advance enables the field of neurodegenerative epilepsy to grow and the studies enable the significance of the term, firing rate, to progress to untold interpretations.

5 Discussion

This discovery of the tau peptide Metallopolymer biomarker for the tauopathies, Parkinson's disease and the epilepsies, is based on the electrochemical definition of the electric charge of the molecule, tau, as electric charge relates to the potential energy that is applied by the potentiostat. A potentiostat is the main tool used in electrochemical and electroanalytical studies, and this instrument manages the voltage or the current applied to an electrochemical cell electrode or diode. In this case, the electrochemical cell is invented by the author, and it operates by producing electrical energy from chemical energy. In this case, the sensor or electrode is a diode, and the diodes include semiconductors. An entire family of sensors are patented and trademarked under the umbrella of the BRODERICK PROBE®, which bears the name of the author's father.

The voltaic signature of the tau peptide Metallopolymer is imaged, and this biomarker of the tau peptide is discovered in real time, and it is LIVE. It is the polymeric metal complex of tau, and it is voltage dependent. It is inferably, one of the cobalt metal complex of tau. However, it is not limited to the cobalt metal complex of tau. The demonstration of tau within our electrochemical empirical design of tau is discussed. The design is centered around electric charge and the solubility of a molecule. In this case, the molecule is tau. The relationship between charge and solubility is dependent on Coulomb's law. The concept of the polarity of a molecule is the key to a healthy brain or an unhealthy brain. The unfolding of the protein from monomer to amyloid state is described as an evolution of configuration of the shape of the molecule, tau [49]. The author is describing the process as an evolution of configuration of electric charge that changes according to solubility. Thus, this chapter brings a novel advance in the study of neurodegeneration. The new way to look at neurodegeneration describes the process of solubility to insolubility of the tau peptide as healthy to unhealthy tau. Therefore, the beginning, monomer, that is soluble and positively charged electrically equates with the non-aggregated state, whereas amyloidosis reflects the insoluble and negatively charged tau equated to neurodegeneration. The tested hypothesis is based on Coulomb's Law.

Coulomb's Law is shown in Fig. 20.

It is worthy of note that tau has a positive overall charge, and in this positive oxidation state, the tau electric charge equates with solubility. It is critically important to share that tau, in this electrochemically positive oxidation voltammetric state, cannot form insoluble tau fibrils/tangles. Therefore, tau may not aggregate into insoluble tau fibrils when tau exists as a positive charge. The electric charge and the

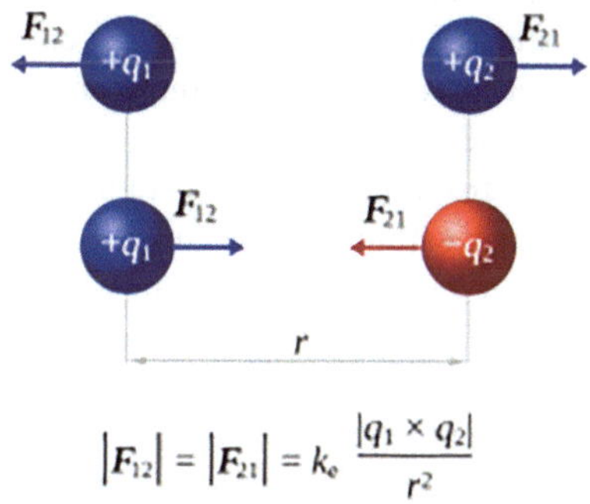

Fig. 20 The magnitude of the electrostatic force ***F*** between two point charges q_1 and q_2 is directly proportional to the product of the magnitudes of charges and inversely proportional to the square of the distance between them. Like charges repel each other, and opposite charges attract each other. (https://en.wikipedia.org/wiki/Coulomb%27s_law. Adapted from Wikipedia under the terms of the Creative Commons Attribution License)

degree of solubility of physiological neuroprotective tau may not inherently shift into pathophysiological and unhealthy tau unless tau is made insoluble by multiple negative charges, that is, for example, polyanions.

It is also important to add that, albeit the tau molecule exists within a positive charge or oxidation potential, there are 441 amino acids within the tau peptide, thereby enabling a negative charge under different circumstances of pH and the like and one may see an overall negative oxidation state for tau. In such a case, the unhealthy tau may cause neurodegeneration. This is the rationale for defining the oxidation state of tau, as we have demonstrated, as positive.

Thus, it is highly suggested that the physiological tau and not the pathophysiological tau, that is the molecule, the ionic compound in this case, that is imaged by the BRODERICK PROBE® nanoprobe biosensor. Indeed, polyanions, such as heparin and other anticoagulants that are heparin-like in electric charge, such as some mucopolysaccharides, that generally promote tau aggregation represent pathophysiological tau and negative ionic compounds [36].

The evidence shows that our discovery is one of neuroprotection since the tau that is seen in this report exists in its positive state as an ionic compound, a cation. It is noteworthy that tau in its negative forms is insoluble and in its negative charge, tau is pathophysiological. We suggest that we see the positive charge of the metal complex of tau. Thus, the physiological form, the healthy form of tau is reported in this chapter. Polyanions are reported by Montgomery [36] and neuroprotective forms of tau are reported by Gorantla et al. [29].

6 Conclusion

The basic tenet of ionic interactions is that the ionic compounds are made of compounds comprised of ions rather than atoms. Ionic compounds are structured so that they form a lattice of oppositely charged particles held by forces called electrostatic. The components of ionic compounds as electrically charged is what makes them soluble in water. It is important to remember that ionic compounds do not form molecules, but instead they exist as lone ions attracted to each other.

Therefore, the conclusion from these original studies may compel one to equate solubility with positive charge without aggregation. The tau ionic compound may well be neuroprotective against neurodegeneration. Neurodegeneration, then, may be associated with a negative charge and insoluble aggregates of neurofibrillary tangles, called tauopathies.

This chapter contributes to patient as well as veterinary theragnostics in the field of neurodegeneration in a novel way with a unique nanotechnology, neuromolecular imaging, using the miniature BRODERICK PROBE® biosensor. Indeed, the studies show that the discernment of neural electroactive waveforms in the patient's or animal's brain may indicate and/or predict the degree to which the patient or animal has presented with these dementia-driven disorders of Parkinson's and the epilepsies. The studies contribute to the assessment of a tauopathy as healthy or unhealthy.

7 Summary

7.1 *Tau in the Epilepsies*

Tauopathies are a group of neurodegenerative diseases in which aggregates of the hyperphosphorylated form of the microtubule-associated protein (MAP), tau, are believed to play a critical role in the pathogenesis. Many of the tauopathies are associated with an increased incidence of development of epilepsy. Deposits of hyperphosphorylated tau have also been documented pathologically in animal models of epilepsy, and in patients with drug-resistant focal epilepsy who have undergone epilepsy surgery. The precise mechanisms by which hyperphosphorylated tau contributes to epileptogenesis are currently uncertain, but there is increasing evidence from animal models that targeting tau-based mechanism, either genetically or pharmacologically, can have both antiseizure and antiepileptogenic effects. Indeed, animal models of epilepsy-associated tauopathies, biomarkers of tau pathology, experimental evidence implicating hyperphosphorylated tau in the pathophysiology of acquired epilepsies, in addition to therapies for tau AD, are described in [35] and critical sensing epilepsy work is contained in [9, 50].

7.2 *Tau in the Disease of Alzheimer*

The tau gene and its reported six tau isoforms have been found in the brain of Alzheimer's patients [39] In tau-related prionopathies, for example, but not limited to mad cow, people and or patients who are affected by this disease experience unusual sensations that include prickling or tingling feelings in the face, hands, legs and feet while sometimes sores appear on the feet for example. Sensations on these body parts can transform into burning sensations and permeate other body parts in a "pins and needles" sensation. Paresthesia can occur, causing a persistent lack of feeling in these body parts, producing clinical numbness.

7.3 *Tau in the Disorder of Parkinson's*

Dugger et al. [26] provides data that

> Dementia is a frequent complication of Parkinson's disease (PD) and that about half of PD dementia (PDD) is hypothesized to be due to progression of the underlying Lewy body pathology into limbic regions and the cerebral cortex while the other half is thought to be due to coexistent Alzheimer's disease (AD). Clinically, however, these are indistinguishable. Since the spread of amyloid plaques to the striatum has been reported to be a sensitive and specific indicator of dementia due to AD, these authors analyzed neuropathologically confirmed cases of PD without dementia (PDND, $N = 31$), PDD without AD (PDD, $N = 31$) and PD with dementia meeting clinicopathological criteria

for AD (PDAD, $N = 40$). The minimum diagnostic criterion for AD was defined as including a clinical history of dementia, moderate or frequent CERAD cortical neuritic plaque density and Braak neurofibrillary stage III-VI. Striatal amyloid plaque densities were determined using Campbell-Switzer and Thioflavine S stains. Striatal plaque densities were significantly higher in PDAD compared to PDD ($p < 0.001$). The presence of striatal plaques was approximately 80% sensitive and 80% specific for predicting AD. In comparison, the presence of cerebral cortex plaques alone was highly sensitive (100%) but had poor specificity (48% to 55%). The results suggest that striatal amyloid imaging may be clinically useful for making the distinction between PDD and PDAD.

The Dugger et.al, work is an important contribution to distinguishing Parkinsonism by amyloid concentration. Positron Emission Tomography in cerebrospinal fluid is new and hopeful. Nuclear magnetic resonance (NMR) for flexible tau forms and Cryo- EM for the most rigid forms or fluid immuno-diagnosis [28] coupled to brain imaging research such as the BRODERICK PROBE® [3, 5–13, 16–19, 23, 24, 30] bring promise. Video microscopy enables another useful tool in this highly active and necessary research arena, both clinical and pre-clinical. Antibody labeling is daily used *antemortem* and *postmortem*; it has brought forth valuable data on tauopathies despite its *postmortem* status [28].

Acknowledgement The author wishes to thank the Broderick Brain Foundation, the F.M. Kirby Foundation, the Center for Advanced Technology, CUNY, and the MacKenzie Foundation for partial support of our laboratory and students during these studies. It is important to note that since this body of work, the development and pioneering of neuromolecular imaging and the BRODERICK PROBE® has spanned over diligent studies yearly, other grants including *the National Institute of Health, National Institute on Drug Abuse*, The Aaron Diamond Foundation, the FACES and PACE Foundations for Epilepsy and the Upjohn Pharmacia Company also deserve honorable mention. Finally, the author gives special thanks to *Eazysense Nanotechnologies Inc.* in New York, the company through which BRODERICK PROBE® nanobiosensors are poised for the market to diagnose and treat brain, spinal cord and skin diseases. Tumor cancers in epilepsy patients such as, but not limited to, glioblastoma is one focus, and discovering sensing algorithms to help the quadriplegic person to walk is another.

References

1. Alzheimer's disease. Symptoms and causes. Mayo Clinic; 2020.
2. Bailey RB. The six kingdoms of life. Thought Co. Nov. 28, 2019.
3. Broderick PA. Cathodic electrochemical current arrangement with telemetric application. US Patent Serial No. 4,883,057, Issued; 1989.
4. Broderick PA. In vivo voltammetric studies on release mechanisms for cocaine with gamma-butyrolactone. Pharmacol Biochem Behav. 1991;40(4):969–75. https://doi.org/10.1016/0091-3057(91)90113-g. PMID: 1816583; PMCID: PMC7133153.
5. Broderick PA. Microelectrodes and their use in Cathodic electrochemical current arrangement with telemetric application. US Patent Serial No. 5,433,710, Issued; 1995.
6. Broderick PA. Microelectrodes and their use in an electrochemical arrangement with telemetric application. US Patent Serial No. 5,938,903, Issued; 1999.
7. Broderick PA. Neuromolecular imaging shows temporal synchrony patterns between serotonin and movement within neuronal motor circuits in the brain. Brain Sci. 2013;3(2):992–1012.

8. Broderick PA. Noninvasive electroactive photonic sensor with polymer memory transduction using organic and inorganic elements as platforms. US Provisional Application No 62/273,693; 2017.
9. Broderick PA. Sensing the epilepsy brain: a unique nanosurgical biomedical device to treat the epilepsy patient. MOJ Clin Med Case Rep. 2018;8(2):60–5.
10. Broderick PA. Noninvasive electroactive photonic protein sensor with polymer photovoltaic optics for memory transduction using organic and inorganic elements as platforms. US Non–Provisional International Application No. PCT/US 16/68879, Issued; 2020.
11. Broderick PA. Identification, diagnosis, and treatment of tau– associated pathologies using electrodes. US Provisional Application No. 63/293,334, December 23, 2021.
12. Broderick PA. Tau peptide signals are seen in Parkinson subjects by Broderick Probe® sensors. MOJ Clin Med Case Rep. 2021;11(2):46–50.
13. Broderick PA. A new approach to tumor cancer with a novel imaging profile for white matter abnormalities including Leukodystrophies: sensing the human brain. Med Res Arch [online]. 2022;10(11) https://doi.org/10.18103/mra.v10i11.3266.
14. Broderick PA. Small sensors for special surgery. MOJ Surg. 2023;11(3):118–20. https://doi.org/10.15406/mojs.2023.11.00236.
15. Broderick PA. Methods of diagnosis and treatment of amyloidosis of the skin using electrodes. (PCT) US Application Serial No. 63/524,250, June 30th, 2024.
16. Broderick PA. NEUROIMAGING: sensing biochemistry in the brain. Singapore: Jenny Stanford Publishing, Pte Ltd.; 2025. https://doi.org/10.1201/9780429029745.
17. Broderick PA, Kolodny EH. Real time imaging of biomarkers in the Parkinson's brain using mini-implantable biosensors. II. Pharmaceutical therapy with bromocriptine. Pharmaceuticals (Basel, Switzerland). 2009;2(3):236–49.
18. Broderick PA, Wenning L. Chapter 2: Neuromolecular imaging in Parkinson's disease. In: Martin CR, Preedy VR, editors. Genetics, neurology, behavior, and diet in parkinson's disease, the neuroscience of Parkinson's disease, vol. 16. Elsevier Academic Press; 2020. p. 251–65.
19. Broderick PA, et al. Identification, diagnosis, and treatment of neuropathologies, Neurotoxicities, tumors, and brain and spinal cord injuries using microelectrodes with microvoltammetry. U S. Patent Serial No. 7,112,319, Issued; 2006.
20. Broderick PA, et al. A clinical trial of an advanced diagnostic biomedical device for epilepsy patients. J Long-Term Effects Med Implants. 2008;18:1050–6934. Vol 18. Crossref- Vol.19, 2009, 103–111.
21. Broderick PA, Doyle WK, Pacia SV, Kuzniecky RI, Devinsky O, Kolodny EH. Conference paper, intraoperative Neuromolecular Imaging (NMI) in comparison with resected epileptogenic tissue. Society for Neuroscience; 2009.
22. Broderick PA, et al. Identification, diagnosis, and treatment of neuropathologies, neurotoxicities, tumors, and brain and spinal cord injuries using electrodes with microvoltammetry. US20120029331A1; 2012.
23. Broderick PA, Wenning L, Li YS. Neuromolecular imaging, a nanobiotechnology for Parkinson's disease: advancing pharmacotherapy for personalized medicine. J Neural Trans (Vienna, Austria). 2016;124(1):57–78.
24. Broderick PA et al. Identification, diagnosis, and treatment of neuropathologies, neurotoxicities, tumors, and brain and spinal cord injuries using electrodes with microvoltammetry. US Patent Serial No. 10,980,460, April 20th, Issued; 2021.
25. Chefer VI, Thompson AC, Zapata A, et al. Overview of brain microdialysis. Curr Protoc Neurosci. 2009;Chapter 7:Unit 7.1. https://doi.org/10.1002/0471142301.ns0701s47.
26. Dugger BN, Serrano GE, Sue LI, et al. Presence of striatal amyloid plaques in Parkinson's disease dementia predicts concomitant Alzheimer's disease: usefulness for amyloid imaging. J Parkinsons Dis. 2012;2(1):57–65.
27. Ferschmann C, Messerschmidt K, Gnörich J, Barthel H, Marek K, Palleis C, Katzdobler S, Stockbauer A, Fietzek U, Finze A, Biechele G, Rumpf JJ, Saur D, Schroeter ML, Rullmann M, Beyer L, Eckenweber F, Wall S, Schildan A, Patt M, Stephens A, Classen J, Bartenstein P, Seibyl J, Franzmeier N, Levin J, Höglinger GU, Sabri O, Brendel M, Scheifele M;

German Imaging Initiative for Tauopathies (GII4T). Tau accumulation is associated with dopamine deficiency in vivo in four-repeat tauopathies. Eur J Nucl Med Mol Imaging. 2024;51(7):1909–22. https://doi.org/10.1007/s00259-024-06637-6. Epub 2024 Feb 17. PMID: 38366196; PMCID: PMC11139736.
28. Fichou Y, Al-Hilaly YK, Devred F, et al. The elusive tau molecular structures: can we translate the recent breakthroughs into new targets for intervention? Acta Neuropathol Commun. 2019;7:31.
29. Gorantla NV, Landge VG, Nagaraju PG, et al. Molecular cobalt (II) complexes for tau polymerization in Alzheimer's disease. ACS Omega. 2019;4(16):16702–14.
30. Haile M, Broderick P, Li Y, et al. Nimodipine reverses the elevation of synaptic striatal dopamine and serotonin during *in vivo* hypoxia. Open Neurosci J. 2011;5:1–7.
31. Hwang K, Vaknalli RN, Addo-Osafo K, Vicente M, Vossel K. Tauopathy and epilepsy comorbidities and underlying mechanisms. Front Aging Neurosci. 2022;14:903973. https://doi.org/10.3389/fnagi.2022.903973. https://www.ncbi.nlm.nih.gov/pmc/articles/PMC10014470/PMID:Neurology. 2018; 90(11): e940–6.
32. Jimenez H, Adrien L, Wolin A, Eun J, Chang EH, Burstein ES, Gomar J, Davies P, Koppel J. The impact of pimavanserin on psychotic phenotypes and tau phosphorylation in the P301L/COMT- and rTg(P301L)4510 mouse models of Alzheimer's disease. Alzheimers Dement (NY). 2022;8(1):e12247. https://doi.org/10.1002/trc2.12247. PMID: 35128032; PMCID: PMC8804623.
33. Koppel J, Jimenez H, Adrien L, H Chang E, Malhotra AK, Davies P. Increased tau phosphorylation follows impeded dopamine clearance in a P301L and novel P301L/COMT-deleted (DM) tau mouse model. J Neurochem. 2019;148(1):127–35. https://doi.org/10.1111/jnc.14593.
34. Linnaeus C. Systemanaturae, sive regna trianaturae systematic eproposita per classes. ordines, genera, & species. Leiden: Haak. Oct.3. 1735; 2013. p. 1–12. https://doi.org/10.5962/bhl.title.68927.
35. Mahoney ER, Dumitrescu L, Seto M, et al. Telomere length associations with cognition depend on Alzheimer's disease biomarkers. Alzheimers Dement. 2019;5(1):883–90.
36. Montgomery KM. Chemical features of polyanions modulate tau aggregation and conformation states. J Am Chem Soc. 2023;145(7), 3926–36. 14, A ... Israel Journal of Chemistry, e202300104.
37. Morris M, Hamto P, Adame A, Devidze N, Masliah E, Mucke L. Age-appropriate cognition and subtle dopamine-independent motor deficits in aged tau knockout mice. Neurobiol Aging. 2013;34(6):1523–9. https://doi.org/10.1016/j.neurobiolaging.2012.12.003. Epub 2013 Jan 16. PMID: 23332171; PMCID: PMC3596503.
38. Mueller RL, Combs B, et al. Tau: a signaling hub protein. Front Mol Neurosci. 2021;14:647054. https://doi.org/10.3389/fnmol.2021.647054. PMID: 33815057; PMCID: PMC8017207.
39. Noble W, Hanger DP, Miller C, et. al. (2013) The importance of tau phosphorylation for neurodegenerative diseases. Front Neurol 4:83.
40. O'Gorman J. In novel electrochemical methods, a thesis presented to Dublin City University in submission for the MSc degree. October 10th, 1998.
41. Oldham K. Semiintegral electroanalysis: analog implementation. Anal Chem. 1973;45(1):39.
42. Prussiner SB. Nobel lecture: prions. Proc Natl Acad Sci. 1998;95(23):13363–83.
43. Prussiner SB, Woerman AL, Mordes DA, et al. Evidence for α-synuclein prisons causing multiple system atrophy in human with parkinsonism. Proc Natl Acad Sci USA. 2015;112(38):E5308–17.
44. Putra M, Puttachary S, Liu G, Lee G, Thippeswamy T. Fyn-tau ablation modifies PTZ-induced seizures and post-seizure hallmarks of early epileptogenesis. Front Cell Neurosci. 2020;14:592374. https://doi.org/10.3389/fncel.592374. PMID: 33363455; PMCID: PMC7752812.
45. Rowley PA, Samsonov AA, Betthauser TJ, Pirasteh A, Johnson SC, Eisenmenger LB. Amyloid and tau PET imaging of Alzheimer disease and other neurodegenerative conditions. Semin Ultrasound CT MR. 2020;41(6):572–83. https://doi.org/10.1053/j.sult.2020.08.011. Epub 2020 Aug 29.

46. Schneider A, Biernat J, von Bergen M, et al. Phosphorylation that detaches tau protein from microtubules (Ser262, Ser214) also protects it against aggregation into Alzheimer paired helical filaments. Biochemistry. 1999;38(12):3549–58.
47. Scholl M, Maass A, Mattson N, et al. Biomarkers for tau pathology. Mol Cell Neurosci. 2019;97:18–33.
48. Trinczek B, Biernat J, Baumann K, et al. Domains of tau protein, differential phosphorylation, and dynamic instability of microtubules. Mol Biol Cell. 1995;2:1887–902.
49. Zeng Y, Yang J, Zhang B, et al. The structure and phase of tau: from monomer to amyloid filament. Cell Mol Life Sci. 2021;78:1873–86. https://doi.org/10.1007/s00018-020-03681-x.
50. Zheng, P, Shultz S, Hovens C, et al. Animal models of acquired epilepsy and tauopathies. In: Models of Seizures and Epilepsy; 2017. p. 1031–41.

The "Hyperactive" Phenotype of Alzheimer's Disease

András Attila Horvath and Anita Kamondi

Abstract Alzheimer's disease (AD) is a leading cause of cognitive decline, with no curative therapy currently available. Emerging research highlights epileptic activity as a modifiable risk factor contributing to disease progression. Animal models and human studies indicate an increased prevalence of epileptic seizures and subclinical epileptiform activity (SEA) in AD, which may exacerbate cognitive impairment. However, significant variability exists in reported SEA prevalence, ranging from 3% to 54%, potentially due to differing definitions, diagnostic methodologies, and biological heterogeneity of AD subtypes. Recent evidence suggests SEA may be more common in individuals with hippocampus-sparing AD, a rapidly progressing variant characterized by distinct neuroimaging and clinical features. Advanced neuroimaging techniques, including MRI and PET, have identified structural and functional brain alterations associated with SEA in AD, providing potential biomarkers for disease phenotyping. Given the role of neuronal hyperexcitability in AD, targeting SEA represents a promising therapeutic strategy. Early clinical trials suggest that levetiracetam may mitigate hyperexcitability and improve cognitive function in SEA-positive individuals. Future research should focus on refining SEA detection methods, integrating neuroimaging biomarkers, and developing personalized treatment approaches to modify AD progression effectively.

Keywords Alzheimer's disease · Epilepsy · Biomarkers · Therapy

---

A. A. Horvath
Neurocognitive Research Center, Gyula Nyírő National Institute of Psychiatry and Addictology, Budapest, Hungary

Department of Anatomy, Semmelweis University, Budapest, Hungary

A. Kamondi (✉)
Neurocognitive Research Center, Gyula Nyírő National Institute of Psychiatry and Addictology, Budapest, Hungary

Department of Neurosurgery and Neurointervention, Semmelweis University, Budapest, Hungary

Department of Neurology, Semmelweis University, Budapest, Hungary

P. A. Broderick (ed.), *Biomedical Imaging in Neurodegeneration*,
https://doi.org/10.1007/978-3-031-95878-6_8

1 Background and Unresolved Issues

Alzheimer's disease (AD) is the leading cause of cognitive decline, with its prevalence increasing rapidly in our society [31]. Although, no curative therapy exists, recent scientific studies have identified modifiable risk factors and comorbid conditions that might serve as potential therapeutic targets [23, 38].

Recently, the pathogenic role of epileptic activity has been proposed as a new potentially modifiable risk factor. First it has been demonstrated in animal models that transgenic AD animals frequently exhibit epileptic seizures and abnormal neural hyperactivity [12, 46]. These findings gain support in human studies, which reveal that patients with AD or mild cognitive impairment (MCI) have an elevated risk for developing epileptic seizures [28, 29, 52]. These seizures contribute to worsened cognitive functions and accelerated disease progression [4, 5, 54], making their detection a high priority. However, most seizures in these patients are non-motor seizures with temporal lobe semiology, which makes their detection difficult especially in individuals with cognitive symptoms [29, 52].

Recent reports also highlight, in the pathogenesis of dementia, the potential significance of isolated epileptiform discharges, which occur without overt epileptic seizures [30, 53]. In neurophysiology studies, these discharges are referred to as subclinical epileptiform activity (SEA) [30, 52, 53]. Traditionally, epileptiform discharges without seizures are considered benign variants in epileptology which might cause transient cognitive changes but are not associated with persistent pathological conditions [9]. These discharges can be detected in healthy individuals and their incidence increases with age [42]. Although the negative impact of SEA on cognitive functions has been suggested in epilepsy studies since the 1990s [1, 8, 26], this phenomenon was not investigated in AD until the study by Vossel and colleagues [52].

While evidence indicates that SEA might correspond to the pathologic imbalance between excitatory and inhibitory circuits resulting in cortical hyperactivation [36], there are numerous discrepancies and inconsistencies in the reported results. As a key question, we need to address why not all individuals living with AD demonstrate SEA and why do we see a large variation across the studies analysing the prevalence of SEA in AD.

2 Variability in the Prevalence of SEA in AD

According to study results, the prevalence of SEA among patients with AD and MCI ranges from 3% to 54% [20, 36]. Most studies indicate a higher prevalence compared to healthy controls (HC), although there are exceptions. A critical consideration is the definition of SEA itself. Early studies did not clarify the neurophysiological definition of SEA [52]. Liedorp et al. [41] referred to the recommendation of the International Federation of Clinical Neurophysiology from

1999, which did not separately define SEA [44]. According to this guideline, terms like epileptiform pattern, discharge, or activity describe transients distinguishable from background activity, characterized by spiky morphology, typically but not exclusively found in interictal EEGs of people with epilepsy. The guideline mentions that such patterns might occur in individuals without epilepsy but does not define SEA as a specific category.

In a decisive guideline article, Noachtar and Rémi defined SEA as paroxysmal EEG graphoelements (spikes or sharp waves) lasting 20–200 ms, disrupting background EEG activity, followed by slow waves [45]. Importantly, these graphoelements do not associate with epileptic seizures. This definition was used later in the studies by Vossel et al. [53] and Horvath et al. [29, 30]. Some recent studies have defined SEA more strictly, focusing on abrupt changes in polarity to emphasize the multiphasic character of spikes and sharp waves. This approach considers a sequence of a minor positive, major negative, and second minor positive component as a marker of epileptiform discharge [11, 39].

In a study by Lam et al. [39], a panel of nine clinical experts evaluated the morphology and existence of SEA in the EEG registrations. SEA was identified in cases where a consensus of more than six experts was achieved. They also conducted supplementary statistical analysis to minimize interpreter variations [35]. In the study by Brunetti et al. [11], a dedicated software was used for spike detection, followed by manual revision of the discharges. The most complex reporting protocol was applied in the study by Babiloni et al. [3], based on two current reference guidelines [24, 37]. SEA was identified when at least four of the following six criteria were met: (1) di- or triphasic waves with sharp or spiky morphology; (2) wave durations different from the ongoing background activity, either shorter or longer; (3) waveform asymmetry with a sharply rising ascending phase and a more slowly decaying descending phase or vice versa; (4) a transient component of SEA followed by an associated slow after-wave; (5) disruption of the background EEG activity surrounding SEA by the presence of SEA; and (6) distribution of the negative and positive EEG potentials on the scalp suggesting a focal cortical source, corresponding to a radial, oblique, or tangential orientation of the (dipole) source.

Beside the differences in the neurophysiological definition of SEA, another possible explanation must be addressed around the large prevalence variability such as the biological heterogeneity of AD.

3 Phenotyping of SEA-Variant AD: A Theoretical Approach

Although pathological forms of both amyloid and tau have been well documented in animal studies to cause neuronal hyperexcitability [34], the question remains whether Aβ or tau is the primary contributor to hyperexcitability in humans. The overwhelming majority of epileptiform discharges in people with AD originate from the temporal cortex [4, 5, 29]. Neuropathological and quantitative neuroimaging studies show that Aβ accumulation begins in the neocortex (temporal, parietal, and/or frontal) and later spreads to the allocortex, brainstem, and eventually the cerebellum [33]. Thus, the high prevalence of epileptiform discharges observed in MCI stage of AD (when amyloid load is high in the temporal cortex) suggests that Aβ is the initiating factor. The primary role of pathological amyloid in inducing hyperexcitability is further supported by neuropathological data demonstrating that tau accumulation starts only after the appearance of amyloid deposits [10, 32]. Furthermore, tau deposition follows a different pathway from amyloid: it begins in the brainstem and spreads through the entorhinal cortex to the neocortical areas of the temporal, parietal, and occipital lobes, and later to the prefrontal areas [10, 32].

Clinicopathological studies have identified four AD subtypes based on tau neurofibrillary tangle (NFT) density [33, 43]: (1) 55–75% of individuals with AD have the typical form, with NFTs equally distributed in the neocortex and hippocampus; (2) 11–17% have the hippocampus-sparing form, with NFTs accumulating predominantly in the association cortices; (3) 14–21% of AD patients have the limbic-predominant type, with NFTs deposited mainly in the hippocampus; and (4) 17% have the so-called minimal atrophy form, characterized by more amyloid than tau pathology. These subtypes exhibit different clinical characteristics. For instance, the hippocampus-sparing type tends to have an earlier onset and a higher prevalence among individuals with higher education levels and is associated with a rapid progression of non-amnestic cognitive symptoms [33, 49].

People with AD who exhibit frequent SEA share similar clinical characteristics with those having the hippocampus-sparing form of AD. These characteristics include a higher frequency of atypical symptoms [53], younger age at disease onset [29], faster progression [4, 5], higher education levels [39], and poorer visuospatial performance [27] compared to individuals with AD without epileptiform activity. These observations suggest that this group might represent the hippocampus-sparing subtype [19]. Supporting this hypothesis, most studies using histopathology or cerebrospinal fluid biomarkers for AD diagnosis stratification found a 13–18% prevalence of AD with epileptiform activity, which is similar to the relative frequency of the hippocampus-sparing subtype [33]. Thus, the density of spiking on the EEG might indirectly reflect the regional distribution of tauopathy in AD.

In each clinicopathological subtype of AD, including classic, limbic-predominant, hippocampus-sparing, and minimal atrophy, the Aβ and p-tau loads differ among cortical structures [33]. The cumulative incidence of dementia among cognitively unimpaired older adults with positive amyloid and tau biomarkers is only 26% after a 5-year follow-up period [50]. People with AD with comparable amyloid plaque burden can still vary in their p-tau levels, which might affect their susceptibility to cortical hyperactivity [25]. These data suggest that the course and effects of neurodegeneration vary among individuals, leading to diverse clinical phenotypes of the disease. The emergence of neuronal hyperexcitability in people with AD likely results from the net effect of the affected functional networks, leading to a wide variability.

4 Phenotyping with Neuroimaging: A Clinical Approach

Cretin and colleagues, based on their previous experiences, have suggested that the extratemporal representation of AD is a red flag for epileptic seizures [17]. Different imaging modalities—especially MRI—play a significant role both in the identification of epilepsy and in the detection of pathological changes in the grey matter areas in AD. In addition, functional MRI (fMRI) plays a crucial role to depict brain networks and to determine their changes.

There are only a few studies focusing on the structural neuroimaging characteristics of AD patients with epileptiform transients, and they do not differentiate between the various stages of dementia. Previous reports have indicated more severe atrophy in AD patients with seizures, without specifying a distinct pattern. This atrophy primarily correlated with the level of cognitive decline and was assessed through visual analysis [2, 7, 52]. In the first systematic neuroimaging study investigating these relations [27], patients with seizures did not show significantly greater atrophy in most lobar structures, except for the parietal lobe. A clear general trend has emerged: epileptic AD patients had lower volumes and thickness in all measured brain structures. The MMSE and ACE scores, which characterize dementia severity, correlated more strongly with the level of atrophy [21, 27]. Specifically, the ACE score correlated with the atrophy of temporal and frontal lobe structures.

The atypical posterior atrophy pattern was supported by functional neuroimaging studies revealing that posterior hypometabolism, as shown by positron emission tomography (PET) in areas such as the temporo-parietal junction, posterior cingulate gyrus, and parietal lobes, might be associated with epileptic seizures in AD [16, 47]. These findings align with reports suggesting that extratemporal hypometabolism or atrophy signals pathological hyperactivity in AD [15, 18].

As a further supporting evidence, a graph network analysis study was performed [40] enrolling 81 patients with AD and 45 healthy controls. Twenty-five patients had epilepsy as well. A 3D T1-weighted brain MRI sequence was used, and cortical as well as subcortical segmentation was performed with FreeSurfer. Graph theory was used to obtain the global brain network. In this study, a significant difference was found in local efficiency, mean clustering coefficient, and path length between AD and healthy controls. Additionally, significant differences were identified between AD patients with and without epilepsy. In those patients who were diagnosed with AD and had epilepsy, the local efficiency, mean clustering coefficient, average degree, and assortative coefficient were lower, while characteristic path length was longer compared to patients without epilepsy. Furthermore, in the intrinsic thalamic network, the mean clustering coefficient was higher and the characteristic path length was shorter in patients with epilepsy than without.

Another recent study investigated the morphological patterns of hippocampal atrophy in patients with mesial temporal lobe epilepsy (MTLE) and AD [22]. Twenty-five patients with AD, 23 MTLE patients, and 48 healthy control subjects were involved. All subjects were scanned with a 3 T MRI scanner, where a conventional 3D T1-weighted sequence was used. A volumetric analysis—with FreeSurfer—and a morphologic analysis—with FSL-FIRST—was applied to compare the hippocampal volume and morphology. Despite that there was no correlation between AD and epilepsy, some key results were found. The mean ipsilateral hippocampal volume of the MTLE patients was 2650 mm^3, while in AD patients that was 1682 mm^3. In MTLE patients, the vertex-based morphologic analysis revealed significant ipsilateral (left) hippocampal atrophy in the anterior region, corresponding to the anatomical head of the hippocampus. The ipsilateral body of the hippocampus also had atrophy. The contralateral (right) hippocampus showed slight deformation in the posterior region. In AD patients, the morphological analysis showed generalized deformation of bilateral hippocampus. Almost the entire hippocampus had a significant atrophy.

The unique patterns of hippocampal atrophy observed in MTLE and AD through volumetric and morphologic analyses could function as a structural marker, shedding light on the correlation between radiological observations and the respective underlying pathologies.

In an 18 fluorodeoxyglucose PET-CT (FDG PET-CT) study, cognitive and imaging characteristics in individuals with probable AD were compared [55]. At the initial assessment, 18% of individuals in the late-onset epilepsy (LOE) cohort with cognitive symptoms demonstrated cognitive impairment within the dementia range, and one individual received a probable AD diagnosis. Cognitive impairment patterns in the LOE group were commonly associated with impaired frontal and left temporal regions, whereas all AD patients displayed parieto-temporal patterns. The AD group demonstrated more severe FDG PET hypometabolism in parietal and occipital lobes, but not in the temporal and frontal lobes. Throughout the 3.0 ± 3.2 years of follow-up, improvements in seizure frequency in the LOE group were linked to enhancement in the total Addenbrooke cognitive examination-III score, and there were no additional instances of conversion to probable AD or group-level cognitive decline.

5 Conclusions and Future Perspective

There is a clear need for proper phenotyping of AD individuals with SEA since pathologic hyperactivity represents a potentially modifiable pathophysiological factor. Some recent studies have already highlighted the relevance of targeting SEA as a disease modification therapy. In a 2021 phase IIa randomized double-blind placebo-controlled crossover clinical trial involving individuals with AD, levetiracetam treatment did not improve overall executive function scores compared to placebo [51]. However, a subgroup analysis revealed significant improvement in the Stroop naming subscale and in a virtual route learning test in participants with epileptiform activity on their EEG, compared to the placebo group. Despite some limitations (small number of participants ($n = 34$) and relatively young average age at onset (57.5 years)), this study suggests that stratifying patient populations based on EEG results is an important strategy when designing drug trials to identify the most effective medication for AD-related epilepsy.

In individuals with MCI due to AD, levetiracetam was found to normalize hippocampal hyperexcitability and improve memory performance [6]. Additionally, a 2023 double-blind placebo-controlled within-subject crossover design study in people with AD showed that levetiracetam increases hippocampal blood flow, as measured by arterial spin-labelling MRI [48]. Currently, a randomized clinical trial is underway to evaluate the efficacy of levetiracetam in treating hyperexcitability in AD, along with a prospective trial on the effect of levetiracetam on neuropsychiatric symptoms related to epilepsy in AD (ClinicalTrials.gov clinicaltrials.gov/study/NCT03875638 [13]; ClinicalTrials.gov clinicaltrials.gov/study/NCT04004702 [14].

Altogether, further studies are needed to support the correlation between pathologic hyperactivity in AD and additional neuroimaging biomarkers. With a potential biomarker setup, we might be able to design individually tailored drug trials based on the presence of SEA as a preselection criterion, and supporting imaging markers could serve as follow-up or outcome measures in a longitudinal study design.

References

1. Aldenkamp AP. Effect of seizures and epileptiform discharges on cognitive function. Epilepsia. 1997;38:52–5. https://doi.org/10.1111/j.1528-1157.1997.tb04520.x.
2. Amatniek JC, Hauser WA, DelCastillo-Castaneda C, Jacobs DM, Marder K, Bell K, et al. Incidence and predictors of seizures in patients with Alzheimer's disease. Epilepsia. 2006;47:867–72. https://doi.org/10.1111/j.1528-1167.2006.00554.x.
3. Babiloni C, Noce G, Bonaventura D, Lizio C, Pascarelli R, Tucci MT, et al. Abnormalities of cortical sources of resting state delta electroencephalographic rhythms are related to epileptiform activity in patients with amnesic mild cognitive impairment not due to Alzheimer's disease. Front Neurol. 2020;11:514136. https://doi.org/10.3389/fneur.2020.514136.
4. Baker J, Libretto T, Henley W, Zeman AAL. Epileptic seizures in Alzheimer's disease. Front Neurol. 2019a;10:1266. https://doi.org/10.3389/fneur.2019.01266.
5. Baker J, Libretto T, Henley W, Zeman A. The prevalence and clinical features of epileptic seizures in a memory clinic population. Seizure. 2019b;71:83–92.
6. Bakker A, et al. Reduction of hippocampal hyperactivity improves cognition in amnestic mild cognitive impairment. Neuron. 2012;74:467–74.
7. Bernardi S, Scaldaferri N, Vanacore N, Trebbastoni A, Francia A, D'Amico A, et al. Seizures in Alzheimer's disease: a retrospective study of a cohort of outpatients. Epileptic Disord. 2010;12:16–21. https://doi.org/10.1684/epd.2010.0290.
8. Binnie CD. Cognitive effects of subclinical EEG discharges. Neurophysiol Clin. 1996;26:138–42.
9. Binnie CD. Cognitive impairment during epileptiform discharges: is it ever justifiable to treat the EEG? Lancet Neurol. 2003;2:725–30. https://doi.org/10.1016/S1474-4422(03)00584-2.
10. Braak H, Braak E. Evolution of the neuropathology of Alzheimer's disease. Acta Neurol Scand. 1996;165:3–12.
11. Brunetti V, D'Atri A, Della Marca G, Vollono C, Marra C, Vita MG. Subclinical epileptiform activity during sleep in Alzheimer's disease and mild cognitive impairment. Clin Neurophysiol. 2020;131:1011–8. https://doi.org/10.1016/j.clinph.2020.02.015.

12. Busche MA, Eichhoff G, Adelsberger H, Abramowski D, Wiederhold KH, Haass C. Clusters of hyperactive neurons near amyloid plaques in a mouse model of Alzheimer's disease. Science. 2008;321:1686–9. https://doi.org/10.1126/science.1162844.
13. ClinicalTrials.gov. clinicaltrials.gov/study/NCT03875638. 2023.
14. ClinicalTrials.gov. clinicaltrials.gov/study/NCT04004702. 2019.
15. Cretin B, Blanc F, Gaultier C, Sellal F. Epileptic amnesic syndrome revealing Alzheimer's disease. Epilepsy Res. 2012;102:206–9. https://doi.org/10.1016/j.eplepsyres.2012.08.002.
16. Cretin B, Di Bitonto L, Blanc F, Magnin E. Left temporal lobe epilepsy revealing left posterior cortical atrophy due to Alzheimer's disease. J Alzheimers Dis. 2015;45:521–6. https://doi.org/10.3233/JAD-141953.
17. Cretin B, Philippi N, Bousiges O, Dibitonto L, Sellal F, Martin-Hunyadi C, et al. Do we know how to diagnose epilepsy early in Alzheimer's disease? Rev Neurol. 2017;173:374–80. https://doi.org/10.1016/j.neurol.2017.03.028.
18. Cretin B, Philippi N, Sellal F, Dibitonto L, Martin-Hunyadi C, Blanc F. Can the syndrome of transient epileptic amnesia be the first feature of Alzheimer's disease? Seizure. 2014;23:918–20. https://doi.org/10.1016/j.seizure.2014.07.008.
19. Cretin B, Sellal F, Philippi N, Bousiges O, Di Bitonto L, Martin-Hunyadi C, Blanc F. Epileptic prodromal Alzheimer's disease, a retrospective study of 13 new cases: expanding the spectrum of Alzheimer's disease to an epileptic variant? J Alzheimers Dis. 2016;52(3):1125–33.
20. Csernus EA, Werber T, Kamondi A, Horvath AA. The significance of subclinical epileptiform activity in Alzheimer's disease: a review. Front Neurol. 2022;13:856500.
21. Csukly G, Siraly E, Fodor Z, Horvath A, Salacz P, Hidasi Z, et al. The differentiation of amnestic type MCI from the nonamnestic types by structural MRI. Front Aging Neurosci. 2016;8:52. https://doi.org/10.3389/fnagi.2016.0005234.
22. Duan Y, Lin Y, Rosen D, Du J, He L, Wang Y. Identifying morphological patterns of hippocampal atrophy in patients with mesial temporal lobe epilepsy and Alzheimer disease. Front Neurol. 2020;11:21.
23. Edwards FAAU. 2019 for Alzheimer's disease: from plaques to neurodegeneration. Trends Neurosci. 2019;42:310–22. https://doi.org/10.1016/j.tins.2019.03.003.
24. Fisher RS, Acevedo C, Arzimanoglou A, Bogacz A, Cross JH, Elger CE. ILAE official report: a practical clinical definition of epilepsy. Epilepsia. 2014;55:475–82. https://doi.org/10.1111/epi.12550.
25. Frisoni GB, Altomare D, Thal DR, Ribaldi F, van der Kant R, Ossenkoppele R, Blennow K, Cummings J, van Duijn C, Nilsson PM, Dietrich PY, Scheltens P, Dubois B. The probabilistic model of Alzheimer disease: the amyloid hypothesis revised. Nat Rev Neurosci. 2022;23(1):53–66.
26. Henriksen O. Education epilepsy: assessment and remediation. Epilepsia. 1990;31:21–5. https://doi.org/10.1111/j.1528-1157.1990.tb05865.x.
27. Horvath A, Kiss M, Szucs A, Kamondi A. Precuneus-dominant degeneration of parietal lobe is at risk of epilepsy in mild Alzheimer's disease. Front Neurol. 2019;10:878.
28. Horváth A, Szucs A, Barcs G, Noebels JL, Kamondi A. Epileptic seizures in Alzheimer disease: a review. Alzheimer Dis Assoc Disord. 2016;30:186–92. https://doi.org/10.1097/WAD.0000000000000134.
29. Horváth A, Szucs A, Hidasi Z, Csukly G, Barcs G, Kamondi A. Prevalence semiology, and risk factors of epilepsy in Alzheimer's disease: an ambulatory EEG study. J Alzheimers Dis. 2018;63:1045–54. https://doi.org/10.3233/JAD-170925.
30. Horvath AA, Papp A, Zsuffa J, Szucs A, Luckl J, Radai F. Subclinical epileptiform activity accelerates the progression of Alzheimer's disease: a long-term EEG study. Clin Neurophysiol. 2021;132:1982–9. https://doi.org/10.1016/j.clinph.2021.03.050.
31. Jack CR, Therneau TM, Weigand SD, Wiste BA, Kopman DS, Vemuri P. Prevalence of biologically vs clinically defined alzheimer spectrum entities using the national institute on aging–Alzheimer's association research framework. JAMA Neurol. 2019;76:1174–83. https://doi.org/10.1001/jamaneurol.2019.1971.

32. Jack CR Jr, Knopman DS, Jagust WJ, Shaw LM, Aisen PS, Weiner MW, Petersen RC, Trojanowski JQ. Hypothetical model of dynamic biomarkers of the Alzheimer's pathological cascade. Lancet Neurol. 2010;9(1):119–28.
33. Jellinger KA. Recent update on the heterogeneity of the Alzheimer's disease spectrum. J Neural Transm. 2022;129:1–24.
34. Jin N, Ziyatdinova S, Gurevicience I, Tanila H. Response of spike-wave discharges in aged APP/PS1 Alzheimer model mice to antiepileptic, metabolic and cholinergic drugs. Sci Rep. 2020;10:11851.
35. Jing J, Herlopian A, Karakis I, Ng M, Halford JJ, Lam A. Interrater reliability of experts in identifying interictal epileptiform discharges in electroencephalograms. JAMA Neurol. 2019;77:49–57. https://doi.org/10.1001/jamaneurol.2019.3531.
36. Kamondi A, Grigg-Damberger M, Löscher W, Tanila H, Horvath AA. Epilepsy and epileptiform activity in late-onset Alzheimer disease: clinical and pathophysiological advances, gaps and conundrums. Nat Rev Neurol. 2024;20:162–82.
37. Kane N, Acharya J, Benickzy S, Caboclo L, Finnigan S, Kaplan PW. A revised glossary of terms most commonly used by clinical electroencephalographers and updated proposal for the report format of the EEG findings. Revision Clin Neurophysiol Pract. 2017;2:170–85. https://doi.org/10.1016/j.cnp.2017.07.002.
38. Kivipelto M, Mangialasche F, Ngandu T. Lifestyle interventions to prevent cognitive impairment, dementia and Alzheimer disease. Nat Rev Neurol. 2018;14:653–66. https://doi.org/10.1038/s41582-018-0070-3.
39. Lam AD, Sarkis RA, Pellerin KR, Jing J, Dworetzky BA, Hoch DB. Association of epileptiform abnormalities and seizures in Alzheimer disease. Neurology. 2020;95:e2259–70. https://doi.org/10.1212/WNL.0000000000010612.
40. Lee DA, Lee HJ, Kim SE, Park KM. Brain networks and epilepsy development in patients with Alzheimer disease. Brain Behav. 2023 Aug;13(8):e3152.
41. Liedorp M, Stam CJ, van der Flier WM, Pijnenburg YAL, Scheltens P. Prevalence clinical significance of epileptiform EEG discharges in a large memory clinic cohort. Dement Geriatr Cogn Disord. 2010;29:432–7. https://doi.org/10.1159/000278620.
42. McBride AE, Shih TT, Hirsch LJ, Video EEG. Monitoring in the elderly: a review of 94 patients. Epilepsia. 2002;43:165–9. https://doi.org/10.1046/j.1528-1157.2002.24401.x.
43. Murray ME, Graff-Radford NR, Ross OA, Petersen RC, Duara R, Dickson DW. Neuropathologically defined subtypes of Alzheimer's disease with distinct clinical characteristics: a retrospective study. Lancet Neurol. 2011;10:785–96.
44. Noachtar S, Binnie C, Ebersole J, Mauguière F, Sakamoto A, Westmoreland B. A glossary of terms most commonly used by clinical electroencephalographers and proposal for the report form for the EEG findings. The international federation of clinical neurophysiology electroencephalogr. Clin Neurophysiol Suppl. 1999;52:21–41.
45. Noachtar S, Rémi J. The role of EEG in epilepsy: a critical review. Epilepsy Behav. 2009;15:22–33. https://doi.org/10.1016/j.yebeh.2009.02.035.
46. Palop JJ, Chin J, Roberson ED, Wang J, Thwin MT, Bien-Ly N. Aberrant excitatory neuronal activity and compensatory remodeling of inhibitory hippocampal circuits in mouse models of Alzheimer's disease. Neuron. 2007;55:697–711. https://doi.org/10.1016/j.neuron.2007.07.025.
47. Picco A, Archetti S, Ferrara M, Arnaldi D, Piccini A, Serrati C, et al. Seizures can precede cognitive symptoms in late-onset Alzheimer's disease. J Alzheimers Dis. 2011;27:737–42. https://doi.org/10.3233/JAD-2011-110896.
48. Press DZ, Musaeus CS, Zhao L, Breton JM, Shafi MM, Dai W, Alsop DC. Levetiracetam increases hippocampal blood flow in Alzheimer's disease as measured by arterial spin labelling MRI. J Alzheimers Dis. 2023;93(3):939–48.

49. Uretsky M, et al. Longitudinal cognitive performance of Alzheimer's disease neuropathological subtypes. Alzheimers Dement. 2021;7:e12201.
50. Vos SJ, Xiong C, Visser PJ, Jasielec MS, Hassenstab J, Grant EA, Cairns NJ, Morris JC, Holtzman DM, Fagan AM. Preclinical Alzheimer's disease and its outcome: a longitudinal cohort study. Lancet Neurol. 2013;12:957–65.
51. Vossel K, Ranasinghe KG, Beagle AJ, La A, Ah Pook K, Castro M, Mizuiri D, Honma SM, Venkateswaran N, Koestler M, Zhang W, Mucke L, Howell MJ, Possin KL, Kramer JH, Boxer AL, Miller BL, Nagarajan SS, Kirsch HE. Effect of levetiracetam on cognition in patients with Alzheimer disease with and without epileptiform activity: a randomized clinical trial. JAMA Neurol. 2021;78(11):1345–54. https://doi.org/10.1001/jamaneurol.2021.3310.
52. Vossel KA, Beagle AJ, Rabinovici GD, Shu H, Lee SE, Naasan G, et al. Seizures and epileptiform activity in the early stages of Alzheimer disease. JAMA Neurol. 2013;70:1158–66. https://doi.org/10.1001/jamaneurol.2013.136.
53. Vossel KA, Ranasinghe KG, Beagle AJ, Mizuiri D, Honma SM, Dowling AF. Incidence impact of subclinical epileptiform activity in Alzheimer's disease. Ann Neurol. 2016;80:858–70. https://doi.org/10.1002/ana.24794.
54. Vossel KA, Tartaglia MC, Nygaard HB, Zeman AZ, Miller BL. Epileptic activity in Alzheimer's disease: causes and clinical relevance. Lancet Neurol. 2017;16:311–22. https://doi.org/10.1016/S1474-4422(17)30044-3.
55. Zhang X, Ahmed R, Thayer Z, Breen N, McMillan J, Fulham M, Nikpour A. Late-onset epilepsy with cognitive symptoms: comparison of cognitive and imaging profiles with probable Alzheimer's disease. Epilepsy Behav. 2023;146:109371.

Application of Proton MRS in the Alzheimer's Disease Continuum: Recent Advances and Future Directions

Firat Kara and Kejal Kantarci

Abstract Proton (^{1}H) magnetic resonance spectroscopy (MRS) has shown great promise to detect and longitudinally monitor early pathological changes across the Alzheimer's disease (AD) spectrum, highlighting its potential in early diagnosis through noninvasive assessment of neurochemical alterations. The application of ^{1}H MRS in evaluating several metabolites offers insights into the biochemical changes in the preclinical, prodromal, and dementia stages of AD, underscoring its value in understanding disease pathophysiology and aiding in the development of preventative therapies.

Keywords Alzheimer's disease · Mild cognitive impairment · Magnetic resonance spectroscopy · MRS

1 Introduction

The number of people with dementia is predicted to increase to 153 million by 2050 [26]. Alzheimer's disease (AD) is the most common cause of dementia, which accounts for 60–80 of all dementia cases. AD dementia continues to have significant long-term physical, psychological, and socioeconomic impacts on patients, primary caregivers, and society [1]. The clinical symptoms include progressive loss of cognitive function, memory loss, and behavioral changes. The core pathological hallmarks of AD are the accumulation of amyloid-beta (Aβ) plaques outside the neurons and the accumulation of neurofibrillary tangles composed of hyperphosphorylated tau inside the neurons, which leads to the degeneration of nerve cells in the brain. These pathologies can be detected through postmortem examination or in vivo using biomarkers. Most AD cases are sporadic or late-onset AD, accounting for over 95% of cases (usually after 65 years). Currently, there is no cure for AD, but several prescribed drugs are approved by the U.S. Food and Drug Administration to treat

F. Kara (✉) · K. Kantarci
Department of Radiology, Division of Neuroradiology, Mayo Clinic, Rochester, MN, USA
e-mail: Kara.Firat@mayo.edu; kantarci.kejal@mayo.edu

P. A. Broderick (ed.), *Biomedical Imaging in Neurodegeneration*,
https://doi.org/10.1007/978-3-031-95878-6_9

symptoms, such as cholinesterase inhibitors, and to modify the disease progression by targeting Aβ pathology using monoclonal antibodies.

Since the publication of the clinical diagnostic criteria for AD in 1984, the diagnostic and research criteria for AD have evolved considerably. In 2018, the National Institute on Aging and Alzheimer's Association (NIA-AA) committee established the first biological definition of AD as a research framework that incorporates neuroimaging and biofluid biomarkers acquired from living people during the AD continuum [52]. This updated research framework of NIA-AA proposed to categorize individuals based on biomarker evidence of AD pathology using amyloid-tau-neurodegeneration (ATN) [52]. In ATN classification, "A" refers to Aβ, "T" to tau, and "N" to neurodegeneration. In the new research framework, AD is now considered a biologically defined continuum that begins with the appearance of brain pathology in asymptomatic individuals and progresses through stages of increasing pathological burden, eventually leading to clinical symptoms [30, 52]. With recent advances in in vivo biomarker fields (i.e., neuroimaging and biofluid markers), the preclinical detection and monitoring progression of AD pathology is possible.

Recent advancements in disease-modifying therapies have underscored the importance of the early and accurate diagnosis of the underlying pathology of AD using surrogate biomarkers. Early diagnosis of AD before irreversible brain damage manifests is crucial for timely intervention, ultimately aiming to enhance outcomes. Furthermore, surrogate biomarkers allow the longitudinal follow-up of disease progression from the earliest stages, which may increase our understanding of the pathophysiology of the disease and accelerate the development of preventative therapies. Noninvasive nuclear magnetic resonance (MR) methods, such as proton (^{1}H) magnetic resonance spectroscopy (MRS), have shown promise in detecting early pathological alterations in the biochemical profile of the brain.

^{1}H MRS is a diagnostic imaging technique that provides information on neurochemical alterations linked to molecular-level pathological processes. As a noninvasive MR method, ^{1}H MRS has been used to monitor the AD continuum, including the preclinical (i.e., cognitively unimpaired), prodromal, and AD dementia stages. ^{1}H MRS can measure the level of proton-containing molecules, such as neurotransmitters and metabolites, using the chemical shift information of molecules. The ^{1}H MRS metabolites related to the AD continuum, which can be reliably quantified up to 3 Tesla (T) magnetic field strengths, include N-acetyl aspartate (NAA), myo-inositol (mIns), choline-containing metabolites (total choline [tCho] = phosphocholine [PCho] + glycerophosphocholine [GPC]), and glutamine + glutamate (Glx). In addition, alterations in glutamate (Glu), glutamine (Gln), glucose (Glc), gamma-aminobutyric acid (GABA), and glutathione (GSH) have also been reported in preclinical, prodromal, and dementia stages at 3 T and higher magnetic field strengths.

There are numerous reports of metabolite changes measured by quantitative ^{1}H MRS in AD continuum (reviewed recently by [37, 41, 44, 65, 88, 115, 118, 140, 145, 156]. The focus of this chapter was to discuss the role of in vivo ^{1}H MRS in detection of pathological changes across the AD continuum in the brain.

2 Cross-Sectional Studies of Quantitative ^{1}H MRS in AD Dementia

NAA, one of the most abundant metabolites in the brain, is synthesized by aspartate and acetyl-CoA in neural mitochondria and is found primarily in neural cells but not in astrocytes [77, 97, 139]. The metabolite resonances at 2.02 ppm is predominantly attributable to NAA and N-acetylaspartylglutamate (NAAG) which are generally ascribed to NAA or total NAA [77, 97, 107, 135, 139]. NAA is considered a marker of neural health, viability, and synaptic integrity [99]. For example, a lower NAA/creatine (Cr) ratio has been associated with loss of synaptic integrity (i.e., SV2 synaptic vesicle protein) [99]. Reduction of NAA levels in AD compared to healthy controls is one of the most frequent findings of ^{1}H MRS. Decreased NAA or NAA/Cr ratio has been observed in AD in a variety of brain regions including the hippocampus, parietal cortex, occipital, temporal, posterior cingulate, and paratrigonal white matter cingulate [4, 17, 25, 28, 39, 47, 50, 57, 61, 65, 93, 95, 110, 128, 129]. Alterations in the NAA/Cr ratio (or NAA) across various brain regions exhibit patterns analogous to Braak staging of the neurofibrillary tangle pathology of AD [15, 16, 61, 136]. In addition, a decrease in NAA/Cr correlates with amyloid plaques, neurofibrillary tangles, and apolipoprotein E4 (APOE e4) carrier status, a well-known genetic risk factor of AD [22, 63, 66]. A recent study using ATN scheme demonstrated that both early AD ($A + T + N-$) and later AD ($A + T + N+$) had lower NAA/Cr compared to controls [22]. Remarkably at early AD stage, a decrease NAA/Cr ratio, which correlated with increased deposition of amyloid and tau on PET, allowed the discrimination of early AD from cognitively unimpaired controls supporting the use of ^{1}H MRS to detect the earliest stages of AD pathophysiology.

Regarded as a glia marker and osmolyte, an increase in mIns (or mIns/Cr ratio) in several anatomical locations has been consistently been found in AD [11, 12, 88, 118]. The spectral peak of mIns is generally ascribed to 3.6 and 3.9 ppm. The majority of the mIns signal originates from free mIns and mIns phosphate [41]. Elevated mIns levels have been linked with glial proliferation in a demyelinating disease, namely multiple sclerosis [12] and associated with the glial marker 18 kDa TSPO [120]. Elevated mIns and mIns/Cr levels have been reported in the hippocampus, occipital, parietal, temporal, posterior cingulate, and paratrigonal white matter in AD [17, 50, 61, 65, 93, 95, 110, 128, 129]. It has been demonstrated that elevation in mIns/Cr ratio occurs at early stage of the disease progression followed by a decrease in NAA/Cr and an increase in Cho/Cr at later stages of the disease [61]. A few studies reported that a decline NAA/mIns ratio is an accurate diagnostic indicator for differentiating AD patients from healthy controls [67, 110, 122]. Furthermore, NAA/mIns is considered as an effective predictor of memory and cognitive ability in patients with mild cognitive impairment (MCI) and AD [68].

Membrane integrity markers such as Cho have also been shown to be altered in AD continuum. Cho is a marker of white matter integrity, cell membrane (phospholipid) turnover, and cellular density [75, 92, 118]. The ^{1}H MRS visible Cho peak, which is generally ascribed to Cho or total Cho at 3.2 ppm, encompasses signals from compounds including primarily phosphocholine (PCho), and glycerophosphocholine (GPC), found in myelin and the cell membrane, with a possible small contribution from the signal from free choline [7, 75, 92, 107]. Some studies have reported elevated levels of Cho or Cho/Cr [34, 61, 65, 78, 79, 90], while others have found no significant change or a decrease in Cho or Cho/Cr in AD when compared with controls [20, 25, 93, 110, 129, 134]. The elevation in Cho in AD might be due the increased demand for acetylcholine synthesis, triggered by the degradation of acetylcholinergic neurons. This leads to increased membrane catabolism, releasing free choline (PCho and GPC) for use as a precursor in the synthesis of the acetylcholine [150]. Treatment of AD with the muscarinic cholinergic M1 agonist xanomeline was associated with a reduction in the brain Cho/Cr ratio [124]. These findings support the notion that compensating for the cellular demand for free choline for acetylcholine synthesis may diminish membrane breakdown, thereby preventing an increase in choline MRS signal in AD [44, 124].

Alterations in Glu + Gln (Glx) have also been reported in various brain regions in AD, but the results are not in agreement with each other. Glu, a predominant excitatory neurotransmitter, is mainly synthesized from Gln within the Glu-Gln cycle [48]. Glu also acts as a metabolic precursor of GABA, the main inhibitory neurotransmitter in the brain. After Gln is released from astrocytes, it is taken by the neuron and converted to Glu [6]. The spectral signals of Glu (at 2.35 ppm) and Gln (at 2.45 ppm) obtained through conventional ^{1}H MRS overlap at relatively low MR field strengths [11]. Therefore, the composite peak was assigned as "Glx (Glu + Gln)". Due to low concentration and poor resolution of Glu and Gln at low field strengths, many studies have reported only the level of Glx (or Glx/Cr) instead of reporting individual levels of Glu or Gln. Some studies have identified a decrease in Glx (or Glx/Cr) in the mid-cingulate cortex [4], occipital cortex [95], posterior cingulate gyrus [34, 36, 47, 128], anterior cingulate cortex [49], and hippocampus [49, 128] in patients with AD compared to healthy controls, but other studies have reported no significant changes in the midfrontal or temporoparietal gray matter or white matter (parietooccipital) [33, 47]. Only few studies reported Glu level separately from pool of Glu + Glu in AD. Separate level of Glu and Gln can be identified reliably by using advance protocols, such as sLASER, in various brain regions at 3 T, while higher field strengths, such as 7 T, allow their reliable spectral separation using conventional protocols [29, 108]. A decrease in Glu (or Glu/Cr) has been observed in regions such as the hippocampus, posterior cingulate gyrus, and dorsolateral prefrontal cortex of AD compared with healthy controls [76, 83, 123, 149]. Conversely, this decrease was not observed in the putamen, indicating region-specific AD-related neurochemical alterations [147]. In a meta-analysis, reduced Glu levels in the posterior cingulate cortex were found in AD compared to healthy controls [76]. Consistent with findings from human studies, a decrease in Glu in the brains of transgenic mouse models of AD mice compared to non-AD mice was also observed [46, 82]. The Glu/Cr ratio has been linked to synaptic vesicle glycoprotein 2A in both the hippocampus and anterior cingulate gyrus in healthy controls on PET with [11C] UCB-J [105]. Consequently, a reduction in Glu/Cr (or Glx/Cr) in AD could indicate synaptic injury resulting from AD pathology. Treatment studies have not been conclusive regarding changes in the level of Glu (or the Glu/Cr ratio). While a galantamine cholinesterase inhibitor study demonstrated an increase in Glu level and Glu/Cr ratio (reference to baseline) in the hippocampus after 4 months of treatment [111], another study using donepezil treatment demonstrated no alteration in the Glu level in the hippocampus of patients [9]. These discrepancies in treatment studies may have occurred because of differences in the cohorts studied, time between baseline and follow-up, or metalogical differences.

The Cr and phosphocreatine (PCr) system play a role in energy metabolism for adenosine triphosphate regeneration and act as an energy buffer [151]. However, ^{1}H MRS measures the total creatine pool since PCr and Cr resonant frequencies overlap at 3 T and lower fields (<3 T). Therefore, their sum as total Cr has been frequently reported [117]. This total sum is generally ascribed to Cr or total Cr, with a primary peak at 3.03 ppm and a second peak at 3.9 ppm [156]. Total Cr is generally used as an internal reference to calculate relative metabolite levels. This approach is based on the principle that Cr and PCr are in chemical equilibrium, leading to the assumption that the total Cr remains constant across various diseases [117]. In various AD studies, no difference in Cr concentration was observed between the brains of patients with AD and controls [95, 110, 129]. However, some studies have reported that Cr levels change with age [18, 85, 125] and may be affected by white matter hyperintensity load [106]. Furthermore, some studies have reported variable Cr concentrations in patients with AD and/or MCI compared to healthy controls [50, 148]. Therefore, it is essential that Cr (i.e., total Cr) levels should be independently assessed for each study before using it as an internal reference [104].

The main essential energy source of the brain is Glc, which resonates in 3.43–3.8 ppm range and 5.2 ppm [42, 118]. Although ^{1}H MRS can detect Glc levels, it has relatively low sensitivity and specificity at 1.5 T. Moreover, Glc resonances overlap with other metabolite resonances, such as taurine and macromolecules, which creates additional challenges in detecting the spectral signal [45]. Furthermore, additional analysis steps are necessary to accurately estimate Glc levels in gray or white matter when the voxel of interest includes cerebrospinal fluid (CSF) [59]. Owing to these challenges, Glc levels have not been widely studied using conventional ^{1}H MRS techniques in AD. Few studies have reported an increase in Glc levels in AD [95, 98]. These results are in line with the postmortem results showing an increase in Glc levels in AD brain specimens [152]. The increase in Glc levels in AD might be related to impaired cellular uptake and metabolism of Glc [98].

Oxidative stress is a significant factor in AD progression. GSH plays a crucial role as an antioxidant in the brain [119]. Brain GSH has a low ^{1}H MRS detectable concentration (1–3 mmol/L) [119, 137]. GSH signals, which have low signal-to-noise ratio, overlap with resonance from other metabolite peak intensities that have higher concentrations [119]. The reliable quantification of detection of GSH requires advanced editing methods. Only few studies have explored the changes in GSH (GSH/Cr) levels in AD using ^{1}H MRS [32, 80, 81]. Depletion of GSH levels in a variety of brain regions, including the hippocampus, frontal cortex, posterior cingulate cortex, and anterior cingulate cortex, has been demonstrated in AD as compared with age-matched cognitively unaffected healthy older participants [80, 81, 131]. In animal studies, a decreased level of GSH in the brain of a transgenic mouse model of AD (B6.Cg-Tg) has also been reported [155]. In healthy individuals without cognitive impairment, a study reported a negative correlation between GSH levels and brain amyloid load in the temporal and parietal regions. [23]. However, a recent meta-analysis in human pathological specimens reported no change in GSH peroxidase and GSH reductase activities and GSH levels [153]. Further research is warranted to understand the interplay between GSH levels, Aβ, and tau pathology [153].

Little research has been conducted using ^{1}H MRS to study GABA levels in the AD continuum, even though GABAergic dysfunction has been reported in the AD continuum [5]. GABA is a primary inhibitory neurotransmitter and neuromodulator of the brain. A lower GABA/Cr ratio was observed in the parietal region of AD patients than in healthy controls [5]. However, no correlation was found between GABA/Cr ratio and memory scores in patients with AD. Unchanged levels of GABA/Cr ratio in the hippocampus and the anterior cerebral cortex among AD, MCI, and normal control groups have also been reported [49]. Further studies focusing on GABA in the AD continuum are warranted.

3 Cross-Sectional ^{1}H MRS Studies in Preclinical and Prodromal Stages of AD

The diagnosis of AD used to be based almost exclusively on information obtained from medical history and neurological examination. [86, 87]. In recent years, there has been a paradigm shift in the understanding of AD, now viewed as a spectrum ranging from preclinical AD to prodromal AD, and to AD dementia, reflecting an evolving perspective on the AD continuum [30, 52]. Within the continuum of AD, accumulation of early neuropathological features such as amyloid and tau pathologies, structural and functional changes, and loss of synaptic function precede the clinical symptoms of AD [74, 142]. Memory impairment is one of the earliest symptoms of AD [132]. The prodromal stage involves amnestic mild cognitive impairment (MCI), which is considered an intermediate stage in the continuum from physiological aging to dementia. In recent years, there has been considerable effort to identify early indicators of pathological changes in this condition. An individual with amnestic MCI experiences a noticeable decline in memory function, yet maintains a relatively unimpaired performance in other non-amnestic cognitive domains, such as language, attention, executive, or visual-spatial processing domains [112–114]. More recently, the concept of amnestic MCI has been expanded to include nonamnestic cognitive domains [112]. In amnestic MCI, where memory impairment is the primary symptom, and nonamnestic MCI, which involves nonmemory cognitive domain impairment, individuals may experience deficits in either a single or multiple cognitive domains simultaneously [112]. Not all individuals with MCI develop AD. Numerous independent studies have consistently shown that individuals with amnestic MCI have higher risk to develop AD later in life [10, 14, 38, 58]. With the development of AD pathological markers, research criteria for preclinical AD have been updated to define cognitively unimpaired individuals with early pathological changes in AD [132]. Biomarkers are currently essential for the diagnosis of preclinical and prodromal stages [52] or individuals at risk for progression to AD (according to IWN criteria) [31]. These groups are of particular interest for preventive interventions aimed at altering disease progression before irreversible brain damage occurs.

Cross-sectional studies reported that amnestic MCI patients tend to have higher mIns/Cr ratios in the posterior cingulate gyrus [61], white matter paratrigonal white matter [17], lower Cho levels in the medial temporal lobe [19], and lower NAA concentrations in the medial temporal lobe [21] compared to healthy controls. Additionally, a decrease in the NAA/Cr ratio (hippocampus, posterior cingulate gyrus) [2, 67], a decrease in NAA concentration (medial temporal lobe, hippocampus, posterior cingulate gyrus, white matter [i.e., parietal, periventricular, and deep white matter]) [19, 116, 121, 148, 149], a decrease in the NAA/mIns ratio, and an increase in the Cho/Cr ratio (posterior cingulate gyrus) [67] were reported in MCI patients compared to healthy controls. In an antemortem MRS study, the earliest stage of AD pathology was associated with an elevation in mIns/Cr, while the decrease in NAA/Cr was associated with later stages of AD pathology, suggesting that alterations in mIns and NAA levels can detect different stages of pathology of the AD continuum [63, 66]. A recent study showed that elevation in mIns levels precedes amyloid deposition in healthy (amyloid-negative, cognitively unimpaired) APOE e4 carriers with normal CSF Aβ42 levels [144]. While these results support the findings in AD, the changes in metabolite concentrations (or ratios), such as a decrease in NAA and an increase in Cho and mIns, are milder in MCI than in AD [69]. Reports on Glx, glutamate, GSH, and GABA are in rare due to their low concentration or requirement of advanced editing and acquisition protocols (e.g., MEGAPRESSS, sLASER) or higher magnetic field systems. Kantarci et al. showed that Glx could be reliably quantified at 3 T compared to 1.5 T due to increased signal to noise ratio and spectral resolution (Fig. 1). The study showed that there was a trend of a decrease in Glx/Cr ratios (posterior cingulate gyrus) from cognitively normal to MCI to AD (i.e., consisted of individuals with only mild to moderate AD) [67]. Supporting these findings, a decrease NAA/Cr, Glx/Cr, NAA/mIns ratios and an increase in mIns/Cr ratio were reported in patients with amnestic MCI compared with healthy controls [43]. Zeydan et al. reported a decrease in Glu/mIns ratio in the posterior cingulate gyrus in amnestic MCI than in health controls, utilizing an advanced modified MRS sLASER protocol at 3 T, which allows reliable quantification of Glu [154]. Oetzchner et al. reported significantly decreased GABA/Cr (anterior cingulate gyrus, posterior cingulate gyrus), Glu/Cr

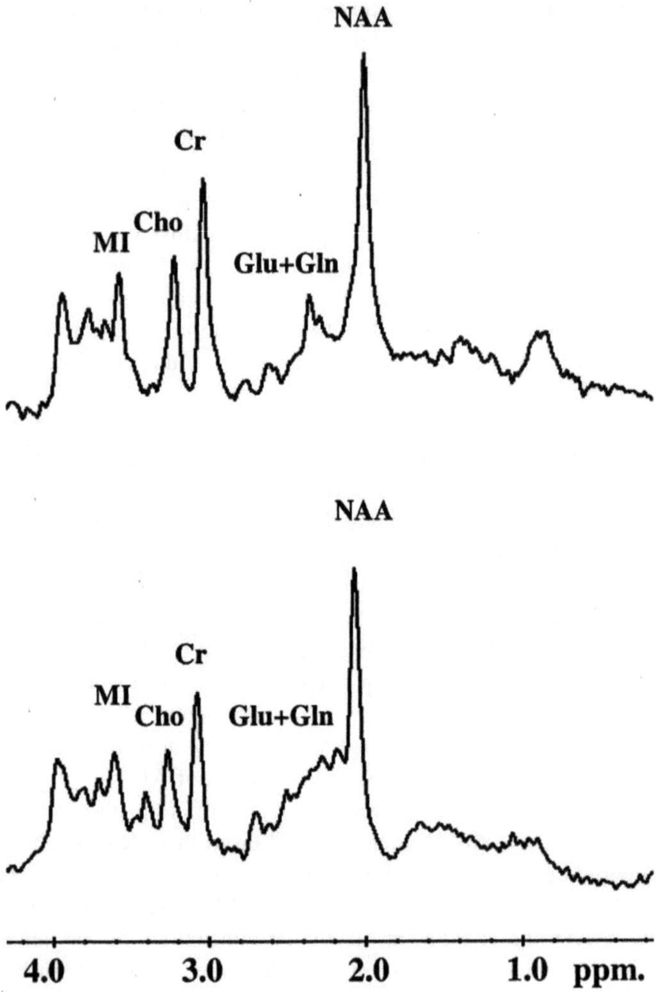

Fig. 1 Examples of ^{1}H MR spectra (2000/30) obtained at 3 T (top) and at 1.5 T (bottom) from the posterior cingulate voxel. Glu + Gln peak resonances are better resolved at 3 T than at 1.5 T. The resonance at 3.4 ppm that is more prominent in the 1.5- than in the 3-T spectra is *scyllo* inositol and/or taurine [8]. This figure is reprinted from "Proton MR spectroscopy in mild cognitive impairment and Alzheimer disease: comparison of 1.5 and 3 T" [67], with permission from AJNR Am J Neuroradiol

(posterior cingulate gyrus), and NAA/Cr (posterior cingulate gyrus) and increased mIns/Cr (anterior cingulate gyrus) in MCI patients compared to healthy controls using 7 T MRS [103]. In accordance with the findings of this study, Won et al. reported a decrease in Glu and NAA in patients with MCI or AD compared to healthy controls at 7 T [149]. Furthermore, a decrease in GABA and Glx in the hippocampus and posterior cingulate gyrus [121] was observed in patients with amnestic MCI compared to cognitively normal elderly controls at 3 T. The study also reported that GABA and Glx levels were similar between APOE e4 carriers and noncarriers, whereas NAA was lower in APOE e4 carriers suggesting APOE e4 genotype has stronger association with NAA [121]. In patients with amnestic MCI, decreased GABA/Cr and GABA levels in the anterior cingulate gyrus and posterior cingulate gyrus were observed when compared to healthy controls [103, 121]. It has been suggested that a reduction in GABA in the posterior cingulate gyrus may serve as a biomarker for cognitive impairment in amnestic MCI, as its level positively correlated with the word learning score [121].

There are inconsistencies in how GSH levels change in MCI, with some studies reporting increases, whereas others reporting decreases or no change. A decreased GSH concentration in the hippocampi of MCI patients compared to healthy old controls has been reported [130] suggesting that GSH levels in specific brain regions could potentially serve as clinically relevant biomarkers for diagnosing MCI and AD [80]. However, another study reported higher GSH levels in both the anterior and posterior cingulate regions of MCI patients than in healthy controls, and this increase was linked to impaired memory consolidation [32]. This finding may be indicative of an early neuroprotective response mechanism that compensates for the increased oxidative stress in MCI [32]. A recent study reported no differences in GSH/Cr levels between patients with MCI and controls [103]. However, there was a positive correlation between PCC GSH/tCr levels and verbal fluency [103] which was in line with results of Duffy [32] suggesting a role of GSH in memory performance. The variation in the results could stem from challenges in measuring low GSH concentrations, differences in methodologies, and variations in study group characteristics, among other factors.

Overall, the most consistently observed results include a decrease in NAA/Cr, NAA, NAA/mIns, and an increase in mIns/Cr in patients with MCI compared to controls [61, 71, 138]. In addition, during the progression from MCI to AD, NAA (and NAA/Cr) consistently decreased in the posterior cingulate and hippocampus, while mIns increased in the posterior cingulate gyrus [76]. Furthermore, a variety of studies demonstrated a decrease in Glx (or Glx/Cr) or Glu (or Glu/Cr) associated with increasing AD pathology at 3 T and higher field strength. These results show the utility of high- (3 T) and ultrahigh-field MRS in detecting MCI-related metabolites with low concentration and/or overlapping spectra.

The potential of ^{1}H MRS to distinguish metabolic alterations between amnestic and nonamnestic MCI has also been demonstrated. For example, while patients with amnestic MCI are characterized by small hippocampal volumes and elevated ratios of mIns/Cr levels, patients with nonamnestic MCI on average have normal hippocampal volumes and ^{1}H MRS metabolite ratios [63, 66]. Interestingly, a few studies have reported no change in NAA/Cr, Cho/Cr, and mIns/Cr ratios of MCI patients in the hippocampus and posterior cingulate gyrus [146] compared to controls. Inconsistent results in MRS studies can be attributed to heterogeneity in MCI types (amnestic and nonamnestic; the former has a higher risk of conversion into AD), magnetic field strength, characteristics of groups, sample size, type of MRS sequences, brain regions, age of participants, preprocessing steps of MRS data, and quality control parameters (e.g., exclusion criteria of metabolites) [60, 67, 108]. For example, there were significant differences in the NAA/Cr ratios between patients with amnestic MCI who converted to AD and those who did not [89]. In summary, these studies suggest that ^{1}H MRS has the potential to detect metabolic alterations not only in AD dementia but also in preclinical and prodromal stages of AD continuum.

Quantitative ^{1}H MRS has demonstrated a potential to predict conversion of cognitively normal individuals to MCI [70] or individuals with MCI to AD dementia [94, 96]. In longitudinal cohort studies, it has been demonstrated that NAA/Cr ratio of the entorhinal cortex, paratrigonal area, posterior cingulate gyrus, and occipital cortex of amnestic MCI patients could predict future conversion to AD [35, 71, 89, 96, 127]. In a population-based study, decreased NAA/mIns ratio in the posterior cingulate gyrus was a strong predictor of MCI in cognitively normal older adults [70]. Individuals with low baseline NAA/mIns in the posterior cingulate gyrus display a steeper cognitive decline trajectory, suggesting that baseline NAA/mIns has the potential to predict cognitive decline trajectories [143]. In line with these findings, in 2-year follow-up study, the NAA/mI ratio in the posterior cingulate gyrus was able to differentiate MCI converters from those MCI that did not develop AD [94].

4 Longitudinal Serial ^{1}H MRS Measurements Correlating with Clinical Disease Progression Across the Entire Severity Spectrum of AD Dementia

While there are numerous cross-sectional ^{1}H MRS studies, there are a limited number of longitudinal MRS studies that investigated metabolic changes in the AD continuum. Adalsteinsson et al. [3] reported a decrease in gray matter NAA over the course of a year in patients with AD [3]. In this study, the percentage change per year in dementia rating scores correlated with a decline in the gray matter NAA. These results contrary to that of Jessen et al. who reported that there were no metabolite ratio difference differences (NAA/Cr and Cho/Cr ratios) in the central region (of the primary motor and sensory cortex) and medial temporal lobe between the first and second measurements, which were conducted 23 months apart [56]. However, changes in NAA between two measurement time points were correlated with cognitive decline. In line with the results of Adalsteinsson et al., Dixon et al. reported a trend toward a decline in the left hippocampal NAA over a year in patients with AD, although it was not statistically significant [28]. However, there was a significant decline in cognitive ability in AD. In line with the results of Jessen et al. [56], Metastasio et al. [89] reported stable NAA, Cho, mIns, and Cr levels in the left and right paratrigonal areas over 1-year interval in stable MCI (those who did not convert to AD), and progressive MCI (those who later developed AD) [89]. However, the baseline NAA/Cr ratios of patients with progressive MCI were lower than those of patients with stable MCI and the controls. Kantarci et al. [69] reported a greater decline in annualized NAA/Cr ratios in the posterior cingulate gyrus of patients with MCI and AD than in controls [69]. The study reported that the Cho/Cr ratio increases in amnestic MCI in those who convert to AD but decreases in the stable MCI group (cognition remains stable). Schott et al. reported a significant decline in both NAA/Cr and NAA/mIns over time in the posterior cingulate gyrus of patients with AD, but only the decrease in NAA/mIns remained statistically significant when compared to controls [126]. In Aβ-positive individuals with MCI, a steepest change in the mIns/Cr and NAA/mIns ratios was observed compared to Aβ-negative individuals with MCI [143]. This study suggests that longitudinal changes in mIns/Cr and NAA/mIns are associated with the underlying Aβ pathology. Supporting these findings, a decline in total NAA/total Cho and total Cr/total Cho in MCI and AD subjects was associated with greater disease severity [24].

Overall longitudinal studies suggest that ^{1}H MRS can be a sensitive noninvasive biomarker for tracking pathological alterations over time across the AD continuum, providing valuable insights into the biochemical underpinnings of AD-related neurodegenerative processes.

5 Correlation of ^{1}H MRS with CSF, Blood, and PET AD Biomarkers and Postmortem Neuropathology

Only a few studies have investigated how the ^{1}H MRS metabolites correlate with premortem and postmortem AD associated pathologies using CSF, blood, amyloid, and tau PET biomarkers and postmortem neuropathology.

5.1 Blood and CSF Biomarkers

The development of blood-based plasma biomarkers is a significant recent advancement in the detection of biological evidence of AD. These biomarkers, which are noninvasive, easily accessible, and cost-effective, play a critical role in detecting AD across the clinical continuum. Key AD-related plasma biomarkers include $A\beta_{42}/A\beta_{40}$, phosphorylated tau (p-tau) proteoforms, neurofilament light chain (NfL), and glial fibrillary acidic protein (GFAP). The $A\beta_{42}/A\beta_{40}$ ratio in plasma has been shown to correlate with CSF AD biomarkers, and amyloid PET [53, 55, 100, 101]. Similarly, p-tau proteoforms correlate with CSF, PET, and postmortem AD neuropathological markers [53–55, 73, 91]. NfL is a marker of neuroaxonal damage, and elevated NfL is associated with higher risk of dementia [40, 84]. Plasma GFAP, a marker of astrocytic activation, is associated with the elevated risk and severity of dementia [141]. All biofluid AD markers in plasma can also be measured in CSF.

However, unlike the collection of plasma biomarkers, obtaining CSF samples is invasive and costly, and it requires specialized training.

Lower hippocampal NAA/Cr in patients with AD was associated with lower CSF Aβ42 levels (which indicates higher Aβ42 deposition in the brain), whereas lower parietal NAA/Cr was associated with higher CSF tau [13]. In two studies, it was reported that elevated posterior cingulate (includes precuneus) mIns/Cr was positively correlated with CSF tau within the AD group [13, 98]. Recently, Spotorno et al. reported an association between elevated mIns and elevated plasma levels of plasma GFAP, but only in APOE ε4 carriers [133]. Furthermore, elevated NfL was negatively correlated with Glu/Cr and NAA/Cr in dorsolateral prefrontal cortex in AD patients [83].

5.2 PET Biomarkers

In cognitively unimpaired older adults, higher mIns/Cr and Cho/Cr ratios correlated with elevated Aβ deposition on PET imaging (11 C Pittsburgh compound B, PiB) [64]. Rieze et al. reported that NAA, GABA, and Glx (posterior cingulate gyrus) were not associated with global and local Aβ deposition on PET with PIB status, and GABA and Glx levels were unrelated to APOE e4 genotype [121]. However, across all participants, GABA and NAA were positively correlated with a word learning score. Kantarci et al. [61] have also reported that mIns but not NAA was correlated with posterior cingulate gyrus Aβ deposition on PET with PIB [64]. Nedelska et al. reported that lower NAA/mIns and higher mIns/Cr at baseline were associated with an increased rate of Aβ deposition on PIB PET [102]. Furthermore, the Glu/mIns ratio was associated with a higher global cortical Aβ deposition on PET with PIB [154]. Only few studies have investigated the association between ^{1}H MRS and both tau and amyloid loads on PET in cognitively unimpaired individuals. An increase in posterior cingulate gyrus tau load on PET with ^{18}F-flortaucipir, but not an elevated Aβ deposition on PET with PIB, was associated with lower NAA/total Cr and Glu/tCr ratios in cognitively unimpaired older adults [72] (Fig. 2). Consistent with these findings, reduced NAA/Cr ratios were associated with an increased overall Aβ and tau burden on PET [22], and a decrease in Glu/tCr in posterior cingulate was associated with tau load on PET with florzolatau in AD [83]. Furthermore, a recent study showed that elevated mIns in the posterior cingulate gyrus was associated with an increase in amyloid and tau loads on PET, but only in APOE e4 carriers [133].

5.3 Postmortem Neuropathology

Local histopathological findings underlying AD-related ^{1}H MRS metabolite alterations have also been studied. NAA/tCr correlates with the postmortem Braak neurofibrillary tangle stage [63, 66] (Fig. 3). In addition, ^{1}H MRS detected early

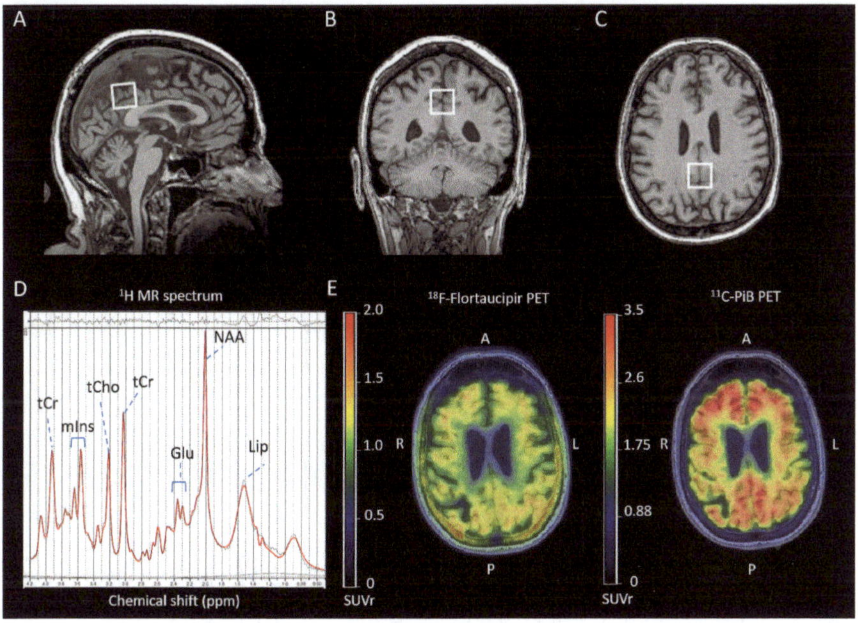

Fig. 2 Sagittal (**a**), coronal (**b**), and transverse (**c**) T1-weighted magnetic resonance images with superimposed posterior cingulate gyrus magnetic resonance spectroscopy volume of interest ($20 \times 20 \times 20$ mm^3). Representative ^{1}H MRS spectra, AV-1451-PET, and PiB-PET of a clinically normal individual. The thick red curve on the representative MRS spectra is the LCModel fit to the data. The thin curve under the spectra is the baseline. The residual (data minus the fit to the data) is shown on the top of the spectra. (**d**) Single voxel proton (^{1}H) magnetic resonance (MR) spectrum acquired from the posterior cingulate gyrus of a cognitively unimpaired participant (age = 81) at 3 T with sLASER sequence. The thick red curve on the representative MR spectrum is the LCModel fit to the data. The thin curve under the spectrum is the fitted baseline. The residual (data minus the fit to the data) is shown at the top of the spectrum. The chemical shift axis is labeled in parts per million (ppm) unit. The Y axis is an intensity scale of each spectral line with no unit. (**e**) The representative cortical flortaucipir PET and PiB PET scans were acquired from the same participant. The participant had low NAA/tCr = 1.20 and Glu/tCr = 0.99, high PCG flortaucipir standard value uptake unit ratio of 1.22, and high PCG PiB standard value uptake unit ratio of 3.06. The PET scans were registered to the T1-weighted MR image and displayed together. We observed flortaucipir uptake in the skull of this participant. The meningeal and bone uptake of flortaucipir is known manifestation of off-target binding. The cause is unknown. In contrast, such off-target uptake in PiB is not seen except with rare cases of bone uptake in diseases with high rates of bone remodeling (e.g., hyperostosis frontalis interna). The representative color scale shows the standardized uptake value ratios Abbreviations: Glu, glutamate; Lip, lipid signal; NAA, N-acetylaspartate; PET, positron emission tomography; PiB, Pittsburgh compound-B; tCr, phosphocreatine + creatine; tCho, phosphocholine + glycerophosphocholine; mIns, myo-inositol. (For interpretation of the references to color in this figure legend, the reader is referred to the Web version of this article.) This figure, originally appearing as Figs. 1 and 2, is reprinted from "1H MR spectroscopy biomarkers of neuronal and synaptic function are associated with tau deposition in cognitively unimpaired older adults", with permission from Elsevier [72].

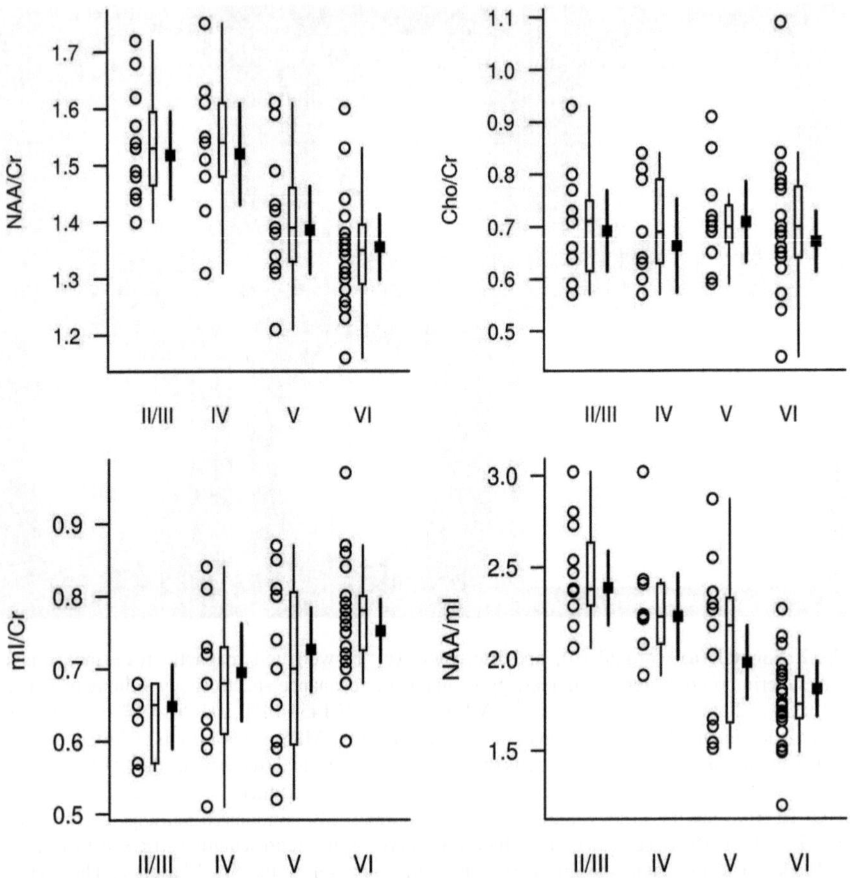

Fig. 3 ^{1}H MR spectroscopic metabolite ratios plotted according to Braak NFT stage (horizontal axis). For each Braak NFT stage diagnosis, individual values, a box plot of the distribution, and the estimated mean and 95% CI (darker lines) for the mean are shown. The mean and CI were derived from ANCOVA models and are assumed for a 78-year-old woman in whom the interval from ^{1}H MR spectroscopy to death is 2 years. (With permission from Radiology [63, 66])

biochemical changes (e.g., NAA/Cr and NAA/mIns) linked to AD neuropathology, including the amyloid burden, synaptic integrity, and tau pathology. The study reported an association between reduced synaptic integrity and a lower NAA/Cr ratio in the posterior cingulate gyrus in AD and control subjects. While a higher pTau burden was associated with lower NAA/Cr and NAA/mI, a higher amyloid burden was associated with elevated mI/Cr and lower NAA/mI ratios [99].

6 Future Directions

The recent presentation of a new draft for Alzheimer's disease (AD) diagnostic criteria at the Alzheimer's Association International Conference 2023 marks a significant step forward in updating the 2018 research framework document. This draft, based on the ATN biomarker classification scheme by Jack et al. [52], aims to incorporate plasma-based biomarkers alongside CSF assays and imaging into revised criteria for biomarker categorization, disease diagnosis, and staging [51, 52] (https://aaic.alz.org/). ^{1}H MRS biomarkers add a crucial neurochemical dimension, and potential enhancing the accuracy of early AD diagnosis. While the associations of some plasma biomarkers, such as GFAP and NfL, with some ^{1}H MRS markers have been studied, the associations between other plasma biomarkers, such as Aβ42/Aβ40, p-tau181, p-tau217, p-tau231, and MRS biomarkers, remain largely unexplored in cross-sectional, serial, and longitudinal studies. Combining ^{1}H MRS biomarkers with plasma biomarkers could provide valuable insights into AD progression, particularly considering the spatial specificity limitations of plasma biomarkers.

The availability of ultrahigh-field MR systems ($\geq$7 T) for clinical imaging presents exciting opportunities. Recently, 7 T MRI systems received U.S. FDA approval for clinical use. ^{1}H MRS at 7 T and higher field strengths offers improved spectral resolution and signal-to-noise ratio, enabling better separation of overlapping spectra and reliable quantification of low-concentration metabolites such as Glu, Glc, Gln, GSH, and GABA. However, the clinical utility of ultrahigh-field MRS in the AD continuum remains largely unexplored.

^{1}H MRS holds promise for tracking disease progression, particularly in clinical trials targeting early predementia pathology. Future research could prioritize investigating the potential of ^{1}H MRS in this regard. Additionally, studies could ensure representation from minority and underrepresented groups, while also considering the impact of biological sex as a variable [62].

Efforts are underway to standardize and optimize ^{1}H MRS methods for both single-center and multicenter studies [108]. Future studies should incorporate consensus recommendations from experts to facilitate multicenter studies and ensure reproducibility of results [27]. Furthermore, adopting recent advancements in ^{1}H MRS, such as the automated volume-of-interest technology that eliminates the requirement for manual voxel placement, in both cross-sectional and longitudinal studies, would enhance the clinical integration of MRS and enhance its use in clinical trials as an outcome measure [109]. Advancements in ^{1}H MRS offer valuable insights into metabolic alterations in the AD continuum. Integrating these advancements into AD research provides a comprehensive approach to understanding disease progression, predicting cognitive decline, and evaluating treatment strategies.

References

1. 2022 Alzheimer's disease facts and figures. Alzheimers Dement. 2022;18(4):700–89. https://doi.org/10.1002/alz.12638.
2. Ackl N, Ising M, Schreiber YA, Atiya M, Sonntag A, Auer DP. Hippocampal metabolic abnormalities in mild cognitive impairment and Alzheimer's disease. Neurosci Lett. 2005;384(1–2):23–8. https://doi.org/10.1016/j.neulet.2005.04.035.
3. Adalsteinsson E, Sullivan EV, Kleinhans N, Spielman DM, Pfefferbaum A. Longitudinal decline of the neuronal marker N-acetyl aspartate in Alzheimer's disease. Lancet. 2000;355(9216):1696–7. https://doi.org/10.1016/s0140-6736(00)02246-7.
4. Antuono PG, Jones JL, Wang Y, Li SJ. Decreased glutamate + glutamine in Alzheimer's disease detected in vivo with (1)H-MRS at 0.5 T. Neurology. 2001;56(6):737–42. https://doi.org/10.1212/wnl.56.6.737.
5. Bai X, Edden RA, Gao F, Wang G, Wu L, Zhao B, Wang M, Chan Q, Chen W, Barker PB. Decreased γ-aminobutyric acid levels in the parietal region of patients with Alzheimer's disease. J Magn Reson Imaging. 2015;41(5):1326–31. https://doi.org/10.1002/jmri.24665.
6. Bak LK, Schousboe A, Waagepetersen HS. The glutamate/GABA-glutamine cycle: aspects of transport, neurotransmitter homeostasis and ammonia transfer. J Neurochem. 2006;98(3):641–53. https://doi.org/10.1111/j.1471-4159.2006.03913.x.
7. Barker PB, Breiter SN, Soher BJ, Chatham JC, Forder JR, Samphilipo MA, Magee CA, Anderson JH. Quantitative proton spectroscopy of canine brain: in vivo and in vitro correlations. Magn Reson Med. 1994;32(2):157–63. https://doi.org/10.1002/mrm.1910320202.
8. Barker PB, Hearshen DO, Boska MD. Single-voxel proton MRS of the human brain at 1.5T and 3.0T. Magn Reson Med. 2001;45(5):765–9. https://doi.org/10.1002/mrm.1104.
9. Bartha R, Smith M, Rupsingh R, Rylett J, Wells JL, Borrie MJ. High field (1)H MRS of the hippocampus after donepezil treatment in Alzheimer disease. Prog Neuro-Psychopharmacol Biol Psychiatry. 2008;32(3):786–93. https://doi.org/10.1016/j.pnpbp.2007.12.011.
10. Bennett DA, Wilson RS, Schneider JA, Evans DA, Beckett LA, Aggarwal NT, Barnes LL, Fox JH, Bach J. Natural history of mild cognitive impairment in older persons. Neurology. 2002;59(2):198–205. https://doi.org/10.1212/wnl.59.2.198.
11. Best JG, Stagg CJ, Dennis A. Chapter 2.5—other significant metabolites: Myo-inositol, GABA, glutamine, and lactate. In: Stagg C, Rothman D, editors. Magnetic resonance spectroscopy. Academic Press; 2014. p. 122–38. https://doi.org/10.1016/B978-0-12-401688-0.00010-0.
12. Bitsch A, Bruhn H, Vougioukas V, Stringaris A, Lassmann H, Frahm J, Brück W. Inflammatory CNS demyelination: histopathologic correlation with in vivo quantitative proton MR spectroscopy. AJNR Am J Neuroradiol. 1999;20(9):1619–27.
13. Bittner DM, Heinze HJ, Kaufmann J. Association of 1H-MR spectroscopy and cerebrospinal fluid biomarkers in Alzheimer's disease: diverging behavior at three different brain regions. J Alzheimers Dis. 2013;36(1):155–63. https://doi.org/10.3233/jad-120778.
14. Bowen J, Teri L, Kukull W, McCormick W, McCurry SM, Larson EB. Progression to dementia in patients with isolated memory loss. Lancet. 1997;349(9054):763–5. https://doi.org/10.1016/S0140-6736(96)08256-6.
15. Braak H, Alafuzoff I, Arzberger T, Kretzschmar H, Del Tredici K. Staging of Alzheimer disease-associated neurofibrillary pathology using paraffin sections and immunocytochemistry. Acta Neuropathol. 2006;112(4):389–404. https://doi.org/10.1007/s00401-006-0127-z.
16. Braak H, Braak E. Neuropathological stageing of Alzheimer-related changes. Acta Neuropathol. 1991;82(4):239–59. https://doi.org/10.1007/BF00308809.
17. Catani M, Cherubini A, Howard R, Tarducci R, Pelliccioli GP, Piccirilli M, Gobbi G, Senin U, Mecocci P. (1)H-MR spectroscopy differentiates mild cognitive impairment from normal brain aging. Neuroreport. 2001;12(11):2315–7. https://doi.org/10.1097/00001756-200108080-00007.

18. Chang L, Ernst T, Poland RE, Jenden DJ. In vivo proton magnetic resonance spectroscopy of the normal aging human brain. Life Sci. 1996;58(22):2049–56. https://doi.org/10.1016/0024-3205(96)00197-X.
19. Chantal S, Braun CM, Bouchard RW, Labelle M, Boulanger Y. Similar 1H magnetic resonance spectroscopic metabolic pattern in the medial temporal lobes of patients with mild cognitive impairment and Alzheimer disease. Brain Res. 2004;1003(1–2):26–35. https://doi.org/10.1016/j.brainres.2003.11.074.
20. Chantal S, Labelle M, Bouchard RW, Braun CM, Boulanger Y. Correlation of regional proton magnetic resonance spectroscopic metabolic changes with cognitive deficits in mild Alzheimer disease. Arch Neurol. 2002;59(6):955–62. https://doi.org/10.1001/archneur.59.6.955.
21. Chao LL, Schuff N, Kramer JH, Du AT, Capizzano AA, O'Neill J, Wolkowitz OM, Jagust WJ, Chui HC, Miller BL, Yaffe K, Weiner MW. Reduced medial temporal lobe N-acetylaspartate in cognitively impaired but nondemented patients. Neurology. 2005;64(2):282–9. https://doi.org/10.1212/01.Wnl.0000149638.45635.Ff.
22. Chen QY, Abrigo J, Liu WT, Han EY, Yeung DKW, Shi L, Au LWC, Deng M, Chen SR, Leung EYL, Ho CL, Mok VCT, Chu WCW. Lower posterior cingulate N-acetylaspartate to creatine level in early detection of biologically defined Alzheimer's disease. Brain Sci. 2022;12(6):Article 722. https://doi.org/10.3390/brainsci12060722.
23. Chiang GC, Mao X, Kang G, Chang E, Pandya S, Vallabhajosula S, Isaacson R, Ravdin LD, Shungu DC. Relationships among cortical glutathione levels, brain amyloidosis, and memory in healthy older adults investigated in vivo with (1)H-MRS and Pittsburgh compound-B PET. AJNR Am J Neuroradiol. 2017;38(6):1130–7. https://doi.org/10.3174/ajnr.A5143.
24. Cho SH, Rhee HY, Oh J, Lee JS, Park S, Shin WC, Paik JW, Ryu CW, Park KC, Jahng GH. Longitudinal functional magnetic resonance spectroscopy study in subjects with mild cognitive impairment and Alzheimer's disease. Curr Alzheimer Res. 2021;18(4):335–46. https://doi.org/10.2174/1567205018666210708145924.
25. Christiansen P, Schlosser A, Henriksen O. Reduced N-acetylaspartate content in the frontal part of the brain in patients with probable Alzheimer's disease. Magn Reson Imaging. 1995;13(3):457–62. https://doi.org/10.1016/0730-725X(94)00113-H.
26. Collaborators GBDDF. Estimation of the global prevalence of dementia in 2019 and forecasted prevalence in 2050: an analysis for the Global Burden of Disease Study 2019. Lancet Public Health. 2022;7(2):e105–25. https://doi.org/10.1016/S2468-2667(21)00249-8.
27. Deelchand DK, Adanyeguh IM, Emir UE, Nguyen T-M, Valabregue R, Henry P-G, Mochel F, Öz G. Two-site reproducibility of cerebellar and brainstem neurochemical profiles with short-echo, single-voxel MRS at 3T. Magn Reson Med. 2015;73(5):1718–25. https://doi.org/10.1002/mrm.25295.
28. Dixon RM, Bradley KM, Budge MM, Styles P, Smith AD. Longitudinal quantitative proton magnetic resonance spectroscopy of the hippocampus in Alzheimer's disease. Brain. 2002;125(Pt 10):2332–41. https://doi.org/10.1093/brain/awf226.
29. Dou W, Kaufmann J, Li M, Zhong K, Walter M, Speck O. The separation of Gln and Glu in STEAM: a comparison study using short and long TEs/TMs at 3 and 7 T. MAGMA. 2015;28(4):395–405. https://doi.org/10.1007/s10334-014-0479-7.
30. Dubois B, Hampel H, Feldman HH, Scheltens P, Aisen P, Andrieu S, Bakardjian H, Benali H, Bertram L, Blennow K, Broich K, Cavedo E, Crutch S, Dartigues JF, Duyckaerts C, Epelbaum S, Frisoni GB, Gauthier S, Genthon R, et al. Preclinical Alzheimer's disease: definition, natural history, and diagnostic criteria. Alzheimers Dement. 2016;12(3):292–323. https://doi.org/10.1016/j.jalz.2016.02.002.
31. Dubois B, Villain N, Frisoni GB, Rabinovici GD, Sabbagh M, Cappa S, Bejanin A, Bombois S, Epelbaum S, Teichmann M, Habert MO, Nordberg A, Blennow K, Galasko D, Stern Y, Rowe CC, Salloway S, Schneider LS, Cummings JL, Feldman HH. Clinical diagnosis of Alzheimer's disease: recommendations of the International Working Group. Lancet Neurol. 2021;20(6):484–96. https://doi.org/10.1016/S1474-4422(21)00066-1.

32. Duffy SL, Lagopoulos J, Hickie IB, Diamond K, Graeber MB, Lewis SJ, Naismith SL. Glutathione relates to neuropsychological functioning in mild cognitive impairment. Alzheimers Dement. 2014;10(1):67–75. https://doi.org/10.1016/j.jalz.2013.01.005.
33. Ernst T, Chang L, Melchor R, Mehringer CM. Frontotemporal dementia and early Alzheimer disease: differentiation with frontal lobe H-1 MR spectroscopy. Radiology. 1997;203(3):829–36. https://doi.org/10.1148/radiology.203.3.9169712.
34. Fayed N, Andrés E, Viguera L, Modrego PJ, Garcia-Campayo J. Higher glutamate+glutamine and reduction of N-acetylaspartate in posterior cingulate according to age range in patients with cognitive impairment and/or pain. Acad Radiol. 2014;21(9):1211–7. https://doi.org/10.1016/j.acra.2014.04.009.
35. Fayed N, Dávila J, Oliveros A, Castillo J, Medrano JJ. Utility of different MR modalities in mild cognitive impairment and its use as a predictor of conversion to probable dementia. Acad Radiol. 2008;15(9):1089–98. https://doi.org/10.1016/j.acra.2008.04.008.
36. Fayed N, Modrego PJ, Rojas-Salinas G, Aguilar K. Brain glutamate levels are decreased in Alzheimer's disease: a magnetic resonance spectroscopy study. Am J Alzheimers Dis Other Dement. 2011;26(6):450–6. https://doi.org/10.1177/1533317511421780.
37. Firbank MJ, Harrison RM, O'Brien JT. A comprehensive review of proton magnetic resonance spectroscopy studies in dementia and Parkinson's disease. Dement Geriatr Cogn Disord. 2002;14(2):64–76. https://doi.org/10.1159/000064927.
38. Flicker C, Ferris SH, Reisberg B. Mild cognitive impairment in the elderly: predictors of dementia. Neurology. 1991;41(7):1006–9. https://doi.org/10.1212/wnl.41.7.1006.
39. Frederick BB, Satlin A, Yurgelun-Todd DA, Renshaw PF. In vivo proton magnetic resonance spectroscopy of Alzheimer's disease in the parietal and temporal lobes. Biol Psychiatry. 1997;42(2):147–50. https://doi.org/10.1016/s0006-3223(97)00242-4.
40. Gafson AR, Barthélemy NR, Bomont P, Carare RO, Durham HD, Julien JP, Kuhle J, Leppert D, Nixon RA, Weller RO, Zetterberg H, Matthews PM. Neurofilaments: neurobiological foundations for biomarker applications. Brain. 2020;143(7):1975–98. https://doi.org/10.1093/brain/awaa098.
41. Gao F, Barker PB. Various MRS application tools for Alzheimer disease and mild cognitive impairment. Am J Neuroradiol. 2014;35(6 suppl):S4–S11. https://doi.org/10.3174/ajnr.A3944.
42. Govindaraju V, Young K, Maudsley AA. Proton NMR chemical shifts and coupling constants for brain metabolites. NMR Biomed. 2000;13(3):129–53. https://doi.org/10.1002/1099-1492(200005)13:3<129::aid-nbm619>3.0.co;2-v.
43. Gozdas E, Hinkley L, Fingerhut H, Dacorro L, Gu M, Sacchet M, Hurd R, Hosseini SMH. H-1-MRS neurometabolites and associations with neurite microstructures and cognitive functions in amnestic mild cognitive impairment. Neuroimage-Clinical. 2022;36:Article 103159. https://doi.org/10.1016/j.nicl.2022.103159.
44. Graff-Radford J, Kantarci K. Magnetic resonance spectroscopy in Alzheimer's disease. Neuropsychiatr Dis Treat. 2013;9:687–96. https://doi.org/10.2147/ndt.S35440.
45. Gruetter R, Garwood M, Uğurbil K, Seaquist ER. Observation of resolved glucose signals in 1H NMR spectra of the human brain at 4 Tesla. Magn Reson Med. 1996;36(1):1–6. https://doi.org/10.1002/mrm.1910360102.
46. Haris M, Nath K, Cai K, Singh A, Crescenzi R, Kogan F, Verma G, Reddy S, Hariharan H, Melhem ER, Reddy R. Imaging of glutamate neurotransmitter alterations in Alzheimer's disease. NMR Biomed. 2013;26(4):386–91. https://doi.org/10.1002/nbm.2875.
47. Hattori N, Abe K, Sakoda S, Sawada T. Proton MR spectroscopic study at 3 Tesla on glutamate/glutamine in Alzheimer's disease. Neuroreport. 2002;13(1):183–6. https://doi.org/10.1097/00001756-200201210-00041.
48. Hertz L. Functional interactions between neurons and astrocytes I. Turnover and metabolism of putative amino acid transmitters. Prog Neurobiol. 1979;13(3):277–323. https://doi.org/10.1016/0301-0082(79)90018-2.
49. Huang DD, Liu D, Yin JZ, Qian TY, Shrestha S, Ni HY. Glutamate-glutamine and GABA in brain of normal aged and patients with cognitive impairment. Eur Radiol. 2017;27(7):2698–705. https://doi.org/10.1007/s00330-016-4669-8.

50. Huang W, Alexander GE, Chang L, Shetty HU, Krasuski JS, Rapoport SI, Schapiro MB. Brain metabolite concentration and dementia severity in Alzheimer's disease: a (1)H MRS study. Neurology. 2001;57(4):626–32. https://doi.org/10.1212/wnl.57.4.626.
51. Jack CR, Andrews JS, Beach TG, et al. Revised criteria for diagnosis and staging of Alzheimer's disease: Alzheimer's Association Workgroup. Alzheimer's Dement. 2024;20:5143–69. https://doi.org/10.1002/alz.13859.
52. Jack CR Jr, Bennett DA, Blennow K, Carrillo MC, Dunn B, Haeberlein SB, Holtzman DM, Jagust W, Jessen F, Karlawish J, Liu E, Molinuevo JL, Montine T, Phelps C, Rankin KP, Rowe CC, Scheltens P, Siemers E, Snyder HM, Sperling R. NIA-AA research framework: toward a biological definition of Alzheimer's disease. Alzheimers Dement. 2018;14(4):535–62. https://doi.org/10.1016/j.jalz.2018.02.018.
53. Janelidze S, Berron D, Smith R, Strandberg O, Proctor NK, Dage JL, Stomrud E, Palmqvist S, Mattsson-Carlgren N, Hansson O. Associations of plasma phospho-Tau217 levels with Tau positron emission tomography in early Alzheimer disease. JAMA Neurol. 2021a;78(2):149–56. https://doi.org/10.1001/jamaneurol.2020.4201.
54. Janelidze S, Mattsson N, Palmqvist S, Smith R, Beach TG, Serrano GE, Chai X, Proctor NK, Eichenlaub U, Zetterberg H, Blennow K, Reiman EM, Stomrud E, Dage JL, Hansson O. Plasma P-tau181 in Alzheimer's disease: relationship to other biomarkers, differential diagnosis, neuropathology and longitudinal progression to Alzheimer's dementia. Nat Med. 2020;26(3):379–86. https://doi.org/10.1038/s41591-020-0755-1.
55. Janelidze S, Teunissen CE, Zetterberg H, Allué JA, Sarasa L, Eichenlaub U, Bittner T, Ovod V, Verberk IMW, Toba K, Nakamura A, Bateman RJ, Blennow K, Hansson O. Head-to-head comparison of 8 plasma amyloid-β 42/40 assays in Alzheimer disease. JAMA Neurol. 2021b;78(11):1375–82. https://doi.org/10.1001/jamaneurol.2021.3180.
56. Jessen F, Block W, Träber F, Keller E, Flacke S, Lamerichs R, Schild HH, Heun R. Decrease of N-acetylaspartate in the MTL correlates with cognitive decline of AD patients. Neurology. 2001;57(5):930–2. https://doi.org/10.1212/wnl.57.5.930.
57. Jessen F, Block W, Träber F, Keller E, Flacke S, Papassotiropoulos A, Lamerichs R, Heun R, Schild HH. Proton MR spectroscopy detects a relative decrease of N-acetylaspartate in the medial temporal lobe of patients with AD. Neurology. 2000;55(5):684–8. https://doi.org/10.1212/wnl.55.5.684.
58. Jicha GA, Parisi JE, Dickson DW, Johnson K, Cha R, Ivnik RJ, Tangalos EG, Boeve BF, Knopman DS, Braak H, Petersen RC. Neuropathologic outcome of mild cognitive impairment following progression to clinical dementia. Arch Neurol. 2006;63(5):674–81. https://doi.org/10.1001/archneur.63.5.674.
59. Joers JM, Deelchand DK, Kumar A, Moheet A, Seaquist E, Henry PG, Öz G. Measurement of hypothalamic glucose under euglycemia and hyperglycemia by MRI at 3T. J Magn Reson Imaging. 2017;45(3):681–91. https://doi.org/10.1002/jmri.25383.
60. Kantarci K. Proton MRS in mild cognitive impairment. J Magn Reson Imaging. 2013;37(4):770–7. https://doi.org/10.1002/jmri.23800.
61. Kantarci K, Jack CR Jr, Xu YC, Campeau NG, O'Brien PC, Smith GE, Ivnik RJ, Boeve BF, Kokmen E, Tangalos EG, Petersen RC. Regional metabolic patterns in mild cognitive impairment and Alzheimer's disease: A 1H MRS study. Neurology. 2000;55(2):210–7. https://doi.org/10.1212/wnl.55.2.210.
62. Kantarci K, Jicha GA. Development of ^{1}H MRS biomarkers for tracking early predementia Alzheimer disease. Neurology. 2019;92(5):209–10. https://doi.org/10.1212/wnl.0000000000006839.
63. Kantarci K, Knopman DS, Dickson DW, Parisi JE, Whitwell JL, Weigand SD, Josephs KA, Boeve BF, Petersen RC, Jack CR Jr. Alzheimer disease: postmortem neuropathologic correlates of antemortem 1H MR spectroscopy metabolite measurements. Radiology. 2008a;248(1):210–20. https://doi.org/10.1148/radiol.2481071590.
64. Kantarci K, Lowe V, Przybelski SA, Senjem ML, Weigand SD, Ivnik RJ, Roberts R, Geda YE, Boeve BF, Knopman DS, Petersen RC, Jack CR Jr. Magnetic resonance spectroscopy, β-amyloid load, and cognition in a population-based sample of cognitively normal older adults. Neurology. 2011;77(10):951–8. https://doi.org/10.1212/WNL.0b013e31822dc7e1.

65. Kantarci K, Petersen RC, Boeve BF, Knopman DS, Tang-Wai DF, O'Brien PC, Weigand SD, Edland SD, Smith GE, Ivnik RJ, Ferman TJ, Tangalos EG, Jack CR Jr. 1H MR spectroscopy in common dementias. Neurology. 2004;63(8):1393–8. https://doi.org/10.1212/01.wnl.0000141849.21256.ac.
66. Kantarci K, Petersen RC, Przybelski SA, Weigand SD, Shiung MM, Whitwell JL, Negash S, Ivnik RJ, Boeve BF, Knopman DS, Smith GE, Jack CR Jr. Hippocampal volumes, proton magnetic resonance spectroscopy metabolites, and cerebrovascular disease in mild cognitive impairment subtypes. Arch Neurol. 2008b;65(12):1621–8. https://doi.org/10.1001/archneur.65.12.1621.
67. Kantarci K, Reynolds G, Petersen RC, Boeve BF, Knopman DS, Edland SD, Smith GE, Ivnik RJ, Tangalos EG, Jack CR Jr. Proton MR spectroscopy in mild cognitive impairment and Alzheimer disease: comparison of 1.5 and 3 T. AJNR Am J Neuroradiol. 2003;24(5):843–9.
68. Kantarci K, Smith GE, Ivnik RJ, Petersen RC, Boeve BF, Knopman DS, Tangalos EG, Jack CR Jr. 1H magnetic resonance spectroscopy, cognitive function, and apolipoprotein E genotype in normal aging, mild cognitive impairment and Alzheimer's disease. J Int Neuropsychol Soc. 2002;8(7):934–42. https://doi.org/10.1017/s1355617702870084.
69. Kantarci K, Weigand SD, Petersen RC, Boeve BF, Knopman DS, Gunter J, Reyes D, Shiung M, O'Brien PC, Smith GE, Ivnik RJ, Tangalos EG, Jack CR. Longitudinal 1H MRS changes in mild cognitive impairment and Alzheimer's disease. Neurobiol Aging. 2007;28(9):1330–9. https://doi.org/10.1016/j.neurobiolaging.2006.06.018.
70. Kantarci K, Weigand SD, Przybelski SA, Preboske GM, Pankratz VS, Vemuri P, Senjem ML, Murphy MC, Gunter JL, Machulda MM, Ivnik RJ, Roberts RO, Boeve BF, Rocca WA, Knopman DS, Petersen RC, Jack CR. MRI and MRS predictors of mild cognitive impairment in a population-based sample. Neurology. 2013;81(2):126–33. https://doi.org/10.1212/WNL.0b013e31829a3329.
71. Kantarci K, Weigand SD, Przybelski SA, Shiung MM, Whitwell JL, Negash S, Knopman DS, Boeve BF, O'Brien PC, Petersen RC, Jack CR Jr. Risk of dementia in MCI: combined effect of cerebrovascular disease, volumetric MRI, and 1H MRS. Neurology. 2009;72(17):1519–25. https://doi.org/10.1212/WNL.0b013e3181a2e864.
72. Kara F, Joers JM, Deelchand DK, Park YW, Przybelski SA, Lesnick TG, Senjem ML, Zeydan B, Knopman DS, Lowe VJ, Vemuri P, Mielke MM, Machulda MM, Jack CR, Petersen RC, Oz G, Kantarci K. H-1 MR spectroscopy biomarkers of neuronal and synaptic function are associated with tau deposition in cognitively unimpaired older adults. Neurobiol Aging. 2022;112:16–26. https://doi.org/10.1016/j.neurobiolaging.2021.12.010.
73. Karikari TK, Pascoal TA, Ashton NJ, Janelidze S, Benedet AL, Rodriguez JL, Chamoun M, Savard M, Kang MS, Therriault J, Schöll M, Massarweh G, Soucy J-P, Höglund K, Brinkmalm G, Mattsson N, Palmqvist S, Gauthier S, Stomrud E, et al. Blood phosphorylated tau 181 as a biomarker for Alzheimer's disease: a diagnostic performance and prediction modelling study using data from four prospective cohorts. The Lancet Neurology. 2020;19(5):422–33. https://doi.org/10.1016/S1474-4422(20)30071-5.
74. Kordower JH, Chu Y, Stebbins GT, DeKosky ST, Cochran EJ, Bennett D, Mufson EJ. Loss and atrophy of layer II entorhinal cortex neurons in elderly people with mild cognitive impairment. Ann Neurol. 2001;49(2):202–13.
75. Lin JC, Gant N. Chapter 2.3—the biochemistry of choline. In: Stagg C, Rothman D, editors. Magnetic resonance spectroscopy. Academic Press; 2014. p. 104–10. https://doi.org/10.1016/B978-0-12-401688-0.00008-2.
76. Liu H, Zhang D, Lin H, Zhang Q, Zheng L, Zheng Y, Yin X, Li Z, Liang S, Huang S. Meta-analysis of neurochemical changes estimated via magnetic resonance spectroscopy in mild cognitive impairment and Alzheimer's disease [systematic review]. Frontiers in aging. Neuroscience. 2021;13. https://www.frontiersin.org/articles/10.3389/fnagi.2021.738971
77. Luyten PR, den Hollander JA. Observation of metabolites in the human brain by MR spectroscopy. Radiology. 1986;161(3):795–8. https://doi.org/10.1148/radiology.161.3.3786735.
78. MacKay S, Ezekiel F, Di Sclafani V, Meyerhoff DJ, Gerson J, Norman D, Fein G, Weiner MW. Alzheimer disease and subcortical ischemic vascular dementia: evaluation by combining MR imaging segmentation and H-1 MR spectroscopic imaging. Radiology. 1996a;198(2):537–45. https://doi.org/10.1148/radiology.198.2.8596863.

79. MacKay S, Meyerhoff DJ, Constans JM, Norman D, Fein G, Weiner MW. Regional gray and white matter metabolite differences in subjects with AD, with subcortical ischemic vascular dementia, and elderly controls with 1H magnetic resonance spectroscopic imaging. Arch Neurol. 1996b;53(2):167–74. https://doi.org/10.1001/archneur.1996.00550020079018.
80. Mandal PK, Saharan S, Tripathi M, Murari G. Brain glutathione levels—a novel biomarker for mild cognitive impairment and Alzheimer's disease. Biol Psychiatry. 2015;78(10):702–10. https://doi.org/10.1016/j.biopsych.2015.04.005.
81. Mandal PK, Tripathi M, Sugunan S. Brain oxidative stress: detection and mapping of anti-oxidant marker 'glutathione' in different brain regions of healthy male/female, MCI and Alzheimer patients using non-invasive magnetic resonance spectroscopy. Biochem Biophys Res Commun. 2012;417(1):43–8. https://doi.org/10.1016/j.bbrc.2011.11.047.
82. Marjanska M, Curran GL, Wengenack TM, Henry PG, Bliss RL, Poduslo JF, Jack CR Jr, Ugurbil K, Garwood M. Monitoring disease progression in transgenic mouse models of Alzheimer's disease with proton magnetic resonance spectroscopy. Proc Natl Acad Sci USA. 2005;102(33):11906–10. https://doi.org/10.1073/pnas.0505513102.
83. Matsuoka K, Hirata K, Kokubo N, Maeda T, Tagai K, Endo H, Takahata K, Shinotoh H, Ono M, Seki C, Tatebe H, Kawamura K, Zhang MR, Shimada H, Tokuda T, Higuchi M, Takado Y. Investigating neural dysfunction with abnormal protein deposition in Alzheimer's disease through magnetic resonance spectroscopic imaging, plasma biomarkers, and positron emission tomography. Neuroimage Clin. 2023;41:103560. https://doi.org/10.1016/j.nicl.2023.103560.
84. Mattsson N, Cullen NC, Andreasson U, Zetterberg H, Blennow K. Association between longitudinal plasma neurofilament light and neurodegeneration in patients with Alzheimer disease. JAMA Neurol. 2019;76(7):791–9. https://doi.org/10.1001/jamaneurol.2019.0765.
85. Maudsley AA, Domenig C, Govind V, Darkazanli A, Studholme C, Arheart K, Bloomer C. Mapping of brain metabolite distributions by volumetric proton MR spectroscopic imaging (MRSI). Magn Reson Med. 2009;61(3):548–59. https://doi.org/10.1002/mrm.21875.
86. McKhann G, Drachman D, Folstein M, Katzman R, Price D, Stadlan EM. Clinical diagnosis of Alzheimer's disease: report of the NINCDS-ADRDA Work Group under the auspices of Department of Health and Human Services Task Force on Alzheimer's disease. Neurology. 1984;34(7):939–44. https://doi.org/10.1212/wnl.34.7.939.
87. McKhann GM, Knopman DS, Chertkow H, Hyman BT, Jack CR Jr, Kawas CH, Klunk WE, Koroshetz WJ, Manly JJ, Mayeux R, Mohs RC, Morris JC, Rossor MN, Scheltens P, Carrillo MC, Thies B, Weintraub S, Phelps CH. The diagnosis of dementia due to Alzheimer's disease: recommendations from the National Institute on Aging-Alzheimer's Association workgroups on diagnostic guidelines for Alzheimer's disease. Alzheimers Dement. 2011;7(3):263–9. https://doi.org/10.1016/j.jalz.2011.03.005.
88. McKiernan E, Su L, O'Brien J. MRS in neurodegenerative dementias, prodromal syndromes and at-risk states: a systematic review of the literature. NMR Biomed. 2023;36(7):e4896. https://doi.org/10.1002/nbm.4896.
89. Metastasio A, Rinaldi P, Tarducci R, Mariani E, Feliziani FT, Cherubini A, Pelliccioli GP, Gobbi G, Senin U, Mecocci P. Conversion of MCI to dementia: role of proton magnetic resonance spectroscopy. Neurobiol Aging. 2006;27(7):926–32. https://doi.org/10.1016/j.neurobiolaging.2005.05.002.
90. Meyerhoff DJ, MacKay S, Constans JM, Norman D, Van Dyke C, Fein G, Weiner MW. Axonal injury and membrane alterations in Alzheimer's disease suggested by in vivo proton magnetic resonance spectroscopic imaging. Ann Neurol. 1994;36(1):40–7. https://doi.org/10.1002/ana.410360110.
91. Mielke MM, Hagen CE, Xu J, Chai X, Vemuri P, Lowe VJ, Airey DC, Knopman DS, Roberts RO, Machulda MM, Jack CR Jr, Petersen RC, Dage JL. Plasma phospho-tau181 increases with Alzheimer's disease clinical severity and is associated with tau- and amyloid-positron emission tomography. Alzheimers Dement. 2018;14(8):989–97. https://doi.org/10.1016/j.jalz.2018.02.013.

92. Miller BL, Chang L, Booth R, Ernst T, Cornford M, Nikas D, McBride D, Jenden DJ. In vivo 1H MRS choline: correlation with in vitro chemistry/histology. Life Sci. 1996;58(22):1929–35. https://doi.org/10.1016/0024-3205(96)00182-8.
93. Miller BL, Moats RA, Shonk T, Ernst T, Woolley S, Ross BD. Alzheimer disease: depiction of increased cerebral myo-inositol with proton MR spectroscopy. Radiology. 1993;187(2):433–7. https://doi.org/10.1148/radiology.187.2.8475286.
94. Mitolo M, Stanzani-Maserati M, Capellari S, Testa C, Rucci P, Poda R, Oppi F, Gallassi R, Sambati L, Rizzo G, Parchi P, Evangelista S, Talozzi L, Tonon C, Lodi R, Liguori R. Predicting conversion from mild cognitive impairment to Alzheimer's disease using brain H-1-MRS and volumetric changes: a two- year retrospective follow-up study. Neuroimage-Clinical. 2019;23:Article 101843. https://doi.org/10.1016/j.nicl.2019.101843.
95. Moats RA, Ernst T, Shonk TK, Ross BD. Abnormal cerebral metabolite concentrations in patients with probable Alzheimer disease. Magn Reson Med. 1994;32(1):110–5. https://doi.org/10.1002/mrm.1910320115.
96. Modrego PJ, Fayed N, Pina MA. Conversion from mild cognitive impairment to probable Alzheimer's disease predicted by brain magnetic resonance spectroscopy. Am J Psychiatry. 2005;162(4):667–75. https://doi.org/10.1176/appi.ajp.162.4.667.
97. Moffett JR, Ross B, Arun P, Madhavarao CN, Namboodiri AM. N-Acetylaspartate in the CNS: from neurodiagnostics to neurobiology. Prog Neurobiol. 2007;81(2):89–131. https://doi.org/10.1016/j.pneurobio.2006.12.003.
98. Mullins R, Reiter D, Kapogiannis D. Magnetic resonance spectroscopy reveals abnormalities of glucose metabolism in the Alzheimer's brain. Ann Clin Transl Neurol. 2018;5(3):262–72. https://doi.org/10.1002/acn3.530.
99. Murray ME, Przybelski SA, Lesnick TG, Liesinger AM, Spychalla A, Zhang B, Gunter JL, Parisi JE, Boeve BF, Knopman DS, Petersen RC, Jack CR Jr, Dickson DW, Kantarci K. Early Alzheimer's disease neuropathology detected by proton MR spectroscopy. J Neurosci. 2014;34(49):16247–55. https://doi.org/10.1523/jneurosci.2027-14.2014.
100. Nabers A, Perna L, Lange J, Mons U, Schartner J, Güldenhaupt J, Saum KU, Janelidze S, Holleczek B, Rujescu D, Hansson O, Gerwert K, Brenner H. Amyloid blood biomarker detects Alzheimer's disease. EMBO Mol Med. 2018;10(5):e8763. https://doi.org/10.15252/emmm.201708763.
101. Nakamura A, Kaneko N, Villemagne VL, Kato T, Doecke J, Doré V, Fowler C, Li Q-X, Martins R, Rowe C, Tomita T, Matsuzaki K, Ishii K, Ishii K, Arahata Y, Iwamoto S, Ito K, Tanaka K, Masters CL, Yanagisawa K. High performance plasma amyloid-β biomarkers for Alzheimer's disease. Nature. 2018;554(7691):249–54. https://doi.org/10.1038/nature25456.
102. Nedelska Z, Przybelski SA, Lesnick TG, Schwarz CG, Lowe VJ, Machulda MM, Kremers WK, Mielke MM, Roberts RO, Boeve BF, Knopman DS, Petersen RC, Jack CR, Kantarci K. H-1-MRS metabolites and rate of beta-amyloid accumulation on serial PET in clinically normal adults. Neurology. 2017;89(13):1391–9. https://doi.org/10.1212/wnl.0000000000004421.
103. Oeltzschner G, Wijtenburg SA, Mikkelsen M, Edden RAE, Barker PB, Joo JH, Leoutsakos JS, Rowland LM, Workman CI, Smith GS. Neurometabolites and associations with cognitive deficits in mild cognitive impairment: a magnetic resonance spectroscopy study at 7 Tesla. Neurobiol Aging. 2019;73:211–8. https://doi.org/10.1016/j.neurobiolaging.2018.09.027.
104. Ongür D, Prescot AP, Jensen JE, Cohen BM, Renshaw PF. Creatine abnormalities in schizophrenia and bipolar disorder. Psychiatry Res. 2009;172(1):44–8. https://doi.org/10.1016/j.pscychresns.2008.06.002.
105. Onwordi EC, Whitehurst T, Mansur A, Statton B, Berry A, Quinlan M, O'Regan DP, Rogdaki M, Marques TR, Rabiner EA, Gunn RN, Vernon AC, Natesan S, Howes OD. The relationship between synaptic density marker SV2A, glutamate and N-acetyl aspartate levels in healthy volunteers and schizophrenia: a multimodal PET and magnetic resonance spectroscopy brain imaging study. Transl Psychiatry. 2021;11(1):393. https://doi.org/10.1038/s41398-021-01515-3.

106. Oppenheimer SM, Bryan RN, Conturo TE, Soher BJ, Preziosi TJ, Barker PB. Proton magnetic resonance spectroscopy and gadolinium-DTPA perfusion imaging of asymptomatic MRI white matter lesions. Magn Reson Med. 1995;33(1):61–8. https://doi.org/10.1002/mrm.1910330109.
107. Öz G, Alger JR, Barker PB, Bartha R, Bizzi A, Boesch C, Bolan PJ, Brindle KM, Cudalbu C, Dinçer A, Dydak U, Emir UE, Frahm J, González RG, Gruber S, Gruetter R, Gupta RK, Heerschap A, Henning A, et al. Clinical proton MR spectroscopy in central nervous system disorders. Radiology. 2014;270(3):658–79. https://doi.org/10.1148/radiol.13130531.
108. Öz G, Deelchand DK, Wijnen JP, Mlynárik V, Xin L, Mekle R, Noeske R, Scheenen TWJ, Tkáč I. Advanced single voxel (1) H magnetic resonance spectroscopy techniques in humans: Experts' consensus recommendations. NMR Biomed. 2020;e4236:e4236. https://doi.org/10.1002/nbm.4236.
109. Park YW, Deelchand DK, Joers JM, Hanna B, Berrington A, Gillen JS, Kantarci K, Soher BJ, Barker PB, Park H, Oz G, Lenglet C. AutoVOI: real-time automatic prescription of volume-of-interest for single voxel spectroscopy. Magn Reson Med. 2018;80(5):1787–98. https://doi.org/10.1002/mrm.27203.
110. Parnetti L, Tarducci R, Presciutti O, Lowenthal DT, Pippi M, Palumbo B, Gobbi G, Pelliccioli GP, Senin U. Proton magnetic resonance spectroscopy can differentiate Alzheimer's disease from normal aging. Mech Ageing Dev. 1997;97(1):9–14. https://doi.org/10.1016/s0047-6374(97)01877-0.
111. Penner J, Rupsingh R, Smith M, Wells JL, Borrie MJ, Bartha R. Increased glutamate in the hippocampus after galantamine treatment for Alzheimer disease. Prog Neuro-Psychopharmacol Biol Psychiatry. 2010;34(1):104–10. https://doi.org/10.1016/j.pnpbp.2009.10.007.
112. Petersen RC. Mild cognitive impairment as a diagnostic entity. J Intern Med. 2004;256(3):183–94. https://doi.org/10.1111/j.1365-2796.2004.01388.x.
113. Petersen RC, Smith GE, Ivnik RJ, Tangalos EG, Schaid DJ, Thibodeau SN, Kokmen E, Waring SC, Kurland LT. Apolipoprotein E status as a predictor of the development of Alzheimer's disease in memory-impaired individuals. JAMA. 1995;273(16):1274–8.
114. Petersen RC, Smith GE, Waring SC, Ivnik RJ, Tangalos EG, Kokmen E. Mild cognitive impairment: clinical characterization and outcome. Arch Neurol. 1999;56(3):303–8. https://doi.org/10.1001/archneur.56.3.303.
115. Piersson AD, Mohamad M, Rajab F, Suppiah S. Cerebrospinal fluid amyloid beta, tau levels, apolipoprotein, and 1H-MRS brain metabolites in Alzheimer's disease: a systematic review. Acad Radiol. 2021;28(10):1447–63. https://doi.org/10.1016/j.acra.2020.06.006.
116. Pilatus U, Lais C, Rochmont Adu M, Kratzsch T, Frölich L, Maurer K, Zanella FE, Lanfermann H, Pantel J. Conversion to dementia in mild cognitive impairment is associated with decline of N-actylaspartate and creatine as revealed by magnetic resonance spectroscopy. Psychiatry Res. 2009;173(1):1–7. https://doi.org/10.1016/j.pscychresns.2008.07.015.
117. Rackayova V, Cudalbu C, Pouwels PJW, Braissant O. Creatine in the central nervous system: from magnetic resonance spectroscopy to creatine deficiencies. Anal Biochem. 2017;529:144–57. https://doi.org/10.1016/j.ab.2016.11.007.
118. Rae CD. A guide to the metabolic pathways and function of metabolites observed in human brain 1H magnetic resonance spectra. Neurochem Res. 2014;39(1):1–36. https://doi.org/10.1007/s11064-013-1199-5.
119. Rae CD, Williams SR. Glutathione in the human brain: review of its roles and measurement by magnetic resonance spectroscopy. Anal Biochem. 2017;529:127–43. https://doi.org/10.1016/j.ab.2016.12.022.
120. Ratai E-M, Alshikho MJ, Zürcher NR, Loggia ML, Cebulla CL, Cernasov P, Reynolds B, Fish J, Seth R, Babu S, Paganoni S, Hooker JM, Atassi N. Integrated imaging of [11C]-PBR28 PET, MR diffusion and magnetic resonance spectroscopy 1H-MRS in amyotrophic lateral sclerosis. NeuroImage: Clinical. 2018;20:357–64. https://doi.org/10.1016/j.nicl.2018.08.007.

121. Riese F, Gietl A, Zölch N, Henning A, O'Gorman R, Kälin AM, Leh SE, Buck A, Warnock G, Edden RA, Luechinger R, Hock C, Kollias S, Michels L. Posterior cingulate γ-aminobutyric acid and glutamate/glutamine are reduced in amnestic mild cognitive impairment and are unrelated to amyloid deposition and apolipoprotein E genotype. Neurobiol Aging. 2015;36(1):53–9. https://doi.org/10.1016/j.neurobiolaging.2014.07.030.
122. Rose SE, de Zubicaray GI, Wang D, Galloway GJ, Chalk JB, Eagle SC, Semple J, Doddrell DM. A 1H MRS study of probable Alzheimer's disease and normal aging: implications for longitudinal monitoring of dementia progression. Magn Reson Imaging. 1999;17(2):291–9. https://doi.org/10.1016/s0730-725x(98)00168-4.
123. Rupsingh R, Borrie M, Smith M, Wells JL, Bartha R. Reduced hippocampal glutamate in Alzheimer disease. Neurobiol Aging. 2011;32(5):802–10. https://doi.org/10.1016/j.neurobiolaging.2009.05.002.
124. Satlin A, Bodick N, Offen WW, Renshaw PF. Brain proton magnetic resonance spectroscopy (1H-MRS) in Alzheimer's disease: changes after treatment with xanomeline, an M1 selective cholinergic agonist. Am J Psychiatry. 1997;154(10):1459–61. https://doi.org/10.1176/ajp.154.10.1459.
125. Saunders DE, Howe FA, van den Boogaart A, Griffiths JR, Brown MM. Aging of the adult human brain: in vivo quantitation of metabolite content with proton magnetic resonance spectroscopy. J Magn Reson Imaging. 1999;9(5):711–6. https://doi.org/10.1002/(sici)1522-2586(199905)9:5<711::aid-jmri14>3.0.co;2-3.
126. Schott JM, Frost C, MacManus DG, Ibrahim F, Waldman AD, Fox NC. Short echo time proton magnetic resonance spectroscopy in Alzheimer's disease: a longitudinal multiple time point study. Brain. 2010;133(11):3315–22. https://doi.org/10.1093/brain/awq208.
127. Seo SW, Lee JH, Jang SM, Kim ST, Chin J, Kim GH, Kim JH, Roh JH, Kim MJ, Kim SH, Na DL. Neurochemical alterations of the entorhinal cortex in amnestic mild cognitive impairment (aMCI): a three-year follow-up study. Arch Gerontol Geriatr. 2012;54(1):192–6. https://doi.org/10.1016/j.archger.2011.04.002.
128. Shiino A, Watanabe T, Shirakashi Y, Kotani E, Yoshimura M, Morikawa S, Inubushi T, Akiguchi I. The profile of hippocampal metabolites differs between Alzheimer's disease and subcortical ischemic vascular dementia, as measured by proton magnetic resonance spectroscopy. J Cereb Blood Flow Metab. 2012;32(5):805–15. https://doi.org/10.1038/jcbfm.2012.9.
129. Shonk TK, Moats RA, Gifford P, Michaelis T, Mandigo JC, Izumi J, Ross BD. Probable Alzheimer disease: diagnosis with proton MR spectroscopy. Radiology. 1995;195(1):65–72. https://doi.org/10.1148/radiology.195.1.7892497.
130. Shukla D, Mandal PK, Mishra R, Punjabi K, Dwivedi D, Tripathi M, Badhautia V. Hippocampal glutathione depletion and pH increment in Alzheimer's disease: an in vivo MRS study. J Alzheimers Dis. 2021;84(3):1139–52. https://doi.org/10.3233/jad-215032.
131. Shukla D, Mandal PK, Tripathi M, Vishwakarma G, Mishra R, Sandal K. Quantitation of in vivo brain glutathione conformers in cingulate cortex among age-matched control, MCI, and AD patients using MEGA-PRESS. Hum Brain Mapp. 2020;41(1):194–217. https://doi.org/10.1002/hbm.24799.
132. Sperling RA, Aisen PS, Beckett LA, Bennett DA, Craft S, Fagan AM, Iwatsubo T, Jack CR Jr, Kaye J, Montine TJ, Park DC, Reiman EM, Rowe CC, Siemers E, Stern Y, Yaffe K, Carrillo MC, Thies B, Morrison-Bogorad M, et al. Toward defining the preclinical stages of Alzheimer's disease: recommendations from the National Institute on Aging-Alzheimer's Association workgroups on diagnostic guidelines for Alzheimer's disease. Alzheimers Dement. 2011;7(3):280–92. https://doi.org/10.1016/j.jalz.2011.03.003.
133. Spotorno N, Najac C, Stomrud E, Mattsson-Carlgren N, Palmqvist S, van Westen D, Ronen I, Hansson O. Astrocytic function is associated with both amyloid-β and tau pathology in non-demented APOE ϵ4 carriers. Brain. Communications. 2022;4
134. Su L, Blamire AM, Watson R, He J, Hayes L, O'Brien JT. Whole-brain patterns of 1H-magnetic resonance spectroscopy imaging in Alzheimer's disease and dementia with Lewy bodies. Transl Psychiatry. 2016;6(8):e877. https://doi.org/10.1038/tp.2016.140.

135. Tallan HH. Studies on the distribution of N-acetyl-L-aspartic acid in brain. J Biol Chem. 1957;224(1):41–5.
136. Therriault J, Pascoal TA, Lussier FZ, Tissot C, Chamoun M, Bezgin G, Servaes S, Benedet AL, Ashton NJ, Karikari TK, Lantero-Rodriguez J, Kunach P, Wang Y-T, Fernandez-Arias J, Massarweh G, Vitali P, Soucy J-P, Saha-Chaudhuri P, Blennow K, et al. Biomarker modeling of Alzheimer's disease using PET-based Braak staging. Nature Aging. 2022;2(6):526–35. https://doi.org/10.1038/s43587-022-00204-0.
137. Trabesinger AH, Weber OM, Duc CO, Boesiger P. Detection of glutathione in the human brain in vivo by means of double quantum coherence filtering. Magn Reson Med. 1999;42(2):283–9. https://doi.org/10.1002/(sici)1522-2594(199908)42:2<283::Aid-mrm10>3.0.Co;2-q.
138. Tumati S, Martens S, Aleman A. Magnetic resonance spectroscopy in mild cognitive impairment: systematic review and meta-analysis. Neurosci Biobehav Rev. 2013;37(10):2571–86. https://doi.org/10.1016/j.neubiorev.2013.08.004.
139. Urenjak J, Williams SR, Gadian DG, Noble M. Specific expression of N-acetylaspartate in neurons, oligodendrocyte-type-2 astrocyte progenitors, and immature oligodendrocytes in vitro. J Neurochem. 1992;59(1):55–61. https://doi.org/10.1111/j.1471-4159.1992.tb08875.x.
140. Valenzuela MJ, Sachdev P. Magnetic resonance spectroscopy in AD. Neurology. 2001;56(5):592–8. https://doi.org/10.1212/wnl.56.5.592.
141. Verberk IMW, Laarhuis MB, van den Bosch KA, Ebenau JL, van Leeuwenstijn M, Prins ND, Scheltens P, Teunissen CE, van der Flier WM. Serum markers glial fibrillary acidic protein and neurofilament light for prognosis and monitoring in cognitively normal older people: a prospective memory clinic-based cohort study. The Lancet Healthy Longevity. 2021;2(2):e87–95. https://doi.org/10.1016/S2666-7568(20)30061-1.
142. Villemagne VL, Burnham S, Bourgeat P, Brown B, Ellis KA, Salvado O, Szoeke C, Macaulay SL, Martins R, Maruff P, Ames D, Rowe CC, Masters CL. Amyloid β deposition, neurodegeneration, and cognitive decline in sporadic Alzheimer's disease: a prospective cohort study. Lancet Neurol. 2013;12(4):357–67. https://doi.org/10.1016/s1474-4422(13)70044-9.
143. Voevodskaya O, Poulakis K, Sundgren P, van Westen D, Palmqvist S, Wahlund LO, Stomrud E, Hansson O, Westman E. Brain myoinositol as a potential marker of amyloid-related pathology: a longitudinal study. Neurology. 2019;92(5):e395–405. https://doi.org/10.1212/wnl.0000000000006852.
144. Voevodskaya O, Sundgren PC, Strandberg O, Zetterberg H, Minthon L, Blennow K, Wahlund LO, Westman E, Hansson O, Swedish Bio FSG. Myo-inositol changes precede amyloid pathology and relate to APOE genotype in Alzheimer disease. Neurology. 2016;86(19):1754–61. https://doi.org/10.1212/wnl.0000000000002672.
145. Wang H, Tan L, Wang HF, Liu Y, Yin RH, Wang WY, Chang XL, Jiang T, Yu JT. Magnetic resonance spectroscopy in Alzheimer's disease: systematic review and meta-analysis [review]. J Alzheimers Dis. 2015;46(4):1049–70. https://doi.org/10.3233/JAD-143225.
146. Wang Z, Zhao C, Yu L, Zhou W, Li K. Regional metabolic changes in the hippocampus and posterior cingulate area detected with 3-Tesla magnetic resonance spectroscopy in patients with mild cognitive impairment and Alzheimer disease. Acta Radiol. 2009;50(3):312–9. https://doi.org/10.1080/02841850802709219.
147. Warsi MA, Molloy W, Noseworthy MD. Correlating brain blood oxygenation level dependent (BOLD) fractal dimension mapping with magnetic resonance spectroscopy (MRS) in Alzheimer's disease. MAGMA. 2012;25(5):335–44. https://doi.org/10.1007/s10334-012-0312-0.
148. Watanabe T, Shiino A, Akiguchi I. Absolute quantification in proton magnetic resonance spectroscopy is useful to differentiate amnesic mild cognitive impairment from Alzheimer's disease and healthy aging. Dement Geriatr Cogn Disord. 2010;30(1):71–7. https://doi.org/10.1159/000318750.
149. Wong D, Atiya S, Fogarty J, Montero-Odasso M, Pasternak SH, Brymer C, Borrie MJ, Bartha R. Reduced hippocampal glutamate and posterior cingulate N-acetyl aspartate in mild cognitive impairment and Alzheimer's disease is associated with episodic memory performance and white matter integrity in the cingulum: a pilot study. J Alzheimers Dis. 2020;73(4):1385–405. https://doi.org/10.3233/JAD-190773.

150. Wurtman RJ, Blusztajn JK, Maire JC. "Autocannibalism" of choline-containing membrane phospholipids in the pathogenesis of Alzheimer's disease-A hypothesis. Neurochem Int. 1985;7(2):369–72. https://doi.org/10.1016/0197-0186(85)90127-5.
151. Wyss M, Kaddurah-Daouk R. Creatine and creatinine metabolism. Physiol Rev. 2000;80(3):1107–213. https://doi.org/10.1152/physrev.2000.80.3.1107.
152. Xu J, Begley P, Church SJ, Patassini S, Hollywood KA, Jüllig M, Curtis MA, Waldvogel HJ, Faull RL, Unwin RD, Cooper GJ. Graded perturbations of metabolism in multiple regions of human brain in Alzheimer's disease: snapshot of a pervasive metabolic disorder. Biochim Biophys Acta. 2016;1862(6):1084–92. https://doi.org/10.1016/j.bbadis.2016.03.001.
153. Zabel M, Nackenoff A, Kirsch WM, Harrison FE, Perry G, Schrag M. Markers of oxidative damage to lipids, nucleic acids and proteins and antioxidant enzymes activities in Alzheimer's disease brain: a meta-analysis in human pathological specimens. Free Radic Biol Med. 2018;115:351–60. https://doi.org/10.1016/j.freeradbiomed.2017.12.016.
154. Zeydan B, Deelchand DK, Tosakulwong N, Lesnick TG, Kantarci OH, Machulda MM, Knopman DS, Lowe VJ, Jack CR Jr, Petersen RC, Öz G, Kantarci K. Decreased glutamate levels in patients with amnestic mild cognitive impairment: an sLASER proton MR spectroscopy and PiB-PET study. J Neuroimaging. 2017;27(6):630–6. https://doi.org/10.1111/jon.12454.
155. Zhang C, Rodriguez C, Spaulding J, Aw TY, Feng J. Age-dependent and tissue-related glutathione redox status in a mouse model of Alzheimer's disease. J Alzheimers Dis. 2012;28(3):655–66. https://doi.org/10.3233/jad-2011-111244.
156. Zhang N, Song X, Bartha R, Beyea S, D'Arcy R, Zhang Y, Rockwood K. Advances in high-field magnetic resonance spectroscopy in Alzheimer's disease. Curr Alzheimer Res. 2014;11(4):367–88. https://doi.org/10.2174/1567205011666140302200312.

Neurodegenerative Disorders: Involvement of the Locus Coeruleus in the Modulation of Diabetic Pain

Angel Nuñez, Alberto Mesa-Lombardo, Nuria García-Magro, and Yasmina B. Martin

Abstract The locus coeruleus (LC) is located in the rostral pons and is characterized by its rich concentration of noradrenergic neurons. It provides the most of noradrenergic innervation for the whole brain, playing a critical role in a wide range of cognitive and physiological functions. The LC plays also an important role in pain control. Recent neuropathological data has shown that LC degeneration may facilitate the occurrence of neurodegenerative diseases such as Alzheimer's disease or chronic neuropathic pain (Martins et al., Anesthesiology 123(3):642–-53, 2015; Galgani and Giorgi, Curr Neurol Neurosci Rep, 23(12):925—936, 2023). Diabetes mellitus is one of the most common serious metabolic disorders in humans. Manifestations of diabetic cerebral disorders have been demonstrated at a neurochemical, electrophysiological, structural, and behavioral level. Animal models of diabetes exhibit alterations in neurotransmission, such as reduced evoked GABA release, increased excitatory glutamate signaling, and reduced dopamine signaling (Broderick and Jacoby, Diabetes, 37(7):956—60, 1988; Perez-Taboada et al., Mov Disord 35(9):1636—1648, 2020). These alterations may explain the increased risk for the development of affective disorders observed in these patients.

This review describes the complex role of the LC and the noradrenergic system within the context of diabetic neuropathic neurodegeneration. Our data show that electrical or chemical stimulation techniques of the LC can reduce diabetic neuropathic pain. Therefore, therapies targeting the LC could be beneficial in attenuating or preventing the development of persistent pain and related comorbidities.

A. Nuñez (✉) · A. Mesa-Lombardo
Department of Anatomy, Histology and Neurosciences, Universidad Autónoma de Madrid, Madrid, Spain
e-mail: angel.nunez@uam.es

N. García-Magro
Facultad de Ciencias de la Salud, Universidad Francisco de Vitoria, Pozuelo de Alarcón, Madrid, Spain

Y. B. Martin
Facultad de Medicina, Universidad Francisco de Vitoria, Pozuelo de Alarcón, Madrid, Spain

P. A. Broderick (ed.), *Biomedical Imaging in Neurodegeneration*,
https://doi.org/10.1007/978-3-031-95878-6_10

Keywords Neuropathic pain · Noradrenergic transmission · Diabetes · Noradrenergic receptor · Neurodegeneration

1 Introduction

The increasing prevalence of diabetes in recent times has contributed to a rise in the number of patients with diabetic complications. Diabetic peripheral neuropathy is an axonopathy that affects at least 50% of patients with diabetes mellitus. It is characterized by sensory loss and pain initially affecting small unmyelinated C fibers, followed by large, myelinated nerve fibers. This condition has a variety of clinical manifestations, including corneal nerve loss, dysesthesia (abnormal sense of touch), and pain [1, 2].

Neuropathic pain is usually caused by chronic, progressive nerve disease, although it can also occur as the result of central or peripheral nervous system injury or infection. In addition, neuropathic pain can be a symptom or complication of several diseases and conditions. These include multiple sclerosis, multiple myeloma, cancer, and diabetes [1, 3]. The efficacy of the current therapies for chronic pain and trigeminal neuropathic pain, in particular, remains low [4]. Neuropathic pain is one of the aspects of the diabetic polyneuropathy affecting 10–26% of people with diabetes and markedly affecting their quality of life, influencing the ability to perform the activities of daily living. Neuropathic pain appears predominantly as hyperalgesia and/or allodynia [5].

The locus coeruleus (LC) is a small, bilaterally paired important nucleus in the central nervous system (CNS) that plays a critical role in a wide range of cognitive and physiological functions. The LC is characterized by its rich concentration of noradrenergic neurons. The name of this nucleus, "Blue Spot," is derived from the Latin, which describes the bluish pigmentation resulting from the presence of neuromelanin. This pigment is associated with the synthesis and storage of noradrenaline (NA), also known as norepinephrine, the primary neurotransmitter produced by the LC neurons. These specialized cells synthesize and release the neurotransmitter NA in ample areas of the brain, which plays an essential role in modulating numerous neuronal processes. The effect of NA on neurons can be excitatory, increasing the activity of the postsynaptic neuron, or inhibitory, decreasing its activity. These opposite effects are produced by the interaction of the NA with different specific receptors (see below).

The ability of the LC to release NA and modulate neuronal activity in key brain areas makes it a central hub to the coordination of a variety of cognitive and emotional processes such as the regulation of attention, alertness, and vigilance. Through its widely distributed projections, the LC exerts significant control over key brain areas such as the thalamus, the prefrontal cortex, and the limbic system. The release of NA by the LC increases neuronal excitability in these regions, enhancing the ability to respond to relevant stimuli and promoting adaptation to the environment. In addition, the LC is thought to play an important role in pain control.

In this case, the main effect of NA is inhibitory. Thus, the descending NA inhibitory projections are considered to be one of the main pathways in the endogenous pain modulation system. For these reasons, the deterioration of LC with age or for pathological conditions, such as diabetes, may contribute to neurodegenerative diseases, particularly Alzheimer' disease, or chronic neuropathic pain [1, 5, 6].

2 Noradrenergic Control of Neuronal Activity

The LC is a relatively small structure with about 3000 neurons in rodents [7, 8] and about 45,000–50,000 cells in the adult human brain [9]. LC neurons are identified by their immunoreactivity for tyrosine hydroxylase and dopamine-β-hydroxylase, the two enzymes critically involved in NA biosynthesis. The ability of the LC to release NA and modulate neuronal activity in different brain areas makes it a central node in the coordination of a variety of cognitive and emotional processes such as the sleep/wake state, arousal, attention, mood, stress reactions, and pain [10–13]. Through its widely distributed projections, the LC exerts significant control over the key areas of the brain involved in these processes, such as the thalamus, the prefrontal cortex, and the limbic system [14–16]. In addition, the LC is also involved in mood regulation and stress response. Alterations in LC function and NA levels in the brain have been associated with psychiatric disorders such as depression, schizophrenia [17], anxiety or stress [18–20], depression [21], and Alzheimer's disease [22], underscoring the importance of a detailed understanding of this nucleus and its functions.

The LC also extends a widespread descending pathway which includes nuclei in the pain pathway such as the spinal cord and the caudalis division of the spinal trigeminal nucleus (Sp5C) [8, 23–27]. Regarding pain modulation, initial experiments have shown that activation of the LC region inhibits nociceptive activity in the dorsal horn neurons [28–31] and increases the spinal content of NA [32]. Thus, the LC inhibitory system appears to play a significant role in nociceptive sensory processing [12, 33]. However, although the inhibitory effect is predominant at spinal levels, some authors have shown that the LC also facilitates pain sensation in supraspinal regions, mainly when the pain is chronic, and by participating in its emotional component [34]. These contradictory results are due to the existence of groups of neurons in the LC that are activated under different conditions and excite specific NA receptors (see below).

3 Cellular Organization Within the LC

The LC was first described in the late eighteenth century by the French anatomist, Félix Vicq d'Azyr. However, it was not until the advent of fluorescence histochemistry in the 1960s that it was discovered that the LC was comprised predominantly of monoaminergic neurons [35, 36]. The LC was initially regarded as a mere homogenous anatomical structure. However, over the years, extensive research has demonstrated its pivotal role in regulating numerous physiological and cognitive functions, establishing it as a significant player in the intricate scenery of the brain [37]. Recent studies have shown that LC has different neuronal subgroups, with clearly segregated projections and with different effects on their projection areas. Excitatory or inhibitory effects have been reported in these areas, mainly due to activation of different NA receptors.

There are several noradrenergic areas, classified from A1 to A7, in the CNS [14]. These nuclei have originally been described in rodents, and most of them have also been identified in primates [38, 39], including humans [40]. The LC, or area A6, is the largest of these groups, and its neurons have extensively branched axons that project throughout the brain providing the main source of NA to the neocortex, hippocampus, amygdala, thalamus, cerebellum, and spinal cord [7].

Traditionally, it was thought that NA neurons in the LC project together to other areas of the brain through a highly branched axonal arborization, responding in a coordinated and uniform manner to relevant stimuli or sensory stimuli [14]. However, there is evidence that LC neurons work as cellular subsets showing different projections to other areas of the brain, spinal, or corticolimbic regions [27, 41–43] or with neurochemical properties [44, 45].

The LC neurons exhibit diverse morphologies [46]. The majority of neurons are multipolar cells: mainly medium-sized cells (~35 μm) with an elongated and polar shape, featuring three or four lengthy and slender dendrites located in the ventral LC [47, 48] and smaller fusiform cells (~20 μm) located in the dorsal LC region [49, 50]. Additionally, the caudal and ventrolateral regions of the LC, which encompass the subcoeruleus area, contain smaller spindle-shaped pigmented neurons. Notably, these caudal noradrenergic cells in the subcoeruleus region have distinct targets in the brainstem and spinal cord, compared with the more anteriorly situated LC neurons [27]. The dendrites of LC neurons are densely innervated and activated by ascending spinothalamic axons that carry somatosensory and nociceptive information [26, 51].

Noradrenergic LC neurons also include a diverse range of neuropeptides, such as galanin, which is the most abundant (80%), neuropeptide Y, which is only expressed in the dorsal portion of the LC, somatostatin, cholecystokinin, and cocaine- and amphetamine-regulated transcripts [52–55]. The release of NA and one of these cotransmitters can influence NA activity at both presynaptic or postsynaptic sites [56, 57].

LC contains many neurotransmitter receptors that, depending on their expression, modify the individual response of LC neurons to similar inputs. It contains nicotinic acetylcholine receptors [58], GABA [59], orexin/hypocretin [60], and opioid receptors [61]. Therefore, LC neurons also delivery NA into the LC that activates two subtypes of G-protein-coupled adrenoreceptor "autoreceptors." The $\alpha 2$ receptor is the most abundant, but LC neurons also express $\alpha 1$ receptors [62]. The $\alpha 2$ autoreceptors exert inhibitory effects: when they are situated in a somatodendritic position, they suppress neuronal firing but when they are located on the nerve terminal membrane, where they are known as the "presynaptic receptors," they inhibit the release of NA [63] (see below).

LC neurons fire in two distinct modes: tonic and phasic discharge patterns [64], and the switch between these modes is regulated by the different behavioral states. Spontaneous tonic activity is characterized by a sustained and highly regular discharge pattern (2–5 Hz) and occurs during arousal and waking states. The phasic discharge pattern is characterized by brief 10–20 Hz bursts of two to three action potentials that are often followed by a sustained suppression of spontaneous activity for 100–200 ms [65, 66]. The LC neurons are silent during rapid eye movement (REM) sleep and have low tonic firing during the slow wave sleep. Their tonic firing rate increases during nonstressed waking, and they exhibit high tonic firing during stress periods [67]. Moreover, the LC neurons change from a tonic to a phasic mode during attentional processes in order to participate in the analysis of task-relevant stimuli. The phasic firing pattern of LC neurons allows focusing on the performance of relevant tasks by filtering and inhibiting irrelevant stimuli and is therefore closely associated with highly accurate behavioral responses [64]. In addition to attentional functions, the LC system also performs a critical role in emotional memory, and in the control of motor, sensory, and autonomic functions.

4 Organization of the Anatomical Projections from the LC

LC projections are widespread and extend throughout different brain areas, giving it a broad and diverse role in the control of brain activity (Fig. 1). The major LC projections include three distinct pathways: the ascending pathway, the cerebellar pathway, and the descending pathway [68].

The ascending pathway from the LC sends projections to structures in the midbrain, including the periaqueductal gray substance, the nucleus raphe dorsalis, and the colliculi [69]. In addition, it projects to the thalamus, particularly the pulvinar/lateral posterior complex, periventricular, anteroventral, ventral posterolateral, and reticular nucleus, as well as the midline, intralaminar, and mediodorsal thalamic nuclei [69]. The most extensive innervation in the cerebral cortex is found in the somatosensory and motor cortices [70] and associated areas, including the prefrontal and parietal cortices [71]. Other targets include the entorhinal cortex and the limbic system (including the amygdala, hippocampus, cingulate, and parahippocampal gyri) and basal forebrain cholinergic groups [70, 72–75]. All these projections are involved in the regulation of attention, alertness, memory, and emotional responses.

The LC also projects to the cerebellum [76], a structure involved in motor control and motor learning. The LC communicates with the cerebellar nuclei and cerebellar cortex via the superior cerebellar peduncle. These projections are thought to be involved in the regulation of the coordination and balance of movement.

The descendent pathway of the LC sends projections into the motor nuclei of the brainstem, such as the dorsal part of the vagal nucleus and the inferior olivary complex, before descending into the spinal cord [77]. This pathway, known as

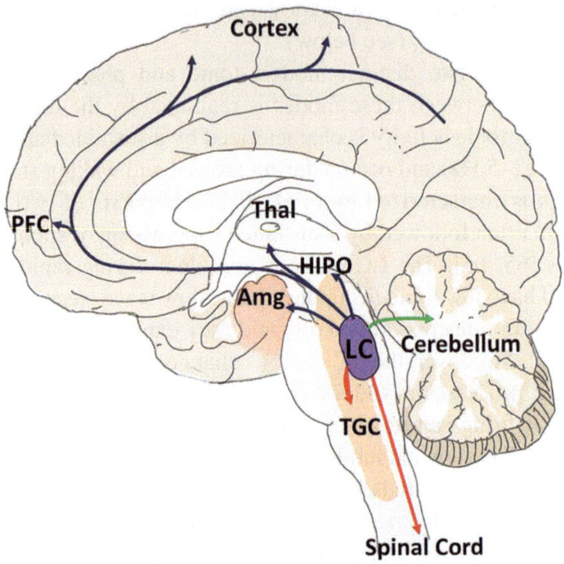

Fig. 1 Major locus coeruleus (LC) projections throughout the central nervous system (CNS). The LC is implicated in a variety of brain functions, including sleep–wake control, sensory perception, attention, or learning. Consequently, LC projects to numerous structures of the CNS: the ascending pathway (black arrows), the cerebellar pathway (green arrow), and the descending pathway (red arrows). The descending noradrenergic (NA) pathway is one of the main modulatory systems of pain sensation. Abbreviations: *Amg* amygdala, *HIPO* hippocampus, *PFC* prefrontal cortex, *TGC* trigeminal complex, *Thal* thalamus

the "locus coeruleus–spinal cord pathway," innervates spinal neurons controlling motor regulation as well as sensory and pain responses. Most of the studies of pain have concentrated on the axonal projections to the spinal dorsal horn. Descending projections to the spinal cord have been described anatomically [77–79]. Later, using adenoviral vector injections into the dorsal horn of the L4 and L5 lumbar spinal showed retrogradely stained neurons in the LC (80%), A5 (12%), and A7 (8%) areas [80]. These neurons were mainly found in the ventral pole of the LC nucleus, where they had a fusiform (70%), multipolar (24%), or round morphology (6%). The descending NA pathway from the LC to the spinal cord is mainly ipsilateral (63%), although a substantial part of the pathway also crosses the midline to innervate the contralateral side of the dorsal horn.

There is an organized cellular topography based on the output targets of the LC, which can be observed along the dorsoventral axis and the anterior–posterior axis. Thus, located in the dorsal region of the LC, there are neurons that project to regions of the forebrain, while in the ventral part the neurons are mostly involved in cerebellar projection and descending to the spinal cord. The LC cells projecting to the hypothalamus are found in the anterior region, while those projecting to the thalamus are located in the posterior region. Additionally, the LC cells projecting to the cortex and amygdala are scattered throughout the LC [45].

The LC receives synaptic inputs from numerous different brain regions where each neuron can receive at least 9–15 inputs with a mostly integrative function [45]. Inputs from the brain stem play a significant role in activating the LC, responding to important sensory stimuli such as tactile, visual, vestibular, and olfactory stimuli [81]. On the other hand, top-down synaptic inputs from the prefrontal cortex (PFC) and the central amygdala can modulate the level of LC activation according to behavioral or emotional states [82–84].

5 Noradrenergic Transmission

The NA is the major and the most important neurotransmitter synthesized by the LC neurons, exerting its action on specific pre- or postsynaptic adrenergic receptors. In addition, NA released from extrasynaptic sites may diffuse in the extracellular fluid and influence neurons, astrocytes, and microvessels via volume transmission. The effect of NA depends on the type of receptors expressed on the postsynaptic neurons.

The NA receptors are divided into two main categories: the α- and β-receptors. Several subtypes of each type are recognized. The α-adrenoceptors can be found as α1A, α1B, α1D, α2A, α2B, α2C, and α2D [85]. The α1A-, α2A-, and α2B-adrenoceptors are found extensively in the supraspinal areas, mostly coinciding with the wide distribution of ascending pathways [86, 87]. The distribution of the α1B subtype is more restricted, and it is generally found in the thalamus, amygdala, and dorsal and median raphe nuclei [87]. Three β-adrenoceptor types (β1, β2, and β3) are described and widely distributed in the CNS [85, 88]. They are known for their role in the regulation of cardiovascular, airway, and other peripheral functions. In addition, these receptors have also been implicated in functions of the CNS, including the regulation of sympathetic tone, learning and memory, mood, and food intake [89]. β-Adrenoreceptors (β-ARs) are also implicated in nociceptive processing [90]. β-AR agonists acting on β2-AR suppress neuropathic allodynia after chronic treatment [91].

In general, the activation of α1 or β-AR produces excitatory neuronal effects while the activation of α2 adrenergic receptors produces inhibitory effects on cell activity [45, 92]. The α2-NA receptors were initially characterized as presynaptic autoreceptors that serve as part of a negative feedback loop to regulate the LC firing discharge and reducing NA release in the LC and downstream areas [93–95]. The activation of presynaptic α2-NA receptors on the primary afferents inhibits voltage-gated Ca^{2+} channels to reduce the release of excitatory neurotransmitters in the spinal cord. Consequently, antagonists acting on α2-NA receptors increase neuronal excitability as well as NA release [95, 96]. The α2-NA receptors are also found in nonadrenergic cells, acting as heteroreceptors and contributing to the regulation of other neurotransmission systems [97].

It has also been shown that the NA receptors are not restricted to presynaptic locations but also have postsynaptic functions. The activation of α2-receptors evokes the hyperpolarization of the membrane potential by G-protein-mediated activation of K+ channels in the spinal cord [98, 99] and in the trigeminal subnucleus caudalis [100]. The α1-NA receptors are G-protein-coupled receptors that evoke depolarization of trigeminal subnucleus caudalis neurons as well as the thalamic and cortical neurons [100, 101]. The α1-NA receptors activate phospholipase C and phosphatidyl inositol intracellular signaling, resulting in the inactivation of protein kinase C and the release of intracellular calcium [102].

Finally, the β-NA receptors are members of a seven-transmembrane family of G-protein-coupled receptors that increase cAMP signaling in neurons [89]. The activation of β1-NA receptors leads to an increase in the inward Na + current through the hyperpolarization-activated cyclic nucleotide-gated channels, which carry a mixed Na+/K+ current [103]. β receptors are also found on glia, where they have multiple actions including reducing the uptake of glutamate [104].

6 Noradrenaline Influences Nociception

A large body of basic research has suggested that pain inhibitory systems, descending from the brain stem, may be crucial in the control of pain. Tricyclic antidepressants, serotonin–norepinephrine reuptake inhibitors or the α2-NA receptor agonist clonidine, which are thought to mimic the effect of noradrenergic analgesic systems, have long become the first-line treatment for clinical neuropathic pain [105].

Nerve injury activates LC neurons [106]. Nociceptive information arrives to the LC through dendritic synapses that are heavily innervated and activated by the ascending spinothalamic tract axons bringing innocuous and nociceptive information from the somatosensory pathways [26, 51]. For example, noxious heat or chemical stimulation increases the firing rate of LC neurons [107]. The expression of c-fos or pCREB proteins is currently used as a marker of neuronal activity. Their expression is also increased in the LC after nerve ligation indicating an increase of neuronal activity [106, 108].

In studies conducted on different species and in various pain models [29–31, 109], it has been demonstrated that LC neurons play an important role in pain modulation by inhibiting the transmission of painful information to higher structures. Electrical stimulation of the LC, which induces spike firing in these neurons, is capable of relieving neuropathic pain by attenuating allodynia and hyperalgesia through an increase of the release of NA in the spinal cord [12]. For example, trains of cathodal pulses of 10–100 Hz frequency, 50–200 μA amplitude, and 0.5 ms width increase NA level within laminae I–VIII of the lumbar spinal cord [110]. Descending projections from the LC to the spinal cord were characterized as pain inhibitory by Jones and Gebhart, who reported that electrical stimulation of the LC reduced the heat stimulus-evoked activity of dorsal horn neurons in anesthetized rodents [29, 57]. Later, experiments using optogenetic stimulation in animals genetically modified to express channelrhodopsin-2 specifically in NA neurons have shown that optogenetic stimulation of the LC neurons produces a significant antinociceptive effect in the thermal hind paw withdrawal test [107].

Furthermore, it has been demonstrated that constriction injury of the sciatic nerve increases tyrosine-hydroxylase (TH) and dopamine β-hydroxylase (DbH) immunoreactive axonal terminals in the ipsilateral L4–L6 dorsal horn, indicating an increase in NA innervation to the lumbar dorsal horn after nerve injury [111]. In agreement with these findings, electrolytic lesions of the LC increase the pain hypersensitivity in animals experiencing pain [112]. The anti-dopamine-beta-hydroxylase-saporin (anti-DbH-saporin) is taken up by noradrenergic nerve endings, resulting in selective ablation of noradrenergic neurons. Intrathecal or intracerebroventricular administration of this neurotoxin increases mechanical hypersensitivity after nerve injury [33, 113, 114]. In summary, all these experiments demonstrate that the NA neurons of the LC are responsible for antinociceptive effects.

Subsequent studies have shown that descending projections release NA [32, 110] that act via α2-NA receptors to inhibit both primary afferents and in postsynaptic neurons [31, 57, 98, 115]. The activation of the α2-NA receptors reduced NMDA-evoked responses in the medullary dorsal horn [116]. Intrathecal injection of the α2-adrenoreceptor antagonist yohimbine produced hyperalgesia on the limb contralateral to tibial nerve transection [117]. In contrast, spinal delivery of α2-NA receptor agonists reduces neuropathic pain in both animals and humans [118, 119]. These findings suggest that stimulation of NA receptors may be a potential target for pain therapy [12].

One of the mechanisms to explain its inhibitory effects has been studied in the presynaptic α2-NA receptors on the primary afferents in the spinal cord. Activation of α2-NA receptors inhibits voltage-gated N-type Ca^{2+} channels to reduce the release of excitatory neurotransmitters. Another mechanism involved in the activation of postsynaptic α2-adrenergic receptors is the opening of the inwardly rectifying K+ channels to hyperpolarize cells, thereby reducing neuronal excitability [98, 99]. Both mechanisms reduce spinal pain transmission [118, 119]. Thus, intrathecal administration of an α2-NA receptor agonist such as clonidine or antidepressants has demonstrated great efficacy in the treatment of pain [12]. Whole cell patch clamp recordings from spinal cord neurons show that the spike-firing frequency in tonic-firing neurons is reduced by clonidine through the control of voltage-gated Na + and K+ channels [120].

In addition, the α2-NA receptors are coupled with a G protein, which reduces the activity of adenylyl cyclase, simultaneously suppressing both the production of cAMP and the activity of protein kinase A (PKA) [121, 122]. These intracellular signaling pathways are involved in the regulation of the number of membrane receptors and neuronal excitability. In fact, the expression of the NMDA receptor subunit 1 in the dorsal horn is increased in rats with chronic constriction injury-induced neuropathy [123].

The caudalis division of the spinal trigeminal nucleus (Sp5C) involved in the relay of orofacial nociceptive information [124–127]. Neuroanatomical experiments

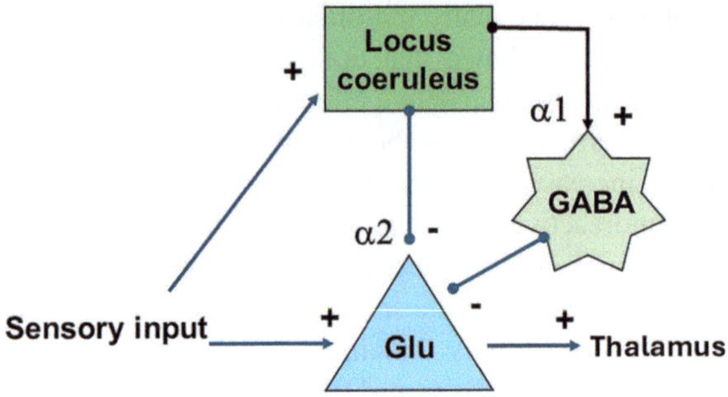

Fig. 2 Locus coeruleus (LC) may inhibit glutamatergic, projecting neurons to the thalamus by means of direct projections, activating α2-NA receptors, or though activation of GABAergic neurons that also synapse with glutamatergic, projecting neurons

have shown reciprocal pathways connecting the LC to trigeminal sensory nuclei [128, 129]. Electrophysiological studies have also shown that stimulation of the LC in different animal species inhibits the spontaneous neuronal activity and evoked responses in Sp5C neurons [130, 131]. In addition, LC stimulation produced the inhibition of neuronal responses to tactile stimulation in the trigeminal main sensory nucleus and the rostral area of the trigeminal spinal-tract nucleus [132]. Electrophysiological recordings made from neurons of the Sp5C nucleus showed that NA inhibits presynaptic glutamate release by acting on the presynaptic α2-NA receptors [133]. Thalamic projecting Sp5C neurons also receive an inhibitory input from GABAergic interneurons that may be activated by collaterals of the same NA axons that synapse in projecting neurons. Thus, LC may also inhibit projecting neurons through these interneurons because in vitro experiments have shown that NA increase the frequency of inhibitory postsynaptic potentials in these neurons [134]. Therefore, LC could exert an inhibitory synergistic effect through activation of α2 NA receptors on projecting trigeminal neurons to the thalamus and through activation of GABAergic neurons that also inhibit projecting neurons, controlling sensory flow in the Sp5C nucleus (Fig. 2).

In agreement with this hypothesis, experiments from our laboratory in rodents have shown LC electrical stimulation (10–100 μA of intensity) induced inhibition of vibrissal responses in the Sp5C neurons when applied at 50 or 100 ms before vibrissal stimulation in control animals [135]. LC-evoked inhibition was decreased by an intraperitoneal injection of the antagonist of the α2-NA receptor, yohimbine, indicating that inhibition was due to the activation of this type of receptor (see Fig. 3). At the same time, projecting Sp5C neurons received an inhibitory input from GABAergic neurons that were probably activated by collaterals of the same NA axons that synapse in projecting neurons. GABAergic neurons were excited by α1 NA receptors because LC-evoked inhibition was blocked by the α1-NA receptor antagonist benoxathian (unpublished results). The participation of α1-NA receptors on the inhibitory interneurons has been previously suggested [136].

Numerous studies have focused on the role of α2-adrenergic receptors in nociceptive sensory processing, but less information is available concerning the role of the β-NA receptors. β2-NA receptors are, however, expressed in areas directly

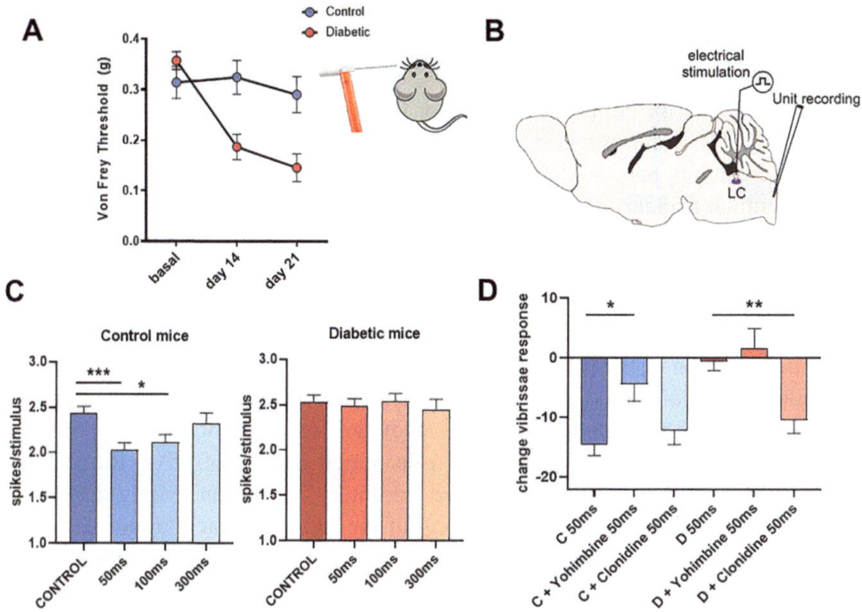

Fig. 3 An animal model of diabetes was used to demonstrate a reduction of LC-mediated inhibition. (**a**) Diabetic mice showed pain in the orofacial area because they had a decrease in the withdrawal threshold to the mechanical stimulation in the vibrissal pad; inset on the right shows a diagram of the stimulus. (**b**) A scheme of the experimental setup is shown. Electrical stimuli were applied to the LC to evaluate its inhibitory action on vibrissal responses in Sp5C neurons. (**c**) Electrical stimulation reduced vibrissal responses in the Sp5C neurons of control mice when was applied at 50 or 100 ms before vibrissal stimulation, indicating that LC exerts a inhibitory action on sensory responses (LC-evoked inhibition). However, the LC-evoked inhibition was reduced in diabetic mice. (**d**) LC-evoked inhibition was decreased by an intraperitoneal injection of the antagonist of the α2-NA receptors, yohimbine, suggesting that this inhibition is due to activation of α2-NA receptors. Accordingly, clonidine increased LC-evoked inhibition in diabetic mice (data modified from [135])

participating in pain pathways [137], and human genetic studies have also confirmed the contribution of β2-NA receptors in chronic pain disorders [138].

Although LC activity has been predominantly associated with inhibition of neuropathic pain, there is growing evidence indicating that descending and ascending projections from the LC may facilitate neuropathic pain [29, 34, 139]. These opposing effects are attributed to the different neuronal LC subpopulations because the antinociceptive effect is more frequently found when the neurons located in the ventral region of the LC are stimulated [107]. These facilitatory effects could be also due to LC projections to the dorsal reticular nucleus (DRt) [106] or via the ascending noradrenergic pathways to the mPFC [46, 64]. The DRt receives noradrenergic innervation from the LC that may contribute to the facilitation of pain transmission at the dorsal horn [140, 141]. Local manipulations to the DRt that decrease or increase noradrenaline levels attenuate or increase neuropathic pain, respectively [106]. A major noradrenergic efferent pathway from the LC is the projection to the mPFC [64], an important cortical area that is activated in patients with chronic pain [142, 143]. In rodents, electrical stimulation of the LC elicits activation of the mPFC neuronal activity through activation of α1-NA receptors [144, 145]. These data suggest, during intense or long-term activation after nerve injury, descending pain-modulatory outflows from the LC to the DRt, spinal cord, and trigeminal nucleus, as well as ascending pain modulatory outputs from the LC to the mPFC become predominantly facilitatory, behaving as a generator of chronic pain [34].

In chronic neuropathic pain models, there are changes in the NAergic innervation of the spinal dorsal horn and in α2-NA receptor sensitivity [111, 117]. The inhibitory descending action of LC predominates following tibial nerve injury when mechanical allodynia, cold allodynia, and heat hyperalgesia appear. Once neuropathic sensitization is established, a diminution of the influence of LC projections is observed, which is accompanied with a decreased density of NA-positive fibers on the dorsal horn [117], indicating that inhibitory NA action mainly acts in the acute phase after injury. The contribution of the LC to the facilitation of chronic neuropathic pain is less well known [34, 136, 139]. In an orofacial pain animal model of trigeminal neuropathy, the elimination of NA neurons through the injection of anti-DbH-saporin either intracerebroventricularly (i.c.v.) or into the Sp5C nucleus results in a reduction of mechanical hypersensitivity, suggesting that LC may play a pain-facilitatory function in the development of chronic neuropathic pain [144]. Therefore, the effect of the LC activation on neuropathic pain may shift over time, beginning with inhibition mediated by descending projections from the ventral region of the LC in acute pain, and then, switching to facilitation mediated by ascending projections from the dorsal LC to the mPFC in chronic pain.

Thus, it is essential to have a good understanding of the functioning of this nucleus because the opposing actions of this system may partly explain the limited effectiveness of tricyclic antidepressants and norepinephrine reuptake inhibitors in the treatment of chronic pain [146]. Taken together, these results confirm the powerful control that the LC has over pain sensation. The gabapentin is often recommended as first-line treatments for the management of neuropathic pain. It has been proposed that gabapentin activates LC neurons increasing NA release, which in turn stimulates α2-adrenoreceptors, leading to analgesia [147, 148]. However, further studies are still needed to distinguish the neuroanatomical and neurochemical characteristics of the pain-reducing and pain-enhancing actions of the LC in the modulation of pain.

7 Diabetic Neuropathy

Diabetes mellitus is one of the most common serious metabolic disorders in humans. Manifestations of diabetic cerebral disorders have been demonstrated at a neurochemical, electrophysiological, structural, and behavioral level. Rodent models of diabetes exhibit alterations in neurotransmission, such as reduced evoked GABA release, increased excitatory glutamate signaling, and reduced dopamine signaling, which suggest specific neuronal deficits. For example, altered monoaminergic functions, such as serotoninergic and dopaminergic transmission, have been observed in animal models of diabetes. These alterations may explain the increased risk for the development of affective disorders observed in these patients [149–151].

In addition, diabetic neuropathy is a common complication of both type 1 and type 2 diabetic patients that show sensory loss, increasing falls, pain and consequently, a reduction in their quality of life [152, 153]. In addition, there is a considerable amount of evidence that supports the notion that the entire neuron, from the perikaryon to the axonal terminal, is affected by diabetes. For example, type 2 diabetes mellitus is associated with neurodegenerative disorders such as Alzheimer's disease. It has been postulated that diabetes could be a novel mechanism of neurodegeneration wherein the Alzheimer's disease pathophysiology can be explained from the perspective of unregulated insulin/IGF (insulin-like growth factor) signaling and improper glucose metabolism [154].

Approximately 30–50% of patients with diabetic neuropathy develop neuropathic pain [5]. Diabetic neuropathy is a neurodegenerative disorder of the peripheral nervous system that preferentially targets sensory axons, autonomic axons, and later, to a lesser extent, motor axons [153, 155]. Patients report spontaneous burning pain or brush-evoked allodynia, when a normally nonnoxious stimulus evokes pain, and paresthesias. It is unknown whether neuropathic pain is due to damage to peripheral axons and their associated Schwann cells, or to neuronal perikarya. A loss of intraepidermal nerve fibers is associated with the presence of neuropathic pain [153]. However, one of the most important reasons for the generation of peripheral neuropathy may be the impairment of neuronal activity caused by diabetes. Lesioned sensory neurons, as occur in diabetic neuropathy, develop hyperexcitability and thus can generate action potential discharges in the absence of a stimulus (spontaneous activity) and can develop an altered stimulus–response function [156–158]. Primary afferent hyperexcitability is a critical point to drive pain because a local anesthetic injection into the nerves innervating the lower limb in patients with neuropathic pain blocks it [159]. The origin of this hyperexcitability may be due to the changes evoked by diabetes in the intrinsic neuronal properties of the axonal terminals. For example, painful diabetic neuropathy is associated with an increased expression of Nav1.8 channels in sensory neurons, facilitating action potential transmission [160]. T-type Ca^{2+} channels have also been implicated in painful diabetic neuropathy through the regulation of the subthreshold excitability of nociceptors. CaV3.2 activity is enhanced in diabetes resulting in dorsal root ganglion neuron hyperexcitability [161]. Therefore, an increase in Ca^{2+} conductance may be responsible for the increase in hyperexcitability. Consequently, reducing Ca^{2+} current decreases pain-related hypersensitivity as has been demonstrated in an animal model of diabetes [162]. In addition, morphological changes in the central neurons also occur in diabetes. In animal models of diabetic neuropathy, spinal neurons show hyperexcitability to peripheral stimuli, which is associated with altered shape, i.e., an increased length and spine head diameter, increased neuronal density, and redistribution of dendritic spines [163, 164]. Changes in glial cells have also been described in diabetic neuropathy.

The balance between hyperexcitability from the periphery and a reduction in inhibition from the brainstem has been suggested to have a role in the development of neuropathic pain. Data from both animal models and patients suggest that spinal disinhibition can contribute to the development of neuropathic pain in diabetic neuropathy [165].

Changes in LC modulatory actions may be also involved in diabetic neuropathy and may contribute to development of chronic pain. To deal with this issue, we applied short-lasting electrical stimulation (0.2 ms duration) in the LC to activate its neurons. A bipolar stimulating electrode was placed in the LC according to the atlas of Paxinos and Franklin (coordinates from Bregma: A: −5.4 mm, L: 0.9 mm lateral, H: 3.5 mm; [166]). Electrical stimulation of the LC induced an inhibition of vibrissal responses in the Sp5C nucleus when they were stimulated with a short delay (50 or 100 ms) after LC electrical stimuli (LC-evoked inhibition). However, the LC-evoked inhibition was reduced in a rodent model of diabetes, which showed pain in the orofacial area [135]. The LC-evoked inhibition was due to the activation of the α2-NA receptors because it was reduced by intraperitoneal injection of the receptor antagonist yohimbine (2 mg/Kg) and increased by the intraperitoneal injection of the receptor agonist clonidine (2 mg/Kg); clonidine effect was clearly observed in diabetic animals in which LC-evoked inhibition was reduced (Fig. 2). Therefore, these experiments suggest that the reduction of the LC-evoked inhibition could favor the development of chronic pain in diabetes. This reduction may be due to a decrease of LC activity.

It is well known that the number of intraepidermal nerve fibers in diabetic subjects is decreased compared with age-matched nondiabetic healthy control subjects, and diabetic subjects have shorter nerves that often end bluntly at the dermal surface of the basement membrane [167]. Therefore, these data suggest that chronic pain in diabetic patients may be due to peripheral nerve degeneration. Our studies cannot rule out this possibility; however, our data showed that the descending inhibitory pathway from LC was reduced in diabetic animals. This reduction may be due to a decrease in the TH+ LC neurons as well as in the TH+ stained fibers in the Sp5C of diabetic mice [135]. In addition, it is also known that TH+ neurons are present in the ganglion although their function is not well established [168–170]. We observed a reduction of TH+ neurons in the ganglion that may be a result of the peripheral nerve degeneration, as has been described above in diabetic patients. However, the fact that in our experiments, both terminals of lamina II, where inputs come from the periphery end, and those of laminae III– IV, whose fibers come from other nuclei, were reduced, strongly suggests that most of the TH+ fibers reduced in diabetic animals come from the LC. Although our findings suggest a reduction in the NA inputs to Sp5C, their α2-NA receptors remained active in Sp5C neurons because the LC-evoked inhibition could be increased in the diabetic mice when the α2-NA receptor agonist clonidine was injected intraperitoneally (see Fig. 3d). These data explain why drugs such as clonidine, which activate α2-NA receptors, may have a therapeutic effect in relieving chronic pain in diabetic patients.

GABAergic and glycinergic neurons modulate the synaptic transmission in the Sp5C and their activity changes after the induction of pain by infraorbital chronic constriction injury [171]. Our preliminary findings suggest that the balance between the direct inhibition evoked by the LC through α2-NA receptors and indirect inhibition evoked by the activation of GABAergic and/or glycinergic neurons could be altered in diabetic animals.

The differential effects of NA activity on neuropathic pain might result from differences in the etiology of the neuropathic pain. In a chronic facial neuropathic pain model, ascending LC NA input to mPFC facilitates pain transmission by the activation of α1-NA receptor [143] (see above). In this context, chronic nerve constriction that induces neuropathic pain increases glutamatergic synaptic transmission and basal extracellular glutamate concentrations in the LC [172, 173]. These findings suggest a facilitatory function of LC in chronic pain, mainly through ascending LC NA pathways, rather than the LC inhibitory descending pathway, which has been described in acute pain models. Trigeminal neuralgia increases the risk of developing depressive, anxiety, or sleep disorders [4]. Animal models with trigeminal neuropathic pain also exhibit anxiety-like behavior in rodents. Trigeminal neuropathic pain produced by chronic constriction injury of the infraorbital nerve was found to cause anxiety-like behavior in rodents evaluated in behavioral tests such as the open field, elevated plus-maze, or light/dark transition tests, while no depressive-like behavior was observed in the forced swimming test [174, 175]. For those reasons, the pain experienced by diabetic patients may be caused by both a reduction in the inhibitory NA projections from the LC to the spinal cord and caudal trigeminal nucleus, and the facilitation of pain due to ascending NA projections to the cerebral cortex.

8 Conclusion and Perspectives

Although an association between diabetes mellitus and neurodegeneration has been recognized for a century, it is often not considered as a complication of this pathology. Cognitive dysfunction, usually present as mild cognitive impairment, can occur with either type 1 or type 2 diabetes mellitus. Although the precise mechanism remains unclear, diabetes can exacerbate neurodegenerative processes. Thus, one promising area of research is the investigation of antidiabetic drugs for the treatment of neurodegenerative diseases such as Alzheimer's or Parkinson's diseases [176].

Results indicate that the LC is an important hub for controlling the sensory and emotional integration of pain. The inhibition caused by LC plays an important role in the inhibition of acute pain. LC neuronal activity may also facilitate pain sensation in chronic pain states. This facilitator-inhibitor balance can be altered in some neuropathies such as diabetes, facilitating the existence of persistent pain. In the case of diabetes, the lack of trophic factors such as insulin and IGF-I could be responsible for the degeneration of the LC [177]. Therefore, therapies aimed at preventing LC deterioration could be beneficial in attenuating or preventing the development of persistent pain and related comorbidities.

References

1. Hebert HL, et al. Risk factors for neuropathic pain in diabetes mellitus. Pain. 2017;158(4):560–8.
2. D'Onofrio L, et al. Small nerve fiber damage and langerhans cells in type 1 and type 2 diabetes and LADA measured by corneal confocal microscopy. Invest Ophthalmol Vis Sci. 2021;62(6):5.
3. Fallon MT. Neuropathic pain in cancer. Br J Anaesth. 2013;111(1):105–11.
4. Gierthmuhlen J, Baron R. Neuropathic pain. Semin Neurol. 2016;36(5):462–8.
5. Abbott CA, et al. Prevalence and characteristics of painful diabetic neuropathy in a large community-based diabetic population in the U.K. Diabetes Care. 2011;34(10):2220–4.
6. Galgani A, Giorgi FS. Exploring the role of locus Coeruleus in Alzheimer's disease: a comprehensive update on MRI studies and implications. Curr Neurol Neurosci Rep. 2023;23(12):925–36.
7. Lindvall O, Bjorklund A. The organization of the ascending catecholamine neuron systems in the rat brain as revealed by the glyoxylic acid fluorescence method. Acta Physiol Scand Suppl. 1974;412:1–48.
8. Loughlin SE, Foote SL, Grzanna R. Efferent projections of nucleus locus coeruleus: morphologic subpopulations have different efferent targets. Neuroscience. 1986;18(2):307–19.
9. Sharma Y, et al. Comparative anatomy of the locus coeruleus in humans and nonhuman primates. J Comp Neurol. 2010;518(7):963–71.
10. Glavin GB. Stress and brain noradrenaline: a review. Neurosci Biobehav Rev. 1985;9(2):233–43.
11. Mitchell HA, Weinshenker D. Good night and good luck: norepinephrine in sleep pharmacology. Biochem Pharmacol. 2010;79(6):801–9.
12. Pertovaara A. The noradrenergic pain regulation system: a potential target for pain therapy. Eur J Pharmacol. 2013;716(1–3):2–7.
13. Borodovitsyna O, Flamini M, Chandler D. Noradrenergic modulation of cognition in health and disease. Neural Plast. 2017;2017:6031478.
14. Dahlstrom A, Fuxe K. Localization of monoamines in the lower brain stem. Experientia. 1964;20(7):398–9.
15. Foote S.L.a.M.J.H. Extrathalamic modulation of cortical function. Ann. Rev Neurosci. 1987;10:67–95.
16. Aston-Jones G. Brain structures and receptors involved in alertness. Sleep Med. 2005;6(Suppl 1):S3–7.
17. Friedman JI, Adler DN, Davis KL. The role of norepinephrine in the pathophysiology of cognitive disorders: potential applications to the treatment of cognitive dysfunction in schizophrenia and Alzheimer's disease. Biol Psychiatry. 1999;46(9):1243–52.
18. Valentino RJ, Van Bockstaele E. Convergent regulation of locus coeruleus activity as an adaptive response to stress. Eur J Pharmacol. 2008;583(2–3)(203):194.
19. Arnsten AF. Stress weakens prefrontal networks: molecular insults to higher cognition. Nat Neurosci. 2015;18(10):1376–85.
20. Reyes BA, et al. Social stress engages neurochemically-distinct afferents to the rat locus Coeruleus depending on coping strategy. eNeuro. 2015;2(6):ENEURO.0042.
21. Delgado PL. Depression: the case for a monoamine deficiency. J Clin Psychiatry. 2000;61(Suppl 6):7–11.

22. Weinshenker D. Functional consequences of locus coeruleus degeneration in Alzheimer's disease. Curr Alzheimer Res. 2008;5(3):342–5.
23. Loughlin SE, Foote SL, Bloom FE. Efferent projections of nucleus locus coeruleus: topographic organization of cells of origin demonstrated by three-dimensional reconstruction. Neuroscience. 1986;18(2):291–306.
24. Mason ST, Fibiger HC. Physiological function of descending noradrenaline projections to the spinal cord: role in post-decapitation convulsions. Eur J Pharmacol. 1979;57(1):29–34.
25. Millan MJ. Descending control of pain. Prog Neurobiol. 2002;66(6):355–474.
26. Panneton WM, Gan Q, Livergood RS. A trigeminoreticular pathway: implications in pain. PLoS One. 2011;6(9):e24499.
27. Westlund KN, DanCoulter J. Descending projections of the locus coeruleus and subcoeruleus/medial parabrachial nuclei in monkey: axonal transport studies and dopamine-beta-hydroxilase immunocytochemistry. Brain Res Rev. 1980;2:235–64.
28. Hodge CJ Jr, et al. Locus coeruleus modulation of dorsal horn unit responses to cutaneous stimulation. Brain Res. 1981;204(2):415–20.
29. Jones SL, Gebhart GF. Quantitative characterization of ceruleospinal inhibition of nociceptive transmission in the rat. J Neurophysiol. 1986;56(5):1397–410.
30. Jones SL. Descending noradrenergic influences on pain. Prog Brain Res. 1991;88:381–94.
31. West WL, Yeomans DC, Proudfit HK. The function of noradrenergic neurons in mediating antinociception induced by electrical stimulation of the locus coeruleus in two different sources of Sprague-Dawley rats. Brain Res. 1993;626(1–2):127–35.
32. Crawley JN, Roth RH, Maas JW. Locus coeruleus stimulation increases noradrenergic metabolite levels in rat spinal cord. Brain Res. 1979;166(1):180–4.
33. Llorca-Torralba M, et al. Noradrenergic locus Coeruleus pathways in pain modulation. Neuroscience. 2016;338:93–113.
34. Taylor BK, Westlund KN. The noradrenergic locus coeruleus as a chronic pain generator. J Neurosci Res. 2017;95(6):1336–46.
35. Carlsson A, et al. Histochemical localization at the cellular level of hypothalamic noradrenaline. Acta Physiol Scand. 1962;54:385–6.
36. Carlsson A, Falck B, Hillarp NA. Cellular localization of brain monoamines. Acta Physiol Scand Suppl. 1962;56(196):1–28.
37. Maeda T. The locus coeruleus: history. J Chem Neuroanat. 2000;18(1–2):57–64.
38. Jacobowitz DM, MacLean PD. A brainstem atlas of catecholaminergic neurons and serotonergic perikarya in a pygmy primate (Cebuella pygmaea). J Comp Neurol. 1978;177(3):397–416.
39. Felten DL, Sladek JR Jr. Monoamine distribution in primate brain V. Monoaminergic nuclei: anatomy, pathways and local organization. Brain Res Bull. 1983;10(2):171–284.
40. Bogerts B. A brainstem atlas of catecholaminergic neurons in man, using melanin as a natural marker. J Comp Neurol. 1981;197(1):63–80.
41. Jones BE, et al. Ascending projections of the locus coeruleus in the rat. I. Axonal transport in central noradrenaline neurons. Brain Res. 1977;127(1):1–21.
42. Chandler DJ, et al. Redefining noradrenergic neuromodulation of behavior: impacts of a modular locus Coeruleus architecture. J Neurosci. 2019;39(42):8239–49.
43. Poe GR, et al. Locus coeruleus: a new look at the blue spot. Nat Rev Neurosci. 2020;21(11):644–59.
44. Chandler DJ, Gao WJ, Waterhouse BD. Heterogeneous organization of the locus coeruleus projections to prefrontal and motor cortices. Proc Natl Acad Sci USA. 2014;111(18):6816–21.
45. Schwarz LA, Luo L. Organization of the locus coeruleus-norepinephrine system. Curr Biol. 2015;25(21):R1051–6.
46. Aston-Jones G, Rajkowski J, Cohen J. Locus coeruleus and regulation of behavioral flexibility and attention. Prog Brain Res. 2000;126:165–82.

47. Chan-Palay V, Asan E. Quantitation of catecholamine neurons in the locus coeruleus in human brains of normal young and older adults and in depression. J Comp Neurol. 1989;287(3):357–72.
48. Patt S, Gerhard L. A Golgi study of human locus coeruleus in normal brains and in Parkinson's disease. Neuropathol Appl Neurobiol. 1993;19(6):519–23.
49. Swanson LW. The locus coeruleus: a cytoarchitectonic, Golgi and immunohistochemical study in the albino rat. Brain Res. 1976;110(1):39–56.
50. Grzanna R, Molliver ME. The locus coeruleus in the rat: an immunohistochemical delineation. Neuroscience. 1980;5(1):21–40.
51. Westlund KN, Craig AD. Association of spinal lamina I projections with brainstem catecholamine neurons in the monkey. Exp Brain Res. 1996;110(2):151–62.
52. Holets VR, et al. Locus coeruleus neurons in the rat containing neuropeptide Y, tyrosine hydroxylase or galanin and their efferent projections to the spinal cord, cerebral cortex and hypothalamus. Neuroscience. 1988;24(3):893–906.
53. Sutin EL, Jacobowitz DM. Neurochemicals in the dorsal pontine tegmentum. Prog Brain Res. 1991;88:3–14.
54. Simpson KL, Waterhouse BD, Lin RC. Origin, distribution, and morphology of galaninergic fibers in the rodent trigeminal system. J Comp Neurol. 1999;411(3):524–34.
55. Devoto P, et al. Co-release of noradrenaline and dopamine in the prefrontal cortex after acute morphine and during morphine withdrawal. Psychopharmacology. 2002;160(2):220–4.
56. Burnstock G. Purinergic receptors and pain. Curr Pharm Des. 2009;15(15):1717–35.
57. Gebhart GF. Descending modulation of pain. Neurosci Biobehav Rev. 2004;27(8):729–37.
58. Egan TM, North RA. Actions of acetylcholine and nicotine on rat locus coeruleus neurons in vitro. Neuroscience. 1986;19(2):565–71.
59. Luque JM, Malherbe P, Richards JG. Localization of GABAA receptor subunit mRNAs in the rat locus coeruleus. Brain Res Mol Brain Res. 1994;24(1–4):219–26.
60. Marcus JN, et al. Differential expression of orexin receptors 1 and 2 in the rat brain. J Comp Neurol. 2001;435(1):6–25.
61. Mansour A, et al. Mu, delta, and kappa opioid receptor mRNA expression in the rat CNS: an in situ hybridization study. J Comp Neurol. 1994;350(3):412–38.
62. Chamba G, et al. Distribution of alpha-1 and alpha-2 binding sites in the rat locus coeruleus. Brain Res Bull. 1991;26(2):185–93.
63. Starke K. Presynaptic autoreceptors in the third decade: focus on alpha2-adrenoceptors. J Neurochem. 2001;78(4):685–93.
64. Aston-Jones G, Cohen JD. An integrative theory of locus coeruleus-norepinephrine function: adaptive gain and optimal performance. Annu Rev Neurosci. 2005;28:403–50.
65. Akaike T. Periodic bursting activities of locus coeruleus neurons in the rat. Brain Res. 1982;239:629–33.
66. Clayton EC, et al. Phasic activation of monkey locus ceruleus neurons by simple decisions in a forced-choice task. J Neurosci. 2004;24(44):9914–20.
67. Aston-Jones G, Chiang C, Alexinsky T. Discharge of noradrenergic locus coeruleus neurons in behaving rats and monkeys suggests a role in vigilance. Prog Brain Res. 1991;88:501–20.
68. Benarroch EE. Locus coeruleus. Cell Tissue Res. 2018;373(1):221–32.
69. Morrison JH, et al. Laminar, tangential and regional organization of the noradrenergic innervation of monkey cortex: dopamine-beta-hydroxylase immunohistochemistry. Brain Res Bull. 1982;9(1–6):309–19.
70. Gaspar P, et al. Catecholamine innervation of the human cerebral cortex as revealed by comparative immunohistochemistry of tyrosine hydroxylase and dopamine-beta-hydroxylase. J Comp Neurol. 1989;279(2):249–71.
71. Morrison JH, et al. Noradrenergic innervation patterns in three regions of medial cortex: an immunofluorescence characterization. Brain Res Bull. 1979;4(6):849–57.
72. Leichnetz GR, et al. Afferent and efferent connections of the cholinoceptive medial pontine reticular formation (region of the ventral tegmental nucleus) in the cat. Brain Res Bull. 1989;22(4):665–88.

73. Radley JJ, Williams B, Sawchenko PE. Noradrenergic innervation of the dorsal medial prefrontal cortex modulates hypothalamo-pituitary-adrenal responses to acute emotional stress. J Neurosci. 2008;28(22):5806–16.
74. Gompf HS, et al. Locus ceruleus and anterior cingulate cortex sustain wakefulness in a novel environment. J Neurosci. 2010;30(43):14543–51.
75. Sadikot AF, Parent A. The monoaminergic innervation of the amygdala in the squirrel monkey: an immunohistochemical study. Neuroscience. 1990;36(2):431–47.
76. Nystrom B, Olson L, Ungerstedt U. Noradrenaline nerve terminals in human cerebral cortices: first histochemical evidence. Science. 1972;176(4037):924–6.
77. Westlund KN, Coulter JD. Descending projections of the locus coeruleus and subcoeruleus/medial parabrachial nuclei in monkey: axonal transport studies and dopamine-beta-hydroxylase immunocytochemistry. Brain Res. 1980;2(3):235–64.
78. Proudfit HK, Clark FM. The projections of locus coeruleus neurons to the spinal cord. Prog Brain Res. 1991;88:123–41.
79. Kwiat GC, Basbaum AI. The origin of brainstem noradrenergic and serotonergic projections to the spinal cord dorsal horn in the rat. Somatosens Mot Res. 1992;9(2):157–73.
80. Howorth PW, Teschemacher AG, Pickering AE. Retrograde adenoviral vector targeting of nociresponsive pontospinal noradrenergic neurons in the rat in vivo. J Comp Neurol. 2009;512(2):141–57.
81. Tabansky I, et al. Molecular profiling of reticular gigantocellularis neurons indicates that eNOS modulates environmentally dependent levels of arousal. Proc Natl Acad Sci USA. 2018;115(29):E6900–9.
82. Sara SJ, Herve-Minvielle A. Inhibitory influence of frontal cortex on locus coeruleus neurons. Proc Natl Acad Sci USA. 1995;92(13):6032–6.
83. Jodo E, Chiang C, Aston-Jones G. Potent excitatory influence of prefrontal cortex activity on noradrenergic locus coeruleus neurons. Neuroscience. 1998;83(1):63–79.
84. McCall JG, et al. CRH engagement of the locus Coeruleus noradrenergic system mediates stress-induced anxiety. Neuron. 2015;87(3):605–20.
85. Bylund DB, et al. International Union of Pharmacology nomenclature of adrenoceptors. Pharmacol Rev. 1994;46(2):121–36.
86. Tavares I, Lima D, Coimbra A. The ventrolateral medulla of the rat is connected with the spinal cord dorsal horn by an indirect descending pathway relayed in the A5 noradrenergic cell group. J Comp Neurol. 1996;374(1):84–95.
87. Day HE, et al. Distribution of alpha 1a-, alpha 1b- and alpha 1d-adrenergic receptor mRNA in the rat brain and spinal cord. J Chem Neuroanat. 1997;13(2):115–39.
88. Nicholas AP, Pieribone VA, Hokfelt T. Cellular localization of messenger RNA for beta-1 and beta-2 adrenergic receptors in rat brain: an in situ hybridization study. Neuroscience. 1993;56(4):1023–39.
89. Hein L, Kobilka BK. Adrenergic receptors from molecular structure to in vivo function. Trends Cardiovasc Med. 1997;7(5):137–45.
90. Gupta M, et al. Mechanism of antinociceptive effect of nimodipine in experimental diabetic neuropathic pain. Methods Find Exp Clin Pharmacol. 2003;25(1):49–52.
91. Yalcin I, et al. Chronic treatment with agonists of beta(2)-adrenergic receptors in neuropathic pain. Exp Neurol. 2010;221(1):115–21.
92. Szabadi E. Functional neuroanatomy of the central noradrenergic system. J Psychopharmacol. 2013;27(8):659–93.
93. Aghajanian GK, VanderMaelen CP. alfa2-adrenoceptor-mediated hyperpolarization of locus coeruleus neurons: intracellular studies in vivo. Science. 1982;215:1394–6.
94. Abercrombie ED, Jacobs BL. Microinjected clonidine inhibits noradrenergic neurons of the locus coeruleus in freely moving cats. Neurosci Lett. 1987;76(2):203–8.
95. Adams LM, Foote SL. Effects of locally infused pharmacological agents on spontaneous and sensory-evoked activity of locus coeruleus neurons. Brain Res Bull. 1988;21(3):395–400.

96. Cedarbaum JM, Aghajanian GK. Noradrenergic neurons of the locus coeruleus: inhibition by epinephrine and activation by the alpha-antagonist piperoxane. Brain Res. 1976;112(2):413–9.
97. Gyires K, et al. Alpha(2)-Adrenoceptor subtypes-mediated physiological, pharmacological actions. Neurochem Int. 2009;55(7):447–53.
98. Sonohata M, et al. Actions of noradrenaline on substantia gelatinosa neurones in the rat spinal cord revealed by in vivo patch recording. J Physiol. 2004;555(Pt 2):515–26.
99. Pan HL, et al. Modulation of pain transmission by G-protein-coupled receptors. Pharmacol Ther. 2008;117(1):141–61.
100. Han SK, et al. Noradrenaline inhibits substantia gelatinosa neurons in mice trigeminal subnucleus caudalis via alpha(2) and beta adrenoceptors. Neurosci Lett. 2007;411(2):92–7.
101. Perez DM. Alpha(1)-adrenergic receptors in neurotransmission, synaptic plasticity, and cognition. Front Pharmacol. 2020;11:581098.
102. Birnbaum SG, et al. Protein kinase C overactivity impairs prefrontal cortical regulation of working memory. Science. 2004;306(5697):882–4.
103. Grzelka K, et al. Noradrenaline modulates the membrane potential and holding current of medial prefrontal cortex pyramidal neurons via beta(1)-adrenergic receptors and HCN channels. Front Cell Neurosci. 2017;11:341.
104. Hansson E, Ronnback L. Receptor regulation of the glutamate, GABA and taurine high-affinity uptake into astrocytes in primary culture. Brain Res. 1991;548(1–2):215–21.
105. Dharmshaktu P, Tayal V, Kalra BS. Efficacy of antidepressants as analgesics: a review. J Clin Pharmacol. 2012;52(1):6–17.
106. Martins I, et al. Increased noradrenergic neurotransmission to a pain facilitatory area of the brain is implicated in facilitation of chronic pain. Anesthesiology. 2015;123(3):642–53.
107. Hickey L, et al. Optoactivation of locus ceruleus neurons evokes bidirectional changes in thermal nociception in rats. J Neurosci. 2014;34(12):4148–60.
108. Brightwell JJ, Taylor BK. Noradrenergic neurons in the locus coeruleus contribute to neuropathic pain. Neuroscience. 2009;160(1):174–85.
109. Miller JF, Proudfit HK. Antagonism of stimulation-produced antinociception from ventrolateral pontine sites by intrathecal administration of alpha-adrenergic antagonists and naloxone. Brain Res. 1990;530(1):20–34.
110. Hentall ID, et al. Temporal and spatial profiles of pontine-evoked monoamine release in the rat's spinal cord. J Neurophysiol. 2003;89(6):2943–51.
111. Ma W, Eisenach JC. Chronic constriction injury of sciatic nerve induces the up-regulation of descending inhibitory noradrenergic innervation to the lumbar dorsal horn of mice. Brain Res. 2003;970(1–2):110–8.
112. Tsuruoka M, Willis WD. Descending modulation from the region of the locus coeruleus on nociceptive sensitivity in a rat model of inflammatory hyperalgesia. Brain Res. 1996;743(1–2):86–92.
113. Hayashida K, et al. Depletion of endogenous noradrenaline does not prevent spinal cord plasticity following peripheral nerve injury. J Pain. 2012;13(1):49–57.
114. Jasmin L, Boudah A, Ohara PT. Long-term effects of decreased noradrenergic central nervous system innervation on pain behavior and opioid antinociception. J Comp Neurol. 2003;460(1):38–55.
115. Hammond DL, Yaksh TL. Antagonism of stimulation-produced antinociception by intrathecal administration of methysergide or phentolamine. Brain Res. 1984;298(2):329–37.
116. Zhang KM, et al. alpha2-adrenoceptors modulate NMDA-evoked responses of neurons in superficial and deeper dorsal horn of the medulla. J Neurophysiol. 1998;80(4):2210–4.
117. Hughes SW, et al. Endogenous analgesic action of the pontospinal noradrenergic system spatially restricts and temporally delays the progression of neuropathic pain following tibial nerve injury. Pain. 2013;154(9):1680–90.
118. Baba H, Shimoji K, Yoshimura M. Norepinephrine facilitates inhibitory transmission in substantia gelatinosa of adult rat spinal cord (part 1): effects on axon terminals of GABAergic and glycinergic neurons. Anesthesiology. 2000;92(2):473–84.

119. Kawasaki Y, et al. Alpha 2 adrenoceptor-mediated presynaptic inhibition of primary afferent glutamatergic transmission in rat substantia gelatinosa neurons. Anesthesiology. 2003;98(3):682–9.
120. Wolff M, et al. Clonidine reduces the excitability of spinal dorsal horn neurones. Br J Anaesth. 2007;98(3):353–61.
121. Stone EA. Central cyclic-AMP-linked noradrenergic receptors: new findings on properties as related to the actions of stress. Neurosci Biobehav Rev. 1987;11(4):391–8.
122. Ibrahim MMH, et al. Norepinephrine regulation of adrenergic receptor expression, 5′ AMP-activated protein kinase activity, and glycogen metabolism and mass in male versus female hypothalamic primary astrocyte cultures. ASN Neuro. 2020;12:1759091420974134.
123. Roh DH, et al. Intrathecal clonidine suppresses phosphorylation of the N-methyl-D-aspartate receptor NR1 subunit in spinal dorsal horn neurons of rats with neuropathic pain. Anesth Analg. 2008;107(2):693–700.
124. Hu JW. Response properties of nociceptive and non-nociceptive neurons in the rat's trigeminal subnucleus caudalis (medullary dorsal horn) related to cutaneous and deep craniofacial afferent stimulation and modulation by diffuse noxious inhibitory controls. Pain. 1990;41(3):331–45.
125. Bae YC, et al. Identification of signal substances in synapses made between primary afferents and their associated axon terminals in the rat trigeminal sensory nuclei. J Comp Neurol. 2000;418(3):299–309.
126. Bereiter DA, Hirata H, Hu JW. Trigeminal subnucleus caudalis: beyond homologies with the spinal dorsal horn. Pain. 2000;88(3):221–4.
127. Martin YB, et al. Neuronal disinhibition in the trigeminal nucleus caudalis in a model of chronic neuropathic pain. Eur J Neurosci. 2010;32(3):399–408.
128. Levitt P, Moore RY. Origin and organization of brainstem catecholamine innervation in the rat. J Comp Neurol. 1979;186(4):505–28.
129. Couto LB, et al. Descriptive and functional neuroanatomy of locus coeruleus-noradrenaline-containing neurons involvement in bradykinin-induced antinociception on principal sensory trigeminal nucleus. J Chem Neuroanat. 2006;32(1):28–45.
130. Sasa M, Takaori S. Influence of the locus coeruleus on transmission in the spinal trigeminal nucleus neurons. Brain Res. 1973;55(1):203–8.
131. Tsuruoka M, et al. Coeruleotrigeminal inhibition of nociceptive processing in the rat trigeminal subnucleus caudalis. Brain Res. 2003;993(1–2):146–53.
132. McBride RL, Sutin J. Noradrenergic hyperinnervation of the trigeminal sensory nuclei. Brain Res. 1984;324(2):211–21.
133. Travagli RA, Williams JT. Endogenous monoamines inhibit glutamate transmission in the spinal trigeminal nucleus of the Guinea-pig. J Physiol. 1996;491(Pt 1):177–85.
134. Grudt TJ, Williams JT, Travagli RA. Inhibition by 5-hydroxytryptamine and noradrenaline in substantia gelatinosa of Guinea-pig spinal trigeminal nucleus. J Physiol. 1995;485(Pt 1):113–20.
135. Mesa-Lombardo A, et al. Locus coeruleus inhibition of vibrissal responses in the trigeminal subnucleus caudalis are reduced in a diabetic mouse model. Front Cell Neurosci. 2023;17:1208121.
136. Pertovaara A. Noradrenergic pain modulation. Prog Neurobiol. 2006;80(2):53–83.
137. Nicholson B. Treatment of painful polyneuropathies. Curr Pain Headache Rep. 2005;9(3):178–83.
138. Diatchenko L, et al. Idiopathic pain disorders—pathways of vulnerability. Pain. 2006;123(3):226–30.
139. Martin WJ, et al. Differential effects of neurotoxic destruction of descending noradrenergic pathways on acute and persistent nociceptive processing. Pain. 1999;80(1–2):57–65.
140. Almeida A, et al. Brain afferents to the medullary dorsal reticular nucleus: a retrograde and anterograde tracing study in the rat. Eur J Neurosci. 2002;16(1):81–95.
141. Lima D, Almeida A. The medullary dorsal reticular nucleus as a pronociceptive Centre of the pain control system. Prog Neurobiol. 2002;66(2):81–108.

142. Apkarian AV, et al. Neural mechanisms of pain and alcohol dependence. Pharmacol Biochem Behav. 2013;112:34–41.
143. Ong WY, Stohler CS, Herr DR. Role of the prefrontal cortex in pain processing. Mol Neurobiol. 2019;56(2):1137–66.
144. Kaushal R, et al. GABA-A receptor activity in the noradrenergic locus coeruleus drives trigeminal neuropathic pain in the rat; contribution of NAalpha1 receptors in the medial prefrontal cortex. Neuroscience. 2016;334:148–59.
145. Marzo A, et al. Unilateral electrical stimulation of rat locus coeruleus elicits bilateral response of norepinephrine neurons and sustained activation of medial prefrontal cortex. J Neurophysiol. 2014;111(12):2570–88.
146. Finnerup NB, Sindrup SH, Jensen TS. Recent advances in pharmacological treatment of neuropathic pain. F1000 Med Rep. 2010;2:52.
147. Hayashida K, et al. Gabapentin acts within the locus coeruleus to alleviate neuropathic pain. Anesthesiology. 2008;109(6):1077–84.
148. Hayashida KI, Obata H. Strategies to treat chronic pain and strengthen impaired descending noradrenergic inhibitory system. Int J Mol Sci. 2019;20(4)
149. Broderick PA, Jacoby JH. Diabetes-related changes in L-tryptophan-induced release of striatal biogenic amines. Diabetes. 1988;37(7):956–60.
150. Broderick PA, Jacoby JH. Serotonergic function in diabetic rats: psychotherapeutic implications. Biol Psychiatry. 1988;24(2):234–9.
151. Perez-Taboada I, et al. Diabetes causes dysfunctional dopamine neurotransmission favoring nigrostriatal degeneration in mice. Mov Disord. 2020;35(9):1636–48.
152. Boulton AJ, et al. Diabetic neuropathies: a statement by the American Diabetes Association. Diabetes Care. 2005;28(4):956–62.
153. Sorensen L, Molyneaux L, Yue DK. The relationship among pain, sensory loss, and small nerve fibers in diabetes. Diabetes Care. 2006;29(4):883–7.
154. Madhusudhanan J, Suresh G, Devanathan V. Neurodegeneration in type 2 diabetes: Alzheimer's as a case study. Brain Behav. 2020;10(5):e01577.
155. Feldman EL, et al. Diabetic neuropathy. Nat Rev Dis Primers. 2019;5(1):41.
156. Suzuki C, et al. Peripheral and central conduction abnormalities in diabetes mellitus. Neurology. 2000;54(10):1932–7.
157. Orstavik K, et al. Abnormal function of C-fibers in patients with diabetic neuropathy. J Neurosci. 2006;26(44):11287–94.
158. Garcia-Perez H, Harlow SD, Denman C. The effect of pelvic pain and urinary incontinence on women's self-rated health in northern Mexico. Int Urogynecol J. 2018;29(2):243–50.
159. Haroutounian S, et al. Primary afferent input critical for maintaining spontaneous pain in peripheral neuropathy. Pain. 2014;155(7):1272–9.
160. Sun W, et al. Reduced conduction failure of the main axon of polymodal nociceptive C-fibres contributes to painful diabetic neuropathy in rats. Brain. 2012;135(Pt 2):359–75.
161. Orestes P, et al. Reversal of neuropathic pain in diabetes by targeting glycosylation of Ca(V)3.2 T-type calcium channels. Diabetes. 2013;62(11):3828–38.
162. Messinger RB, et al. In vivo silencing of the Ca(V)3.2 T-type calcium channels in sensory neurons alleviates hyperalgesia in rats with streptozocin-induced diabetic neuropathy. Pain. 2009;145(1–2):184–95.
163. Tan AM, et al. Maladaptive dendritic spine remodeling contributes to diabetic neuropathic pain. J Neurosci. 2012;32(20):6795–807.
164. Tan XJ, et al. Reduction of dendritic spines and elevation of GABAergic signaling in the brains of mice treated with an estrogen receptor beta ligand. Proc Natl Acad Sci USA. 2012;109(5):1708–12.
165. Marshall AG, et al. Spinal disinhibition in experimental and clinical painful diabetic neuropathy. Diabetes. 2017;66(5):1380–90.
166. Paxinos G, Franklin KBJ. The mouse brain in stereotaxic coordinates. San Diego: Academic Press; 2003.

167. Shun CT, et al. Skin denervation in type 2 diabetes: correlations with diabetic duration and functional impairments. Brain. 2004;127(Pt 7):1593–605.
168. Lallemend F, Ernfors P. Molecular interactions underlying the specification of sensory neurons. Trends Neurosci. 2012;35(6):373–81.
169. Usoskin D, et al. Unbiased classification of sensory neuron types by large-scale single-cell RNA sequencing. Nat Neurosci. 2015;18(1):145–53.
170. Brumovsky PR. Dorsal root ganglion neurons and tyrosine hydroxylase—an intriguing association with implications for sensation and pain. Pain. 2016;157(2):314–20.
171. Garcia-Magro N, et al. Modulation of mechanosensory vibrissal responses in the trigeminocervical complex by stimulation of the greater occipital nerve in a rat model of trigeminal neuropathic pain. J Headache Pain. 2020;21(1):96.
172. Rohampour K, et al. Peripheral nerve injury potentiates excitatory synaptic transmission in locus coeruleus neurons. Brain Res Bull. 2017;130:112–7.
173. Suto J, Satou K. Further characterization of diabetes mellitus and body weight loss in males of the congenic mouse strain DDD.Cg-A(y.). J Vet Med Sci. 2015;77(2):203–10.
174. McIlwrath SL, et al. Manganese-enhanced MRI reveals changes within brain anxiety and aversion circuitry in rats with chronic neuropathic pain- and anxiety-like behaviors. NeuroImage. 2020;223:117343.
175. Gambeta E, et al. Anxiety- but not depressive-like behaviors are related to facial hyperalgesia in a model of trigeminal neuropathic pain in rats. Physiol Behav. 2018;191:131–7.
176. Santiago JA, et al. Diabetes: a tipping point in neurodegenerative diseases. Trends Mol Med. 2023;29(12):1029–44.
177. Nuñez A, et al. The neurobiology of insulin-like growth factor I: from neuroprotection to modulation of brain states. Mol Psychiatry. 2023;28(8):3220–30.

Potential Therapeutic Strategy for Neurodegenerative Disorders: Insights from Natural Cannabis Compounds and Analogues

Lei Tian, Bingxing Zhang, Xiuding Yang, Xiaopei Zhai, Yue Gao, Yunfei Zhang, Xiaolin Xie, Dezhu Zhang, and Chengyuan Liang

Abstract Natural compounds are being explored as a promising strategy for treating neurodegenerative disorders. This study describes natural cannabinoids from *Cannabis sativa*, including (−)-cannabidiol (CBD) analogues with therapeutic potential. The underlying molecular mechanisms of CBD as a therapeutic candidate for epilepsy, Alzheimer's disease (AD), and Parkinson's disease (PD) were comprehensively clarified. CBD acts as an endogenous agonist for cannabinoid receptors, indirectly suppressing glutamate release and stimulating cannabinoid receptor 1 and 2 receptors. This mechanism makes CBD effective in treating epilepsy and PD while promoting neuroprotection through various signal transduction pathways. CBD also shows promise as a therapeutic candidate for AD by preventing glycogen synthase kinase 3β hyperphosphorylation induced by β-amyloid. Furthermore, this chapter incorporates brain imaging evidence related to CBD exposure in both healthy subjects and patients at a clinical high risk (CHR) for psychosis. By examining the impact of CBD on human brain function through neuroimaging studies, this chapter aims to provide a mechanistic understanding of CBD's potential role as a therapeutic strategy with a neuroprotective effect.

Keywords Neurodegenerative disorders · Cannabinoids · Therapeutic strategies · Molecular mechanisms · Neuroimaging

L. Tian · B. Zhang · X. Yang · Y. Zhang · C. Liang (✉)
Xi'an Key Laboratory for Antiviral and Antimicrobial-Resistant Bacteria Therapeutics Research, Shaanxi University of Science & Technology, Xi'an, China

School of Biological and Pharmaceutical Sciences, Shaanxi University of Science & Technology, Xi'an, China

X. Zhai
Department of Pharmaceutics, School of Pharmacy, Air Force Medical University, Xi'an, China

Y. Gao
College of Pharmacy, Jinan University, Guangzhou, China

X. Xie · D. Zhang
Shaanxi Panlong Pharmaceutical Group Co., Ltd., Xi'an, China

P. A. Broderick (ed.), *Biomedical Imaging in Neurodegeneration*,
https://doi.org/10.1007/978-3-031-95878-6_11

1 Introduction

Cultivated for over 6000 years, *Cannabis* is a traditional crop with significant historical importance, utilised for its fibres, seeds, and stalks in various industries (Fig. 1) [1–3]. It exhibits notable morphological diversity, leading to taxonomic debates [4, 5]. *Cannabis* is classified into three subspecies: *Cannabis sativa*, *Cannabis indica*, and *Cannabis ruderalis* [6]. *Cannabis indica*, with broad leaves and a bushy structure, grows 2–4 feet tall [7, 8]. *Cannabis ruderalis*, typically branchless and low in tetrahydrocannabinol (THC), is often symbiotic with other plants [8]. In comparison, *Cannabis sativa*, the most prevalent subspecies in Western countries, is a towering, non-branching plant with thin leaves that can grow up to 5–18 feet tall. And it is the most well-studied subspecies [9–11]. Recent studies have unveiled that *Cannabis* has diverse properties, including anti-inflammatory [12–14], analgesic [14, 15], anti-allergic [16], antibacterial [17, 18], antihypertensive [19], and antithrombotic effects [20]. Unofficially approved *Cannabis* has been widely used clinically for antitumor applications and for treating neurological diseases [21, 22].

Following Raphael Mechoulam's pioneering isolation of THC [23], a multitude of other phytocannabinoids was subsequently discovered, including cannabidiol (CBD), cannabidivarin (CBDV), tetrahydrocannabivarin (THCV), cannabigerol (CBG), and cannabichromene (CBC). During a substantial timeframe, research on the psychoactive THC overwhelmingly dominated the field. However, due to the strong hallucinogenic properties of THC, the most commonly occurring phytocannabinoid in *Cannabis*, its applications have been limited, and its legality remains restricted in many countries and regions [24, 25]. In contrast, CBD inhibits the uptake of the neurotransmitter anandamide (AEA) and blocks AEA hydrolysis, thereby antagonising the aversive hallucinogenic effects of THC [26]. This chapter focusses on natural cannabinoids from *Cannabis sativa*, particularly (−)-CBD analogues, and their therapeutic potential. It extensively covers the molecular mechanisms underlying CBD's efficacy in treating epilepsy, Alzheimer's disease (AD), and Parkinson's disease (PD). Additionally, it integrates neuroimaging evidence of CBD exposure in healthy subjects and patients at clinical high risk (CHR) for psychosis. The aim is to elucidate the mechanistic basis for CBD's potential utility as a therapeutic intervention with neuroprotective properties.

2 Natural Cannabinoids from *Cannabis Sativa*

Recent research on *Cannabis sativa* has identified over 565 compounds, including more than 120 natural cannabinoids and their derivatives [28]. These cannabinoids, unique for their C_{21} terpenophenolic skeleton, are classified into 11 structural types (Table 1). Types **1–10** display consistent modifications on their terpenophenolic skeletons (Fig. 2), while type **11**, noted for its structural diversity, includes 30 different skeletons (Fig. 3, **11-1~11-30**).

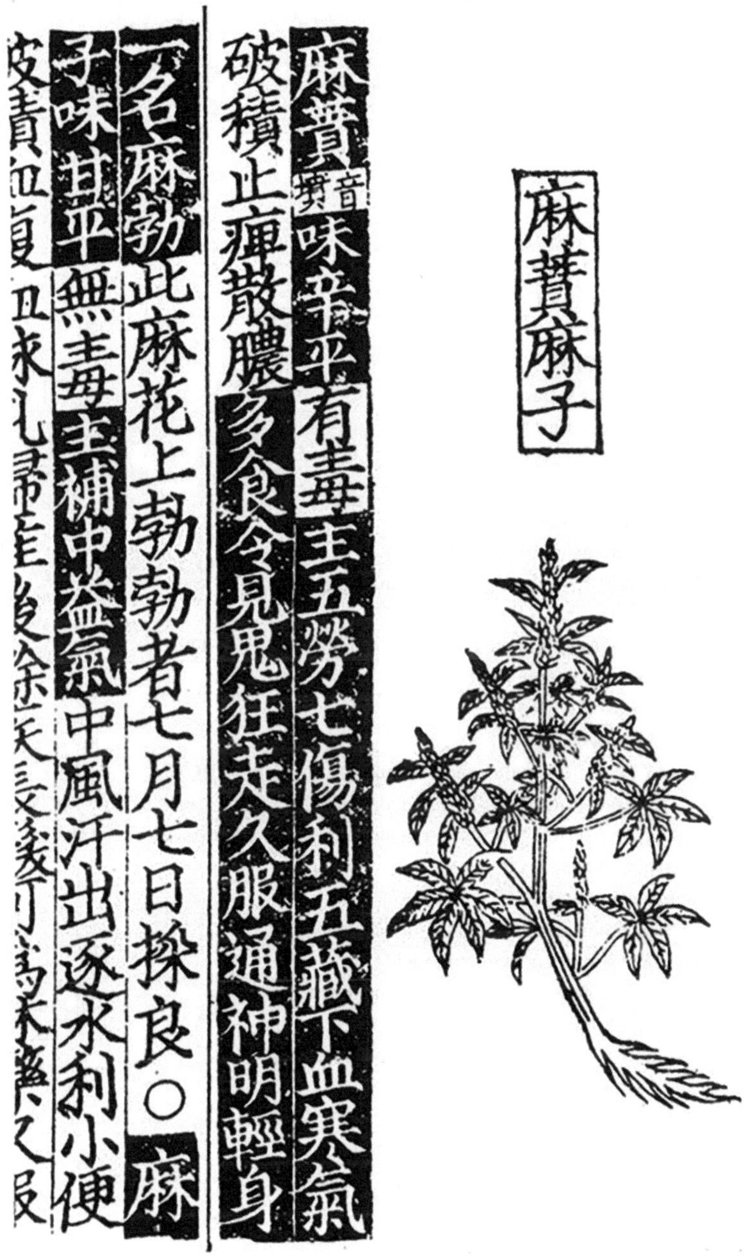

Fig. 1 The depicted illustration of the *Cannabis* plant originates from ancient China and is accompanied by a text detailing its various functions. This particular image was sourced from the *Compendium of Materia Medica*, a text that dates back to the sixteenth century AD during the Ming Dynasty [27]

Table 1 11 general types of natural cannabinoids from *Cannabis sativa* [27]

Type	Cannabinoids	Ref.
1	(−)-Δ9-*trans*-tetrahydrocannabinol (Δ^9-THC)	[29]
2	(−)-Δ8-*trans*-tetrahydrocannabinol (Δ^8-THC)	[30]
3	Cannabigerol (CBG)	[31]
4	Cannabidiol (CBD)	[32]
5	Cannabielsoin (CBE)	[33]
6	Cannabichromene (CBC)	[34]
7	Cannabinodiol (CBND)	[35]
8	Cannabicyclol (CBL)	[36]
9	Cannabitriol (CBT)	[37]
10	Cannabinol (CBN)	[34]
11	**Miscellaneous types**	[28]
11-1	Dehydrocannabifuran (DCBF-C_5)	[38]
11-2	Cannabifuran (CBF-C_5)	[39]
11-3	8-Hydroxy-isohexahydrocannabivirin (OH-iso-HHCV-C_3)	[40]
11-4	Cannabichromanone-C_5 (CBCN-C_5)	[41]
11-5	Cannabichromanone-C_3 (CBCN-C_3)	[41]
11-6	10-Oxo-$\Delta^{6a(10a)}$-tetrahydrocannabinol (OTHC)	[39]
11-7	Cannabicitran	[42]
11-8	(−)-Δ^9-*cis*-(6a*S*,10a*R*)-tetrahydrocannabinol (*cis*-Δ^9-THC)	[28]
11-9	Cannabicoumaronone (CBCON-C_5)	[43]
11-10	Cannabiripsol (CBR)	[44]
11-11	Cannabitetrol (CBTT)	[45]
11-12	(±)-Δ7-*cis*-Isotetrahydrocannabivarin-C_3 (*cis*-iso-Δ^7-THCV)	[32]
11-13	(−)-Δ7-*trans*-(1*R*,3*R*,6*R*)-Isotetrahydrocannabivarin-C_3 (*trans*-iso-Δ^7-THCV)	[32]
11-14	(−)-Δ7-*trans*-(1*R*,3*R*,6*R*)-Isotetrahydrocannabinol-C_5 (*trans*-Iso-Δ^7-THC)	[32]
11-15	Cannabichromanones B	[41]
11-16	Cannabichromanones C	[41]
11-17	Cannabichromanones D	[41]
11-18	(−)-(7*R*)-Cannabicoumarononic acid	[42]
11-19	4-Acetoxy-2-geranyl-5-hydroxy-3-*n*-pentylphenol	[46]
11-20	2-Geranyl-5-hydroxy-3-*n*-pentyl-1,4-benzoquinone	[42]
11-21	5-Acetoxy-6-geranyl-3-*n*-pentyl-1,4-benzoquinone	[46]
11-22	Cannabimovone (CBM)	[47]
11-23	Cannabioxepane (CBX)	[48]
11-24	10α-Hydroxy-$\Delta^{9,11}$-hexahydrocannabinol	[42]
11-25	9β,10β-epoxy hexahydrocannabinol	[42]
11-26	9α-Hydroxy hexahydrocannabinol	[48]
11-27	7-Oxo-9α-hydroxy hexahydrocannabinol	[48]
11-28	10α-Hydroxy hexahydrocannabinol	[48]
11-29	10a*R*-Hydroxy hexahydrocannabinol	[48]
11-30	9α-Hydroxy-10-oxo-$\Delta^{6a,10a}$-tetrahydrocannabinol	[48]

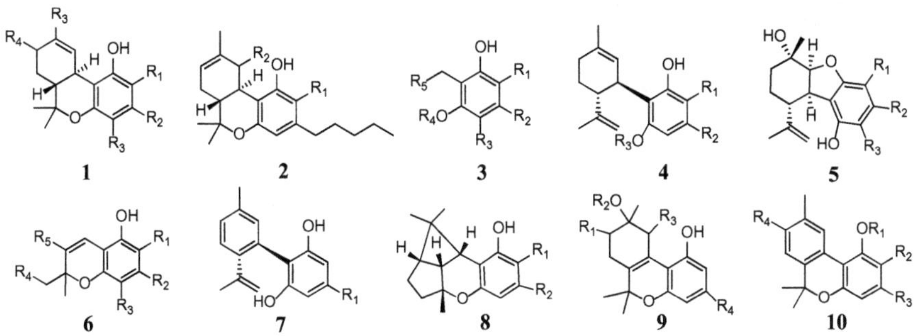

Fig. 2 Skeletons of cannabinoid types **1–10** in Table 1 [27]

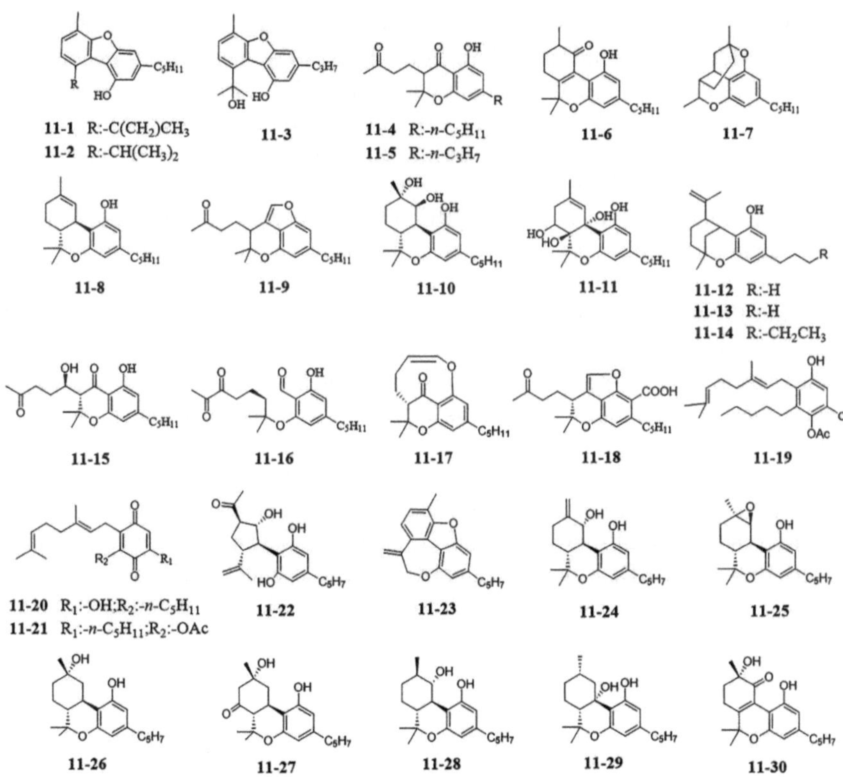

Fig. 3 Structures of type **11** miscellaneous cannabinoids in Table 1 [27]

3 CBD and Its Analogues

3.1 Overview of CBD

In 1940, Adams et al. first isolated CBD (CAS: 13956-29-1) from *Cannabis* [49]. Mechoulam's team later elucidated CBD's structure in 1963 [50]. Jones, in 1977, determined CBD's crystal structure, identifying two optical isomers [51] (Fig. 4). The isomers differ in the stereochemistry linking the terpene core to the resorcinol moiety, with the terpene and aromatic rings almost perpendicular. Natural *Cannabis* plants contain (−)-CBD; (+)-CBD is synthesised organically. Recent studies [52, 53] show all naturally occurring CBD-type cannabinoids have a (−)-trans-(*1R,6R*) configuration, likely causing negative optical rotation.

CBD, historically used unofficially for treating Alzheimer's disease (AD) [54, 55], epilepsy [55, 56], and various neurological disorders, shows promising neurological effects in preclinical studies. In the late 1800s, medical professionals in the United States and United Kingdom noted *Cannabis* extracts reduced epileptic seizures [57]. Iván et al. [58] found CBD decreased rat seizures from afferent stimuli, proving to be an effective, specific anticonvulsant with fewer neurotoxic effects than traditional drugs [59]. It also reduces post-discharge duration and amplitude in the central nervous system (CREB) [60]. Dose-dependent anticonvulsant and antiepileptic effects of CBD were confirmed in various seizure models [61]. The National Institute of Neurological Disorders and Stroke's Epilepsy Therapy Screening Program (2017) found CBD's antiseizure properties in corneal-ignited mice and acute seizure models [62, 63]. Epidiolex®, the first CBD drug approved by the FDA in June 2018 for treatment-resistant epilepsy, is effective for Dravet and Lennox-Gastaut syndromes [64, 65]. Its approval was based on a double-blind trial with 120 patients suffering from drug-resistant seizures [66, 67].

Further studies highlight CBD's anticonvulsant, antinausea, and analgesic effects [68]. It also shows superior antioxidant neuroprotective potency and cytoprotective properties in normal cells, exhibiting effects akin to capsaicin in rheumatoid arthritis models [69–71]. CBD, a dual antagonist of GPR18 and GPR55, is a potential treatment for endometriosis and other diseases [72, 73]. CBD uniquely inversely

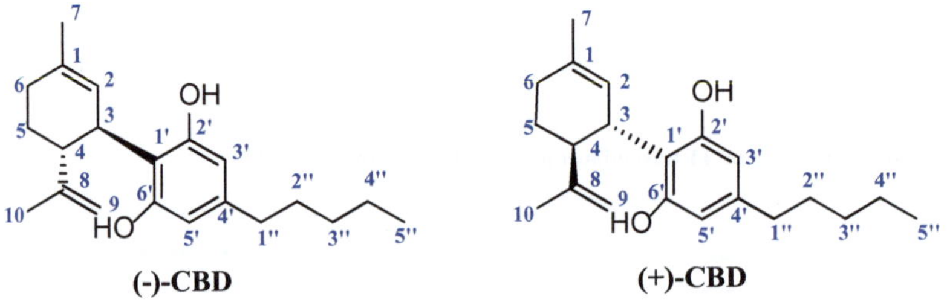

Fig. 4 The chemical structures of the CBD's two optical isomers [27]

agonises GPR12 [74]. A structure–activity relationship (SAR) analysis reveals that CBD's inverse agonism on GPR12 depends on dihydroxyl groups at C-2′ and C-6′ of its benzene ring, influencing its affinity and function. Modifying the C-4′ side chain or removing the hydroxyl group reduces cAMP accumulation via GPR12 [74]. CBD demonstrates efficacy against multiple cancers and inhibits inducible nitric oxide synthase in mouse mammary cancer cells ($IC_{50} = 1.80\ \mu M$), antagonises TRPM8 in human colon adenocarcinoma cells ($IC_{50} = 3.73\ \mu M$), and induces apoptosis in human prostate carcinoma cells ($IC_{50} = 5.40\ \mu M$) [75–77]. Additionally, CBD enhances gamma radiation sensitivity in glioblastoma cells by affecting TRAIL/TRAIL-R2 and TNF/TNFR1 pathways, offering oncological therapeutic potential [78]. CBD also reduces intracellular lipid levels, indicating its use in treating metabolic syndrome, hepatic steatosis, and obesity [79].

3.2 *Classification Based on the (−)-CBD Analogue Structure*

(−)-CBD, a non-psychoactive phytocannabinoid, has spurred research into its analogues due to chemical and metabolic instability and poor solubility. These analogues, including natural, synthetic, and metabolite variants, are divided into three categories based on modifications to the (−)-CBD terpineol skeleton: terpene-modified, benzene ring-modified, and dual-modified (terpene and benzene ring). Table 2 lists these compounds, and Figs. 5–7 illustrate their structures.

3.2.1 (−)-CBD Terpene Moiety-Modified Analogues

Table 2 lists terpene moiety-modified (−)-CBD analogues, compounds **12-24** (Fig. 5). Hydroxyl metabolites **12-14** represent the initial mammalian metabolism stages of (−)-CBD, beginning with hydroxylation at C-7 to form (−)-7-OH-CBD (**12**), which further hydroxylates at C-6 to yield (−)-6,7-di-OH-CBD (**14**). Bill et al. [80] identified **12** as the primary oxidative metabolite in rat liver supernatant. Hydroxylation at C-6 of (−)-CBD leads to (−)-6α-OH-CBD/(−)-6β-OH-CBD (**13**). Human liver microsomes (HLMs), through enzymes CYP2C19 and CYP3A4, are responsible for **13** and **14**'s production [81]. Another key metabolite is (−)-CBD-7-oic acid (**15**), derived from **12**'s further oxidation [82, 83]. Compound **12**, with strong CB1 (Ki = 5.3 ± 0.5 nM) and CB2 (Ki = 101.0 ± 5.1 nM) affinity, may be beneficial for epilepsy and convulsions and inhibits nicotine metabolism by CYP450 ($IC_{50} = 1.2 \pm 0.44\ \mu M$) [84].

Compounds **16-21**, derivatives of (−)-CBD, include hydrogenated (−)-CBD products **16-18** [85]. (−)-CBD was hydrogenated at C-1 or C-8 positions to form compounds **16** and **17** and at both C-8 and C-1 for compound **18**. Mark et al. [86] found that **17** had anticonvulsant effects, comparable antibacterial and antioxidant properties to (−)-CBD, and was biologically safe with low toxicity to human skin fibroblasts [87]. **18** was shown to inhibit nitric oxide release in macrophages and

Table 2 Three types of (−)-CBD analogues [27]

No.	Compound name	Categories	Compound type
12	(−)-7-OH-CBD	(−)-CBD terpene moiety-modified analogues	Metabolite
13	(−)-6α-OH-CBD/(−)-6β-OH-CBD		
14	(−)-6,7-di-OH-CBD		
15	(−)-CBD-7-oic acid		
16	(−)-1,2-H_2-CBD		Synthetic derivative
17	(−)-8,9-H_2-CBD		
18	(−)-H_4-CBD		
19	(−)-8,9-H_2-6-OH-CBD		
20	(−)-8,9-H_2-7-OH-CBD		
21	(−)-8,9-H_2-7-fluoro-CBD		
22	(−)-1,2-dihydroxycannabidiol		Natural compound
23	3,4-dehydro-1,2-dihydroxycannabidiol		
24	(−)-hexocannabitriol		
25	(−)-1″-OH-CBD	(−)-CBD benzene ring-modified analogues	Metabolite
26	(−)-2″-OH-CBD		
27	(−)-3″-OH-CBD		
28	(−)-4″-OH-CBD		
29	(−)-5″-OH-CBD		
30	(−)-CBD-C_3, acid		
31	(−)-CBD-dimethyl ether		
32	(−)-CBD-monomethyl ether (CBDM-C5)		Natural compound
33	(−)-Cannabidiorcol (CBD-C_1)		
34	(−)-Cannabidivarin (CBDV-C_3)		
35	(−)-Cannabidiol-C_4 (CBD-C_4)		
36	(−)-Cannabidiolic acid (CBDA-C_5)		
37	(−)-Cannabidivarinic acid (CBDVA-C_3)		
38	(−)-CBD-DMH		Synthetic derivative
39	(−)-6″-azidohex-2″-yne-CBD		
40	(−)-4′-fluoro-CBD		
41	Abnormal-CBD		
42	(−)-1″-N-acetylcyclobutylamine-Cannabidiorcol (CBD-C_1)		
43	(−)-3″,7-di-OH-CBD	(−)-CBD analogues with simultaneous modifications in both the benzene ring and the terpene moiety	Metabolite

(continued)

Table 2 (continued)

No.	Compound name	Categories	Compound type
44	(−)-4″,6-di-OH-CBD		
45	(−)-4″,7-di-OH-CBD		
46	(−)-5″,6-di-OH-CBD		
47	(−)-2″,6,7-tri-OH-CBD		
48	(−)-4″,6,7-tri-OH-CBD		
49	(−)-5″,6,7-tri-OH-CBD		
50	(−)-1″-OH,7-COOH-CBD		
51	(−)-2″-OH,7-COOH-CBD		
52	(−)-3″-OH,7-COOH-CBD		
53	(−)-4″-OH,7-COOH-CBD		
54	(−)-5″-OH,7-COOH-CBD		
55	(−)-1,2-H_2-CBD-DMH		Synthetic derivative
56	(−)-8,9-H_2-CBD-DMH		
57	(−)-H_4-CBD-DMH		
58	(−)-7-OH-CBD-DMH		
59	(−)-1-COOH-CBD-DMH		
60	(−)-8,9-H_2-CBD-diacetate		
61	(−)-8,9-H_2-6-OH-CBD-diacetate		
62	(−)-8,9-H_2-7-OH-CBD-diacetate		
63	(−)-10-hydroxy-CBD-diacetate		
64	(−)-10-fluoro-CBD-diacetate		

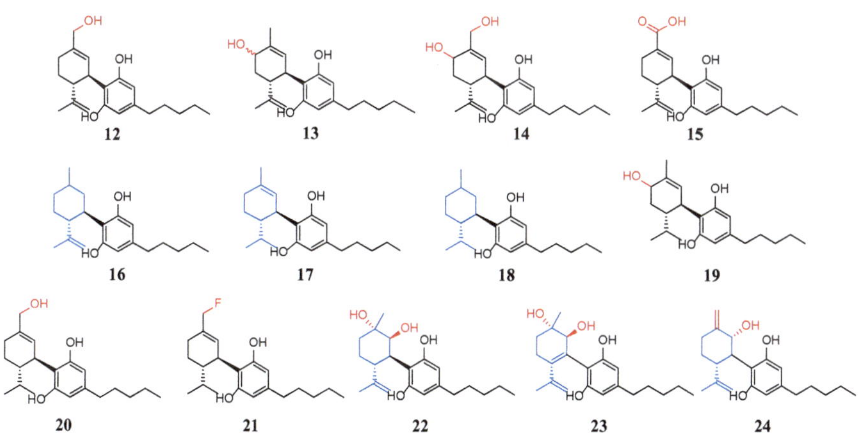

Fig. 5 Structures of (−)-CBD terpene moiety-modified analogues **12-24** [27]

had a significant CB1 binding affinity (Ki value of 145 nM) [87]. Compounds **19** and **20**, hydroxylated at C-6 and C-7, and hydrogenated at C-8 and C-9, were

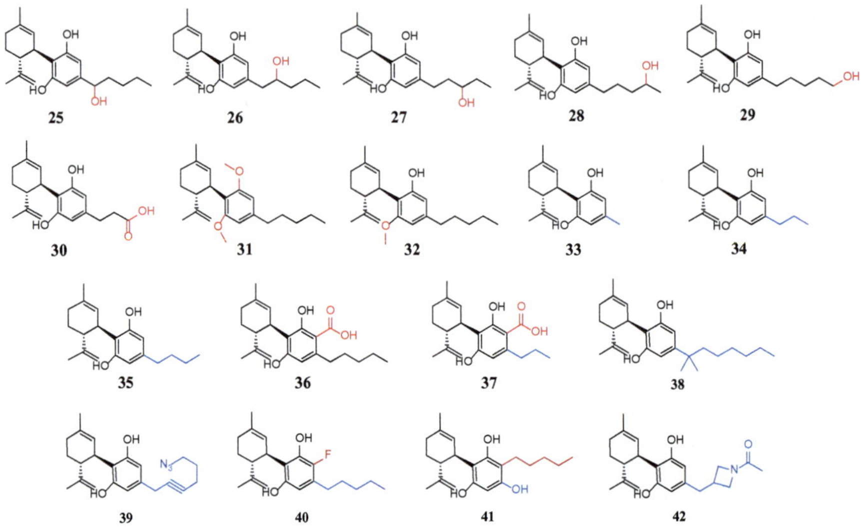

Fig. 6 Structures of (−)-CBD benzene ring-modified analogues **25-42** [27]

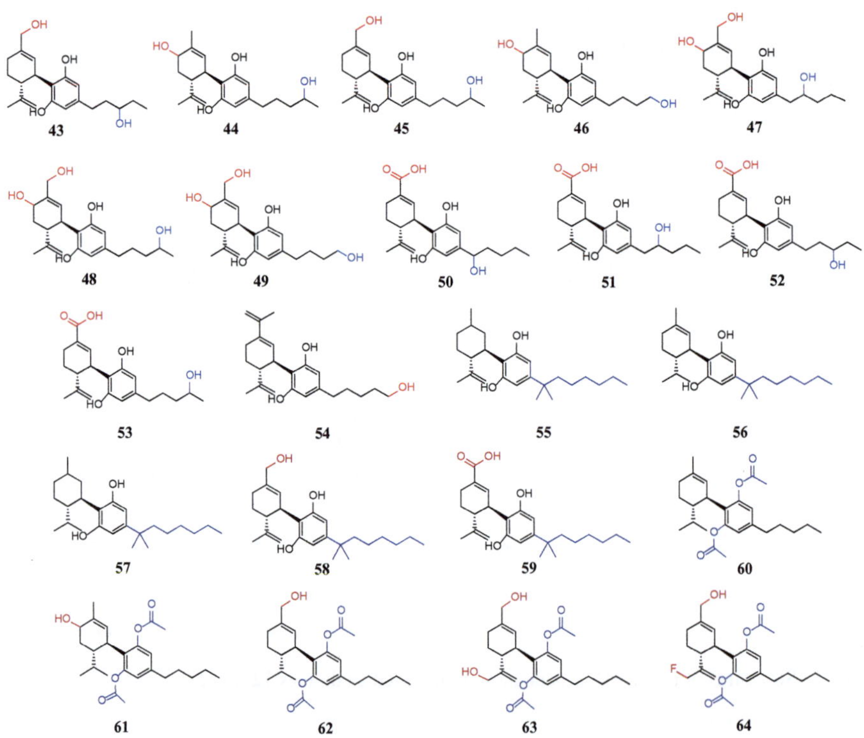

Fig. 7 Structures of (−)-CBD analogues **43-64** with simultaneous modifications in the terpene moiety and the benzene ring [27]

explored by Ewa et al. [88], with compound **20** demonstrating anti-inflammatory effects in mouse encephalitic T cells. Compound **20** shows a higher CB1 binding capacity than (−)-CBD. In addition, interest in fluorinated CBD derivatives, like antidepressant compound **21** with a fluorine-substituted C-7 methyl group, has grown within medicinal chemistry [89]. Giuseppina et al. [90] identified compounds **22-24**, with **24** showing promise in regulating the Nrf2 pathway more effectively than **22** and **23**, suggesting its potential for preventing degenerative diseases.

3.2.2 (−)-CBD Benzene Ring-Modified Analogues

Table 2 details (−)-CBD analogues **25-42**, with modifications to the benzene ring (Fig. 6). Compounds **25-31** represent monohydroxy metabolites of (−)-CBD's pentyl side chain, produced enzymatically. ElSohly et al. [32] summarised six natural CBD analogues (**32-37**) extracted from *Cannabis*, including (−)-CBD-monomethyl ether (**32**), (−)-cannabidiorcol (**33**), (−)-cannabidivarin (**34**), (−)-cannabidiol-C4 (**35**), (−)-cannabidiolic acid (**36**), and (−)-cannabidivarinic acid (**37**). These analogues, especially **31** and **32**, showed reduced inhibition of human CYP450 compared to (−)-CBD. Satoshi et al. [91] reported decreased effectiveness of (−)-CBD analogues with increased methylations in the resorcinol moiety. Notably, short-alkyl chain analogue **33** activates cannabinoid receptor GPR55, enhancing TNF-α and IL-12 production [92]. Compound **34**, with varied pharmacological activities, antagonises TRPM8 receptors and inhibits Caco-2 cell viability [76]. Yedukondalu et al. [93] demonstrated its strong antiproliferative activity against various cancer cells (Table 3). It also disrupts the Wnt/β-catenin signalling pathway in HepG2 cells. GW Pharmaceuticals explored **34**'s therapeutic potential in focal seizures and ASD in a phase II trial (NCT03202303) [91]. Krejčí et al. isolated the first *Cannabis* acid (**36**) in 1955 [94]. Compound **36** inhibits the migration of MDA-MB-231 mastadenoma cells by modulating cAMP-dependent protein kinase A (PKA) and RhoA [95] and reduces cancer cell mobility via RhoA activation [96]. It also inhibits the COX-2 enzyme, reducing inflammation, and enhances 5-HT_{1A} receptor activation, showing promise for appetite disorders and acute nausea [97, 98].

Compounds **38-42** are synthetic (−)-CBD derivatives. Compound **38**, (−)-CBD-dimethylheptyl (DMH), induces apoptosis in the human acute myeloid leukaemia HL-60 cell line and shows weak CB2 binding activity, as noted by Gallily et al [99] It also modulates LPS-induced proinflammatory genes and glutathione homeostasis genes in BV-1 microglial cells, demonstrating anti-inflammatory effects and microglial protection [100]. Compound **39**, with a 6′-azido-2″-hexyne substitution, is a potent CB1 antagonist for studying CB1 receptor function and structure [101]. Compound **40** (PECS-101), a fluorinated (−)-CBD derivative, shows potential as an anti-anxiolytic, antidepressant, and antipsychotic agent and may alleviate PTX-induced mechanical and cold allodynia via PPARγ activation in macrophages [102]. Nicole et al. suggest its fluorination enhances analgesic effects through CB1 and CB2 receptors [103]. Compound **41**, a synthetic cannabinol, binds to GPR55,

Table 3 (−)-Cannabidivarin and (−)-cannabidiolic acid under trial [27]

Drug name	Highest phase	Mechanism of action	Pharmacological activity	Method	Parameter	Ref.
(−) Cannabidivarin (**34**)	Phase II	Transient receptor potential cation channel subfamily M member 8 (TRPM8) antagonist	Cytotoxicity @ 24 h (Caco-2 human colon adenocarcinoma cells)	MTT assay	IC50 = 10.1 μM	[76]
		Wnt/β-catenin inhibitor	Cytotoxicity @ 72 h (A549 human non-small-cell lung carcinoma cells)	MTT assay	IC_{50} = 39 μM	
			Cytotoxicity @ 72 h, (HepG2 human hepatoblastoma cells)	MTT assay	IC_{50} = 64 μM	
			Cytotoxicity @ 72 h (OVCAR human ovarian carcinoma cells)	MTT assay	IC_{50} = 57 μM	[93]
			Cytotoxicity @ 72 h (MCF7 human breast adenocarcinoma cells, hormone-dependent)	MTT assay	IC_{50} = 53 μM	
(−) Cannabidivarin (**34**)	Phase II	Wnt/β-catenin inhibitor	Cytotoxicity @ 72 h (PC3 human prostate adenocarcinoma cells)	MTT assay	IC_{50} = 44 μM	
			Cytotoxicity @ 72 h (SHSY5Y human dopaminergic neuroblastoma cells)	MTT assay	IC_{50} = 55 μM	[93]
			Cytotoxicity @ 72 h (HCT116 human colon carcinoma cells)	MTT assay	IC_{50} = 13 μM	
		CB1 receptor inverse agonist	Protection against nausea and vomiting prophylaxis (rats, male)	Conditioned gaping assay	MED = 0.200 g/kg i.p.	[107]
(−)-Cannabidiolic acid (**36**)	Preclinical	PKA inhibitor	Inhibition of cell migration (MDAMB231 cells)	Transwell chamber assay	MIC≤5.00 μM	[95]
		5-HT1A receptor agonist	Protection against nausea and vomiting prophylaxis (mouse)	[35S]-GTPgammaS binding assay	Kb = 1.80 μM	[97]
		/	Protection against nausea and vomiting prophylaxis (rats, male)	Conditioned gaping assay	MED = 2.50 μg/kg	[98]

reducing apoptosis in pancreatic β-cell lines, suggesting utility in type 2 diabetes mellitus (T2DM) treatment [104]. William et al. [105] developed compound **42** (KLS-13019) for treating hepatic encephalopathy, with 50 times the efficacy and over 400 times the safety of (−)-CBD, and improved oral bioavailability. A recent preclinical study indicates its potential in treating chemotherapy-induced peripheral neuropathy (CIPN) by modulating the mitochondrial sodium–calcium exchanger-1 (mNCX-1) [106].

3.2.3 (−)-CBD Analogues with Simultaneous Modifications in Both the Terpene Moiety and the Benzene Ring

Compounds **43-64** (Fig. 7) encompass a range of (−)-CBD analogues that feature alterations to the terpene moiety and benzene ring. Compounds **43-46** are bihydroxy metabolites of (−)-CBD, metabolised mainly at the C-6/C-7 and benzene ring's pentyl side chain. Trihydroxy metabolites **47-49** of (−)-CBD are also included. Derivatives **50-54** result from varying the side-chain hydroxylation positions of **15**, specifically at C-1″ to C-5″ of the benzene ring.

Compounds **55** and **56** are partial hydrogenation products of (−)-CBD-DMH **35**, catalysed by PtO2, with **56** being the major epimer. Compound **57** stands out for its ability to suppress TNF-α production, offering stronger anti-inflammatory effects than (−)-CBD [85]. Compound **58** is a potent anandamide uptake inhibitor and a potential cannabinoid receptor antagonist, effectively inhibiting mouse vas deferens contractions without involving TRPV1, opioid, CB1, CB2, or α2 adrenergic receptors [108]. (−)-1- COOH-CBD-DMH (**59**) is distinguished by a 1,2-dimethylheptyl group at C-4′ and a carboxyl group replacing the C-1 methyl, preventing vanilloid receptor type 1 (VR1) activation by cannabinoids [109]. It shows anti-inflammatory effects in an otitis model. Compounds **58** and **59** hold therapeutic potential for cystic fibrosis and inflammatory bowel disease, attributed to their anti-inflammatory and antidiarrheal properties [110].

3.3 Neuroprotective Mechanism of CBD

3.3.1 The Neuroprotective Effects of CBD in the Treatment of Epilepsy

The endocannabinoid system (ECS), a complex endogenous signalling network, consists of cannabinoid receptors, fatty acid ligands, and biosynthesis/degradation proteins for endocannabinoids [111–114]. It includes two primary receptors, CB1 and CB2, both G-protein-coupled receptors (GPCRs) expressed in various tissues [115–118]. CB1 receptors predominantly exist in the CNS, especially in the substantia nigra, pale globus pallidus, cerebellar molecular layers, hippocampus, and cerebral cortex. CB2 receptors are found mainly in peripheral areas like the spleen and immune cells [119, 120]. Recent studies indicate CB2's presence in CNS and peripheral nervous systems, impacting neurophysiology [121, 122]. CB1 receptors, crucial in antiepileptic drug development, show potential in animal models [122–125].

Endocannabinoids have been shown to mitigate neuronal hyperexcitation symptoms [126, 127]. In a multiple sclerosis (MS) model, CB1 activation significantly reduced spasms and tremors [128]. Wallace et al. found endocannabinoids more effective than phenobarbital and phenytoin sodium in reducing seizures in a pilocarpine-induced epilepsy rat model [129]. Seizure duration and cessation are regulated by CB1 activation by endocannabinoids. Another study revealed endocannabinoids' seizure-preventing role by examining hippocampal changes in kainic acid (KA)-induced seizures in mice [130]. Glutamic acid, a primary CNS stimulant, can cause excitotoxicity and seizures when excessively transmitted. CB1 activation suppresses presynaptic glutamate release, reducing excitotoxicity [131]. CBD indirectly activates CB1, offering neuroprotection against acute excitotoxicity (Fig. 8). Endocannabinoids, acting as retrograde messengers in depolarisation-induced inhibition (DSI), help reduce seizures [132]. CBD increases AEA and 2-arachidonoylglycerol (2-AG) levels by inhibiting fatty acid amide hydrolase (FAAH), an enzyme metabolising lipid amides [133]. Activating presynaptic CB1 receptors suppresses calcium channels, reducing glutamate release and having anticonvulsant effects. Additionally, endocannabinoids activate G-protein-gated inwardly rectifying K^+ channels (GIRKs), stabilising membrane potential and reducing epileptiform activity, thus preventing seizures [134].

CBD potentially promotes neurite outgrowth by indirectly activating signal transduction pathways via CB1 and CB2 receptor stimulation [135, 136]. Recent studies show that CB1 receptor activation on neuronal somata enhances extracellular signal-regulated kinase (ERK) activity, leading to brain-derived neurotrophic factor (BDNF) expression [137]. BDNF is linked to long-term potentiation in the mature hippocampus [138]. CB1 activation also initiates the inositol triphosphate 3 (IP3)/mitogen-activated protein kinase (MAPK) pathway through Gi/o-βγ transduction, activating protein kinase C (PKC) and increasing critical transcription factor CREB expression, promoting neurite growth [139]. Similarly, CB2 receptors in the CNS, associated with Gi/o protein, activate adenylate cyclase via Gi/o-α transduction, reducing cAMP levels and activating the PI3K/Akt and ERK pathways, further supporting neurite growth [140]. Neurite outgrowth, a complex process, is influenced by nerve growth factors, neurotransmitters, and electrophysiological activity [141]. CBD stimulates CB1 receptors through endocannabinoids, triggering a protective response against excitatory neuronal toxicity. These insights suggest CBD's potential as a therapeutic agent for neurological disorders marked by hyperexcitability and excitotoxicity.

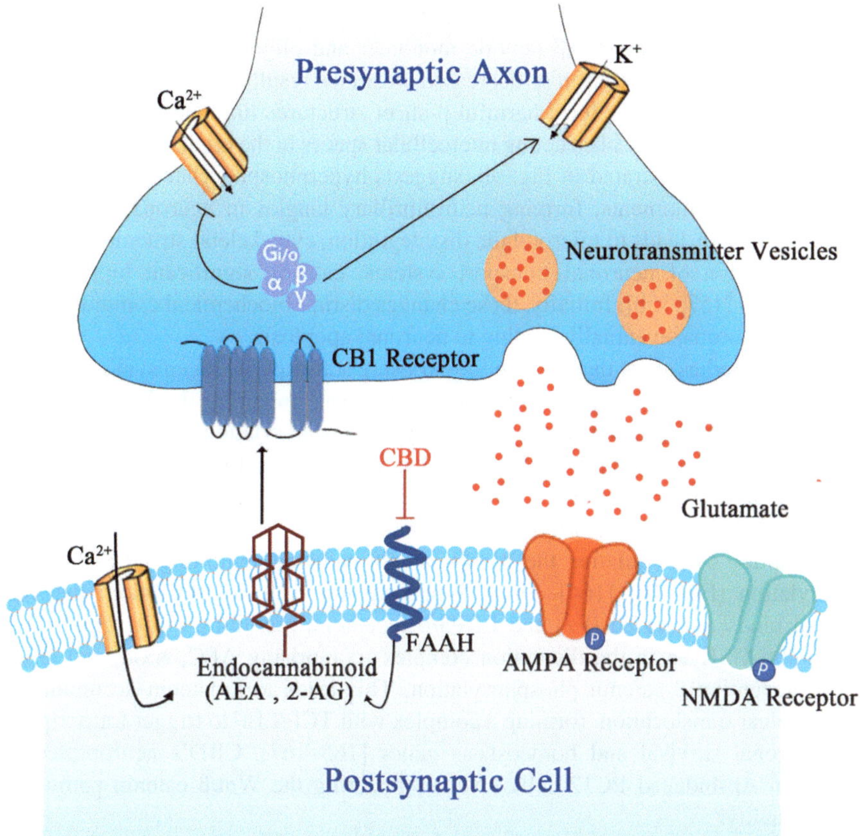

Fig. 8 As a fatty acid amide hydrolase (FAAH) inhibitor, CBD increases the AEA and 2-AG concentrations to activate retrograde signals [27]. Abbreviations: *2-AG* 2-arachidonoylglycerol, *AEA* anandamide, *AMPA* alpha-amino-3-hydroxy-5-methyl-4-isoxazole propionic acid, *CB1* cannabinoid receptor type 1, *FAAH* fatty acid amide hydrolase, *NMDA* N-methyl-D-aspartate

3.3.2 The Neuroprotective Properties of CBD Against β-Amyloid-Induced Alzheimer's Disease

Alois Alzheimer, a German psychiatrist and pathologist, first described AD in 1906 as a gradual, insidious neurodegenerative disorder with a hidden onset and progressive decline [142]. Neuropathological studies have linked AD to neurofibrillary tangles and plaques in the brain, caused by β-amyloid (Aβ) aggregation and tau protein dysregulation, impairing microtubules in brain cells [143–146]. The Aβ hypothesis is central to AD aetiology, triggering a cascade of pathological events harmful to neuronal populations [147, 148]. Aβ peptide aggregates accumulate and deposit in the brain, initiating and progressing AD [149, 150]. As indicated in Fig. 9a, neurons process amyloid precursor protein (APP) via α-, β-, and γ-secretases

[151, 152]. An imbalance in Aβ40–42 fragment production and removal by β- and γ-secretases leads to Aβ peptide monomer and oligomer buildup [152–154]. The increased Aβ generation or decreased clearance results in large aggregated Aβ plaques. These peptides form harmful β-sheet structures like neuritic plaques and neurofibrillary tangles, obstructing intercellular spaces in the brain parenchyma. The tau hypothesis, illustrated in Fig. 9b, suggests hyperphosphorylated tau associates with other tau filaments, forming neurofibrillary tangles in neuronal cell bodies [155, 156]. This leads to microtubule disintegration, cytoskeletal structure collapse, and disruption of neuronal transport systems, causing significant hippocampal neuron loss [157, 158]. Initially, these changes disrupt biochemical communication between neurons, eventually leading to neuronal apoptosis.

Recent studies show that Aβ peptide exposure activates glycogen synthase kinase 3β (GSK-3β) in hippocampal and cortical neurons, impairing the Wnt signalling pathway and contributing to tau protein hyperphosphorylation and neurofibrillary tangles in AD brains [159–161]. CBD has been shown to mitigate Aβ peptide toxicity in PC12 cells via the Wnt/β-catenin pathway [162, 163]. Specifically, CBD enhances T cell factor/lymphocyte enhancement factor (TCF/LEF)-mediated transcription by modulating the Wnt/β-catenin pathway, inhibiting GSK-3β phosphorylation (Fig. 10) [164]. Upon Wnt/β-catenin pathway activation, the Wnt ligand binds to Fizzled (Fz) and LRP5/6 receptors, recruiting Dishevelled (Dvl) to inhibit GSK-3β in the disruption complex (comprising APC, Axin, and GSK-3β), preventing β-catenin phosphorylation. This leads to β-catenin accumulation and nuclear translocation, forming a complex with TCF/LEF to trigger transcription of neuronal survival and homeostasis genes [165–167]. CBD's neuroprotective effect in Aβ-induced PC12 cells involves restoring the Wnt/β-catenin pathway's functionality.

CBD also impedes neuronal apoptosis (Fig. 10). Following Aβ-induced adrenergic receptor stimulation, caspase 8 and 10 activation leads to BID cleavage, t-BID formation, and Bax and Bak insertion into mitochondrial membranes, releasing cytochrome C [168, 169]. Cytochrome C binds to apoptotic protease-activating factor-1 (Apaf-1) and caspase 9, forming an apoptosome that activates effector caspase 3, a key apoptosis stage. In Aβ-treated PC12 cells, CBD upregulates pro-caspase 3 and downregulates caspase 3, suppressing proteins in the apoptosis pathway, offering neuroprotective benefits [170, 171]. Additionally, CBD counters Aβ-induced Ca^{2+} overload in neurons, aligning with its antioxidant properties [172, 173].

3.3.3 The Neuroprotective Effects of CBD in the Treatment of Parkinson's Disease

PD presents motor and nonmotor symptoms like tremors, rigidity, and bradykinesia [174, 175]. Key to PD's neuropathology is the degeneration of dopaminergic neurons in the substantia nigra pars compacta, impacting dopamine secretion and acetylcholine regulation, crucial for motor control [176–178]. The primary treatment strategy for PD aims to boost dopamine levels or stimulate dopaminergic

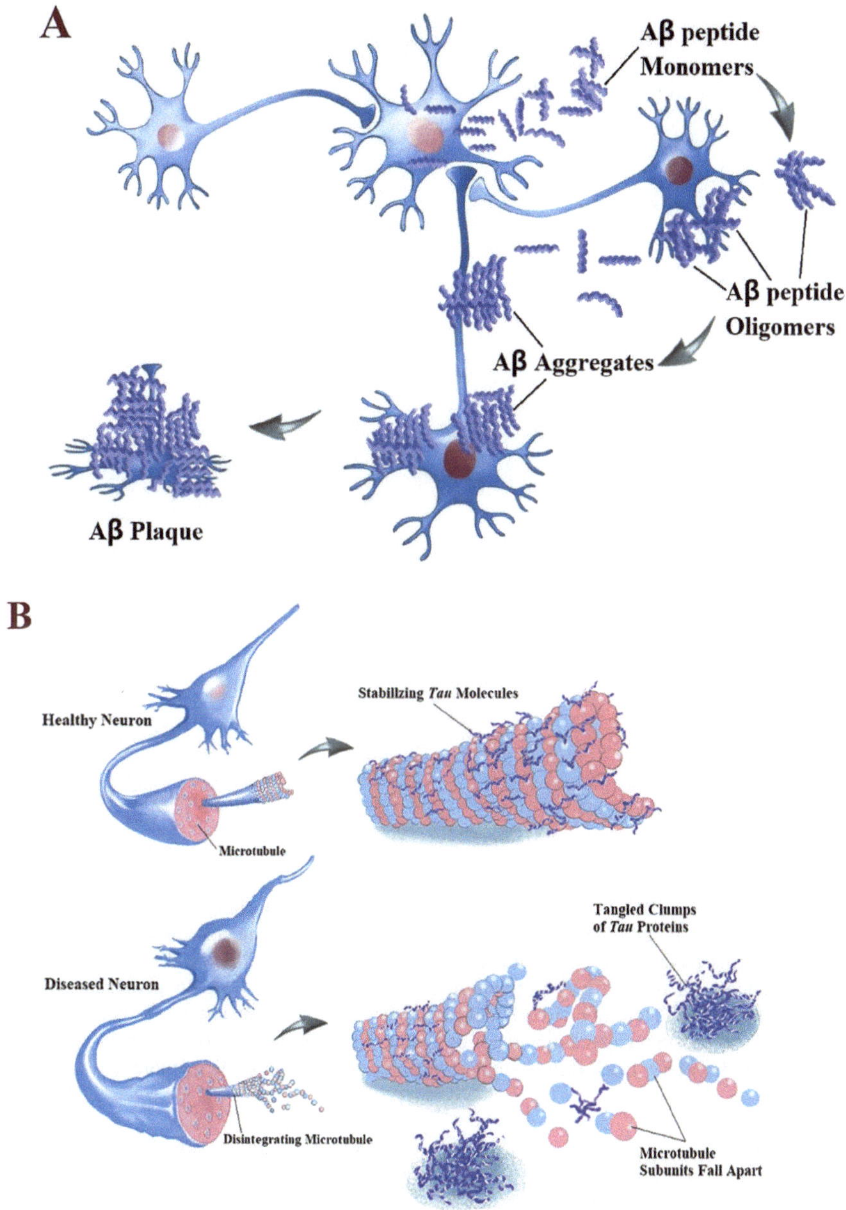

Fig. 9 (**a**) β-amyloid hypothesis model; (**b**) *tau* hypothesis model [27]

neurons. Studies have linked the endocannabinoid system (ECS) to PD, with cannabinoids believed to improve motor function by interacting with the ECS [179]. CBD's action in PD treatment essentially involves activating CB1, CB2,

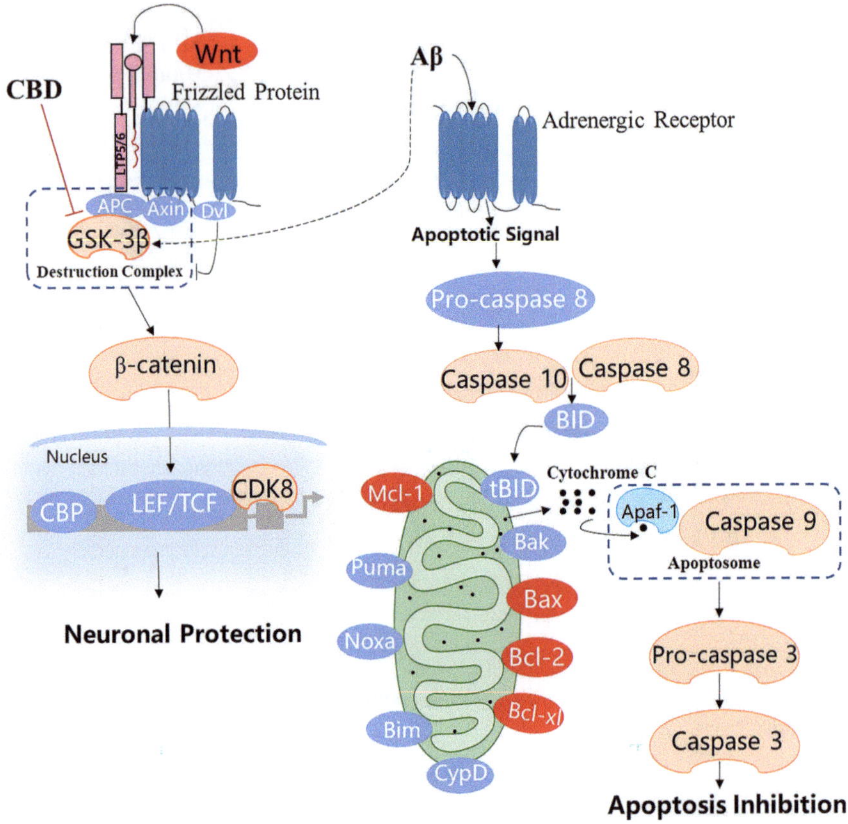

Fig. 10 Two signalling pathways related to the neuroprotective mechanism of CBD [27]. Abbreviations: *APC* adenomatosis polyposis coli, *Apaf-1* apoptotic protease-activating factor-1, *Axin* axis inhibition protein; *Bak* Bcl-2 homologous antagonist, *Bax* Bcl-2-associated X protein, *Bcl-2* B cell lymphoma 2, *Bcl-xl* B cell lymphoma-extra large, *BID* BH3 interacting domain death agonist, *Bim* Bcl-2 interacting mediator of cell death, *CBD* cannabidiol, *CBP* CREB-binding protein, *CDK8* cyclin-dependent kinase 8, *CypD* cyclophilin D, *Dvl* Dishevelled, *GSK-3β* glycogen synthase kinase 3β, *Mcl-1* stands for: myeloid cell leukemia, *Noxa* phorbol-12-myristate-13-acetate-induced protein 1, *Puma* p53 upregulated modulator of apoptosis, *tBID* truncated BH3 interacting domain death agonist, *TCF/LEF* T cell factor/lymphocyte enhancement factor, *Wnt* wingless/integrated

and TRPV-1 receptors and inhibiting the FAAH enzyme [109, 133, 180–182]. CB1 receptors, key in modulating dopaminergic neuron synaptic activity within the nigrostriatal pathway, are abundant in the basal ganglia (BG), which is crucial for motor control. CB2 receptors, also present in this pathway, may significantly affect psychomotor behaviour, anxiety, and depression [183]. Focus on two primary endocannabinoids (eCBs), AEA and 2-AG, has emerged [184–186]. eCBs, residing within postsynaptic neurons, facilitate retrograde signalling. As Fig. 11 illustrates, diacylglycerol (DAG) is converted to 2-AG by diacylglycerol lipase α (DAGL-α) [187, 188]. 2-AG then binds to CB1 or CB2 receptors via the endocannabinoid

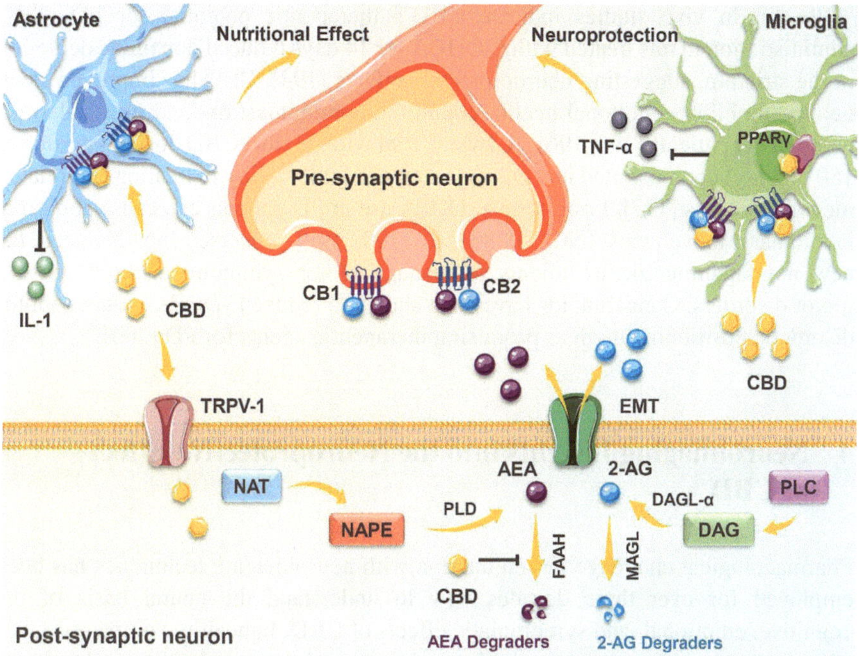

Fig. 11 The mechanism of action of CBD on Parkinson's disease. Abbreviations: *2-AG* 2-arachidonoylglycerol, *AEA* anandamide, *CBD* cannabidiol, *CB1* cannabinoid receptor type 1, *CB2* cannabinoid receptor type 2, *DAG* diacylglycerol, *DAGL-α* diacylglycerol lipase α, *EMT* endocannabinoid membrane transporter, *FAAH* fatty acid amide hydrolase, *MAGL* monoacylglycerol lipase, *NAPE* N-acyl-phosphatidylethanolamine, *NAT* N-acyl transferase, *PLC* phospholipase C, *PPARγ* peroxisome proliferator-activated receptor γ, *TRPV1* transient potential receptor V1

membrane transporter (EMT), ensuring normal neuronal signalling and motor function, and is degraded by monoacylglycerol lipase (MAGL). AEA, produced from N-acetyl-physidylethanolamine (NAPE) by phospholipase D (PLD), also binds to these receptors and is degraded by FAAH [187, 189]. CBD increases AEA levels by inhibiting FAAH enzymes and transporters, activating CB1 or CB2 receptors [190, 191]. CBD stimulates TRPV-1 receptors, increasing Ca^{2+} levels with neuroprotective effects for PD.

In addition, microglia, the brain's innate immune cells, astrocytes, and central nervous system glial cells play roles in PD. Astrocytes support neurons and release neurotrophins, but the abnormal release of proinflammatory factors from these cells can exacerbate neuronal damage, contributing to PD. Cannabinoids regulate astrocyte activity via CB2 receptors or a combination of CB1 and CB2 receptors [192, 193]. CBD binds to astrocyte CB1 and CB2 receptors, promoting their nutritional function and anti-inflammatory mediator release, repairing damaged neurons. It also inhibits the microglial release of neurotoxic factors like IL-1. Moreover, CBD activates PPARγ, suppressing the inflammatory response.

Recent in vivo studies indicate CBD's therapeutic potential for PD. CBD administration in rats treated with 6-OHDA for 14 days reduced dopamine depletion in the striatum, suggesting neuroprotective effects [194]. CBD has improved sleep quality, mobility, emotional health, cognition, communication, and physical well-being in PD patients [195, 196]. A 4-week trial with flexible CBD dosing (starting at 150 mg/day) demonstrated its efficacy, safety, and tolerability in treating PD-related mental disorders [197]. Long-term L-DOPA use in PD patients can cause abnormal involuntary movements (NCT00888186) [198, 199], underscoring the need for new nondopaminergic treatments to manage motor symptoms without causing motor disorders. Cannabinoids have been shown to reduce L-DOPA-induced motor disorders, positioning them as promising therapeutic agents for PD [200].

4 Neuroimaging Insights into the Neuroprotective Effect of CBD

Pharmacological challenge in conjunction with neuroimaging techniques has been employed for over three decades now to understand the neural basis of the cognitive, emotional, and symptomatic effects of CBD. In healthy volunteers, CBD administration in a resting state enhanced frontostriatal connectivity and reduced limbic activity. Additionally, CBD increased blood flow in the left parahippocampal and fusiform gyrus but decreased it in the left posterior cingulate cortex and a cluster including the left amygdala, hippocampus, uncus, and hypothalamus [201]. Moreover, CBD's influence on brain activity was task-specific during different cognitive paradigms. During recall tasks, activation in the hippocampus was modulated and linked to CBD administration [202]. In tasks involving auditory processing, CBD increased activation relative to the placebo in the parahippocampus [203], while a reduction in activation was observed during the processing of fearful faces [204]. Additionally, during the processing of fearful faces, there was a decrease in connectivity between frontolimbic regions, specifically the anterior cingulate cortex and amygdala. Conversely, during salience processing, enhanced connectivity was observed in frontolimbic striatal regions, which include the inferior frontal gyrus, dorsal striatum, and posterior hippocampus [204].

CBD also affects the brain networks of patients who have a psychiatric disorder. It has been observed that CBD reduces cerebral blood flow in the limbic regions, which includes the hippocampus, parahippocampal, and inferior temporal gyrus, in individuals with social anxiety [205]. This is similar to the reduced frontolimbic activity in nonclinical subjects after consuming CBD [201]. However, the immediate neurological effects of CBD may differ between psychiatric patients and healthy people. Subjects at CHR for psychosis displayed intermediate brain activity in memory and reward processing regions following CBD administration compared to placebo and healthy controls [206]. This pattern was also observed in patients with psychosis during memory tasks [206]. Furthermore, the activation of the insula was found to be positively correlated with both positive psychotic symptoms and salience perception. CBD was observed to reduce the heightened activation in the left insula/parietal operculum and was linked to a general deceleration in reaction time [207]. It indicates a possible mechanism for its antipsychotic effects by regulating motor response and normalising motivational salience.

Along with intensive research, it was found that THC and CBD, both derived from *Cannabis*, exhibit distinct and frequently contrasting effects on neural networks in healthy volunteers. These effects are task-specific: THC and CBD show opposite influences in emotional processing within the frontotemporal regions, verbal memory in the frontostriatal areas, response inhibition in frontolimbic striatal circuits, and auditory/visual processing in the temporo-occipital areas [208]. CBD's role appears to mitigate the adverse effects of THC. For instance, the modulation of striatal activation during verbal recall tasks may reflect an increased ease of response as subjects become familiar with the recall cues. THC attenuates this striatal activation, possibly disrupting the neural correlate of eased recall responses. Conversely, CBD might facilitate this process [209]. Furthermore, THC's modulation of striatal and amygdalar activation, correlated with induced psychotic and anxiety symptoms, suggests its role in the acute onset of these symptoms following cannabis use. The opposite effect of CBD in these regions hints at its potential therapeutic value in treating psychosis and anxiety symptoms [209].

5 Conclusion

Historically, *Cannabis* has been used in traditional herbal medicine to treat various conditions like otalgia, arthritis, and convulsions. While *Cannabis* use has been linked to increased psychosis risk, isolated CBD shows promise in treating epilepsy and neurodegenerative disorders like AD due to its neuroprotective properties. Current CBD research reveals three key findings. First, CBD indirectly stimulates CB1 receptors in neuronal damage, reducing glutamate transmission and aiding epilepsy treatment through ECS retrograde signalling involving cannabinoid receptors, endocannabinoids, and related enzymes. Second, CBD acts on ECS and glial cells, showing therapeutic efficacy in PD. It reduces neurotoxic factor release from glial cells, enhances astrocyte nutrition, and offers neuroprotection without causing L-dopamine-induced movement disorders, making it a promising PD drug. Third, CBD's potential in AD treatment involves inhibiting Aβ-induced neurofibrillary tangles through Wnt/β-catenin pathway upregulation and attenuating caspase 3 expression, impeding neuronal apoptosis. Additionally, CBD's antiproliferative properties across tumour cell lines suggest its potential in chemotherapy.

This chapter outlines CBD's molecular signalling pathways and targets, but several issues remain unresolved. A comprehensive investigation into other *Cannabis* bioactive constituents, including lignans, alkaloids, terpenoids, stilbenoids, and flavonoids, is pending. Exploring natural analogues, metabolites, and synthetic CBD derivatives with higher biological activity and lower psychotoxicity is promising for epilepsy, cancer, AD, PD, and neuropathic pain management. Comprehensive studies are required to explore how CBD regulates nerve growth factors, neurotransmitters, and electrophysiological processes. Neuroimaging techniques can offer more evidence about brain imaging to help clarify the neurocognitive mechanisms that relieve psychotic symptoms with CBD. It can also provide a mechanistic understanding of the potential role of CBD as a therapeutic strategy for neurodegenerative disorders.

References

1. Li HL. An archaeological and historical account of Cannabis in China. Econ Bot. 1973;28:437–48.
2. Russo EB. Cannabis and cannabinoids. J Cannabis Ther. 2001;1:103–32.
3. Amaducci S, Pelatti F, Bonatti PM. Fibre development in hemp (Cannabis sativa L.) as affected by agrotechnique. J Ind Hemp. 2005;10:31–48.
4. Chopra GS. Man and marijuana. Int J Addict. 1969;4:215–47.
5. Curran HV, et al. Keep off the grass? Cannabis, cognition and addiction. Nat Rev Neurosci. 2016;17:293–306.
6. Gloss D. An overview of products and bias in research. Neurotherapeutics. 2015;12:731–4.
7. Cannabis: from cultivar to Chemovar II—A metabolomics approach to cannabis classification. Cannabis Cannabinoid Res. 2016;1:202–15.
8. Messina F, Rosati O, Curini M, Marcotullio MC. Cannabis and bioactive cannabinoids, Studies in Natural Products Chemistry 45, 17–57. Elsevier; 2015.
9. Hillig KW. Genetic evidence for speciation in Cannabis (Cannabaceae). Genet Resour Crop Evol. 2005;52:161–80.
10. Pollastro F, Minassi A, Fresu GL. Cannabis Phenolics and their bioactivities. Curr Med Chem. 2018;25:1160–85.
11. Kovalchuk I, et al. The genomics of cannabis and its close relatives. Annu Rev Plant Biol. 2020;71:713–39.
12. Liu WM, Fowler DW, Dalgleish AG. Cannabis-derived substances in cancer therapy—an emerging anti-inflammatory role for the cannabinoids. Curr Clin Pharmacol. 2010;5:281–7.
13. Machado Bergamaschi M, Helena Costa Queiroz R, Waldo Zuardi A, Alexandre S, Crippa J. Safety and side effects of Cannabidiol, a Cannabis sativa constituent. Curr Drug Saf. 2011;6:237–49.
14. Amin MR, Ali DW. Pharmacology of medical Cannabis, Recent Advances in Cannabinoid Physiology and Pathology 151–165. Springer; 2019.
15. Roger G, Pertwee T, Gibson M, Stevenson LA, Ross RA. O-1057, a potent water-soluble cannabinoid receptor agonist with antinociceptive properties. Br J Pharmacol. 2000;129:1577–84.
16. Nuutinen T. Medicinal properties of terpenes found in Cannabis sativa and Humulus lupulus. Eur J Med Chem. 2018;157:198–228.
17. Appendino G, et al. Antibacterial cannabinoids from Cannabis sativa: a structure-activity study. J Nat Prod. 2008;71:1427–30.
18. Maya AF, et al. Uncovering the hidden antibiotic potential of Cannabis. ACS Infect Dis. 2020;6:338–46.
19. Kicman A, Toczek M. The effects of Cannabidiol, a non-intoxicating compound of cannabis, on the cardiovascular system in health and disease. Int J Mol Sci. 2020;21:6740.

20. Coetzee C, Levendal R-A, Venter MVD, Frost CL. Anticoagulant effects of a Cannabis extract in an obese rat model. Phytomedicine. 2007;14:333–7.
21. Rocha FCM, dos Santos Júnior JG, Stefano SC, da Silveira DX. Systematic review of the literature on clinical and experimental trials on the antitumor effects of cannabinoids in gliomas. J Neuro-Oncol. 2014;116:11–24.
22. Carter GT, Abood ME, Aggarwal SK, Weiss MD. Cannabis and amyotrophic lateral sclerosis: hypothetical and practical applications, and a call for clinical trials. Am J Hosp Palliat Care. 2010;27:347–56.
23. Gaoni Y, Mechoulam R. Isolation, structure, and partial synthesis of an active constituent of hashish. J Am Chem Soc. 1964;86:1646–7.
24. Ibarra-Lecue I, et al. Chronic Cannabis promotes pro-hallucinogenic signaling of 5-HT2A receptors through Akt/mTOR pathway. Neuropsychopharmacology. 2018;43:2028–35.
25. Santos NA, d., et al. Evaluating the selectivity of colorimetric test (Fast Blue BB salt) for the cannabinoids identification in marijuana street samples by UV–Vis, TLC, ESI(+)FT-ICR MS and ESI(+)MS/MS. Forensic Chem. 2016;1:13–21.
26. Izzo AA, et al. Inhibitory effect of cannabichromene, a major non-psychotropic cannabinoid extracted from Cannabis sativa, on inflammation-induced hypermotility in mice. Br J Pharmacol. 2012;166:1444–60.
27. Li H, et al. Overview of cannabidiol (CBD) and its analogues: structures, biological activities, and neuroprotective mechanisms in epilepsy and Alzheimer's disease. Eur J Med Chem. 2020;192:112163.
28. Elsohly MA, Radwan MM, Gul W, Chandra S, Galal A. Phytochemistry of Cannabis sativa L. Phytochemistry of Cannabis sativa L. Springer International Publishing; 2017.
29. Ahmed SA, Ross SA, Slade D, Radwan MM, Elsohly MA. Cannabinoid Ester constituents from high-potency Cannabis sativa. J Nat Prod. 2008;71:1119.
30. Gul W, et al. Determination of 11 cannabinoids in biomass and extracts of different varieties of cannabis using high-performance liquid chromatography. J AOAC Int. 2015;98:1523–8.
31. Hanuš LO. Discovery and isolation of anandamide and other endocannabinoids. Cannabinoids in Nature and Medicine, Zurich: Verlag Helvetica Chimica Acta, Wiley VCH 4; 2009. p. 251–66.
32. ElSohly MA, Slade D. Chemical constituents of marijuana: the complex mixture of natural cannabinoids. Life Sci. 2005;78:539–48.
33. Shani A, Mechoulam R. Cannabielsoic acids: isolation and synthesis by a novel oxidative cyclisation. Tetrahedron. 1974;30:2437–46.
34. Poklis JL, Thompson CC, Long KA, Lichtman AH, Poklis A. Disposition of cannabichromene, cannabidiol, and Δ^9-tetrahydrocannabinol and its metabolites in mouse brain following marijuana inhalation determined by high-performance liquid chromatography-tandem mass spectrometry. J Anal Toxicol. 2010;34:516–20.
35. Ch LRJJ, Bercht CAL, Ooyen RV, Spronck HJW. Cannabinodiol: conclusive identification and synthesis of a new cannabinoid from Cannabis sativa. Phytochemistry. 1977;16:595–7.
36. Brenneisen R. Chemistry and analysis of phytocannabinoids and other cannabis constituents, Marijuana and the Cannabinoids 17–49. Springer; 2007.
37. Obata Y, Ishikawa Y. Studies on the constituents of hemp plant (Cannabis sativa L.). J Agric Chem Soc Jpn. 1965;30:619–20.
38. Radwan M, et al. Isolation and characterisation of new cannabis constituents from a high potency variety. Planta Med. 2008;74:267–72.
39. Friedrich-Fiechtl J, Spiteller G. Neue cannabinoide-1. Tetrahedron. 1975;31:479–87.
40. Radwan MM, Wanas AS, Chandra S, ElSohly MA. Natural cannabinoids of cannabis and methods of analysis, Cannabis sativa L.-Botany and Biotechnology 161–182. Springer; 2017.
41. Ahmed SA, et al. Structure determination and absolute configuration of cannabichromanone derivatives from high potency Cannabis sativa. Tetrahedron Lett. 2008;49:6050–3.
42. Lewis MM, Yang Y, Wasilewski E, Clarke HA, Kotra LP. Chemical profiling of medical Cannabis extracts. Acs Omega. 2017;2:6091–103.

43. Appendino G, Chianese G, Taglialatela-Scafati O. Cannabinoids: occurrence and medicinal chemistry. Curr Med Chem. 2011;18:1085–99.
44. Boeren EG, Elsohly MA, Turner CE. Cannabiripsol: A novel Cannabis constituent. Experientia. 1979;35:1278–9.
45. Elsohly HN, Boeren EG, Turner CE, Elsohly MA. Constituents of cannabis sativa L. XXIIII: Cannabitetrol, a new polyhydroxylated cannabinoid. Cannabinoids Chemical Pharmacologic & Therapeutic Aspects, 89–96. Elsevier; 1984.
46. Radwan MM, et al. Non-cannabinoid constituents from a high potency Cannabis sativa variety. Phytochemistry. 2018;69:2627–33.
47. Taglialatela-Scafati O, Pagani A. Cannabimovone, a cannabinoid with a rearranged Terpenoid skeleton from hemp. Eur J Org Chem. 2010;2010:2067–72.
48. Ahmed SA, et al. Minor oxygenated cannabinoids from high potency Cannabis sativa L. Phytochemistry. 2015;117:194–9.
49. Adams R, Hunt M, Clark JH, Structure of Cannabidiol. III. Reduction and cleavage. J Am Chem Soc. 1940;62:735–7.
50. Mechoulam R, Shvo Y. Hashish—I: the structure of Cannabidiol. Tetrahedron. 1963;19:2073–8.
51. Jones PG, Falvello L, Kennard O, Sheldrick GM, Mechoulam R. Cannabidiol. Acta Crystallogr. 1977;33:3211–4.
52. Vollner L, Bieniek D, Korte F. Hashish. XX. Cannabidivarin, a new hashish constituent. Tetrahedron Lett. 1969;3:145–7.
53. Wood JS, Gordon WH, Morgan JB, Williamson RT. Cannabicitran: its unexpected racemic nature and potential origins. Chirality. 2023;35:540–8.
54. Pattnaik F, et al. Cannabis: chemistry, extraction and therapeutic applications. Chemosphere. 2022;289:133012.
55. Montero-Oleas N, Arevalo-Rodriguez I, Nuñez-González S, Viteri-García A, Simancas-Racines D. Therapeutic use of Cannabis and cannabinoids: an evidence mapping and appraisal of systematic reviews. BMC Complementary Medicine and Therapies. 2020;20:12.
56. dos Santos RG, Hallak JEC, Crippa. Neuropharmacological effects of the main phytocannabinoids: a narrative review. Cannabinoids and Neuropsychiatric Disorders 29–45. Springer; 2021.
57. Friedman D, Sirven JI. Historical perspective on the medical use of Cannabis for epilepsy: ancient times to the 1980s. Epilepsy Behav. 2017;70:298–301.
58. Izquierdo I, Orsingher OA, Berardi AC. Effect of cannabidiol and of other Cannabis sativa compounds on hippocampal seizure discharges. Psychopharmacologia. 1973;28:95–102.
59. Consroe PF, Wolkin AL. Cannabidiol—antiepileptic drug comparisons and interactions in experimentally induced seizures in rats. J Pharmacol Exp Ther. 1977;201:26–32.
60. Turkanis SA, Smiley KA, Borys HK, Olsen DM, Karler R. An electrophysiological analysis of the anti-convulsant action of Cannabidiol on limbic seizures in conscious rats. Epilepsia. 1979;20:351–63.
61. Jones NA, et al. Cannabidiol exerts anti-convulsant effects in animal models of temporal lobe and partial seizures. Seizure. 2012;21:344–52.
62. Klein BD, et al. Evaluation of cannabidiol in animal seizure models by the epilepsy therapy screening program (ETSP). Neurochem Res. 2017;42:1939–48.
63. Pernici CD, et al. Development of an anti-seizure drug screening platform for Dravet syndrome at the NINDS contract site for the Epilepsy Therapy Screening Program. Epilepsia. 2021;62:1665–76.
64. Gray RA, Whalley BJ. The proposed mechanisms of action of CBD in epilepsy. Epileptic Disord. 2020;22:S10–5.
65. Newton M, Newton DW. Cannabidiol or CBD oil: help, Hope, and hype for psychiatric and neurologic conditions. J Am Psychiatr Nurses Assoc. 2020;26:447–57.
66. Devinsky O, et al. Trial of cannabidiol for drug-resistant seizures in the Dravet syndrome. N Engl J Med. 2017;376:2011–20.

67. Thiele EA, et al. Cannabidiol in patients with seizures associated with Lennox-Gastaut syndrome (GWPCARE4): a randomised, double-blind, placebo-controlled phase 3 trial. Lancet. 2018;391:1085–96.
68. Solinas M, Cinquina V, Parolaro D. Cannabidiol and cancer—an overview of the preclinical data. Molecular Considerations and Evolving Surgical Management Issues in the Treatment of Patients with a Brain Tumor 399–412. IntechOpen; 2015.
69. Tura M, Mandrioli M, Gallina Toschi T. Preliminary study: comparison of antioxidant activity of Cannabidiol (CBD) and α-tocopherol added to refined olive and sunflower oils. Molecules. 2019;24:3485.
70. Hampson AJ, et al. Neuroprotective antioxidants from marijuanaa. Ann N Y Acad Sci. 2000;899:274–82.
71. Rodríguez Mesa XM, et al. Therapeutic prospects of cannabinoids in the immunomodulation of prevalent autoimmune diseases. Cannabis Cannabinoid Res. 2021;6:196–210.
72. Ferro R, et al. GPR55 signalling promotes proliferation of pancreatic cancer cells and tumour growth in mice, and its inhibition increases effects of gemcitabine. Oncogene. 2018;37:6368.
73. Ryberg E, et al. The orphan receptor GPR55 is a novel cannabinoid receptor. Br J Pharmacol. 2007;152:1092–101.
74. Brown KJ, Laun AS, Song Z-H. Cannabidiol, a novel inverse agonist for GPR12. Biochem Bioph Res Co. 2017;493:451–4.
75. De Petrocellis L, et al. Non-THC cannabinoids inhibit prostate carcinoma growth in vitro and in vivo: pro-apoptotic effects and underlying mechanisms. Br J Pharmacol. 2013;168:79–102.
76. Borrelli F, et al. Colon carcinogenesis is inhibited by the TRPM8 antagonist cannabigerol, a Cannabis-derived non-psychotropic cannabinoid. Carcinogenesis. 2014;35:2787–97.
77. Murase R, et al. Targeting multiple cannabinoid anti-tumour pathways with a resorcinol derivative leads to inhibition of advanced stages of breast cancer. Br J Pharmacol. 2014;171:4464–77.
78. Ivanov VN, Wu J, Hei TK. Regulation of human glioblastoma cell death by combined treatment of cannabidiol, γ-radiation and small molecule inhibitors of cell signaling pathways. Oncotarget. 2017;8:74068.
79. Silvestri C, et al. Two non-psychoactive cannabinoids reduce intracellular lipid levels and inhibit hepatosteatosis. J Hepatol. 2015;62:1382–90.
80. Martin B, et al. Identification of monohydroxylated metabolites of cannabidiol formed by rat liver. J Pharm Pharmacol. 1976;28:275–9.
81. Jiang R, Yamaori S, Takeda S, Yamamoto I, Watanabe K. Identification of cytochrome P450 enzymes responsible for metabolism of cannabidiol by human liver microsomes. Life Sci. 2011;89:165–70.
82. Tchilibon S, Mechoulam R. Synthesis of a primary metabolite of cannabidiol. Org Lett. 2000;2:3301–3.
83. Hanuš LO, et al. Enantiomeric cannabidiol derivatives: synthesis and binding to cannabinoid receptors. Org Biomol Chem. 2005;3:1116–23.
84. Nasrin S, et al. Inhibition of nicotine metabolism by Cannabidiol (CBD) and 7-Hydroxycannabidiol (7-OH-CBD). Chem Res Toxicol. 2023;36:177–87.
85. Ben-Shabat S, Hanuš LO, Katzavian G, Gallily R. New cannabidiol derivatives: synthesis, binding to cannabinoid receptor, and evaluation of their anti-inflammatory activity. J Med Chem. 2006;49:1113–7.
86. Mascal M, et al. Synthetic, non-intoxicating 8,9-dihydrocannabidiol for the mitigation of seizures. Sci Rep. 2019;9:7778.
87. Wu Q, Guo M, Zou L, Wang Q, Xia Y. 8,9-dihydrocannabidiol, an alternative of cannabidiol, its preparation, antibacterial and antioxidant ability. Molecules. 2023;28:445.
88. Kozela E, et al. HU-446 and HU-465, derivatives of the non-psychoactive cannabinoid cannabidiol, decrease the activation of encephalitogenic T cells. Chem Biol Drug Des. 2016;87:143–53.
89. Breuer A, et al. Fluorinated cannabidiol derivatives: enhancement of activity in mice models predictive of anxiolytic, antidepressant and antipsychotic effects. PLoS One. 2016;11:e0158779.

90. Chianese G, et al. A Nrf-2 stimulatory hydroxylated Cannabidiol derivative from hemp (Cannabis sativa). J Nat Prod. 2022;85:1089–97.
91. Pretzsch CM, et al. Effects of cannabidivarin (CBDV) on brain excitation and inhibition systems in adults with and without Autism Spectrum Disorder (ASD): a single dose trial during magnetic resonance spectroscopy. Transl Psychiatry. 2019;9:313.
92. Morales P, Jagerovic N. Synthetic and natural derivatives of cannabidiol, Cannabinoids and Sleep: Molecular, Functional and Clinical Aspects 11–25. Springer; 2021.
93. Nalli Y, et al. Analysing the role of cannabinoids as modulators of Wnt/β-catenin signaling pathway for their use in the management of neuropathic pain. Bioorganic Med Chem Lett. 2019;29:1043–6.
94. Pertwee RG, et al. Cannabidiolic acid methyl ester, a stable synthetic analogue of cannabidiolic acid, can produce 5-HT1A receptor-mediated suppression of nausea and anxiety in rats. Br J Pharmacol. 2018;175:100–12.
95. Takeda S, et al. Cannabidiolic acid, a major cannabinoid in fiber-type Cannabis, is an inhibitor of MDA-MB-231 breast cancer cell migration. Toxicol Lett. 2012;214:314–9.
96. Yamazaki D, Kurisu S, Takenawa T. Regulation of cancer cell motility through actin reorganisation. Cancer Sci. 2005;96:379–86.
97. Bolognini D, et al. Cannabidiolic acid prevents vomiting in S uncus murinus and nausea-induced behaviour in rats by enhancing 5-HT1A receptor activation. Br J Pharmacol. 2013;168:1456–70.
98. Rock EM, Connolly C, Limebeer CL, Parker LA. Effect of combined oral doses of Δ^{9}-tetrahydrocannabinol (THC) and cannabidiolic acid (CBDA) on acute and anticipatory nausea in rat models. Psychopharmacology. 2016;233:3353–60.
99. Gallily R, et al. Γ-Irradiation enhances apoptosis induced by cannabidiol, a non-psychotropic cannabinoid, in cultured HL-60 myeloblastic leukemia cells. Leuk Lymphoma. 2003;44:1767–73.
100. Juknat A, Kozela E, Kaushansky N, Mechoulam R, Vogel Z. Anti-inflammatory effects of the cannabidiol derivative dimethylheptyl-cannabidiol—studies in BV-2 microglia and encephalitogenic T cells. J Basic Clin Physiol Pharmacol. 2016;27:289–96.
101. Thomas A, et al. 6″-Azidohex-2″-yne-cannabidiol: a potential neutral, competitive cannabinoid CB1 receptor antagonist. Eur J Pharmacol. 2004;487:213–21.
102. Silva NR, et al. The Cannabidiol analog PECS-101 prevents chemotherapy-induced neuropathic pain via PPARγ receptors. Neurotherapeutics. 2022;19:434–49.
103. Silva NR, et al. Antinociceptive effects of HUF-101, a fluorinated cannabidiol derivative. Prog Neuro-Psychopharmacol Biol Psychiatry. 2017;79:369–77.
104. Vong CT, Tseng HHL, Kwan YW, Lee SM-Y, Hoi MPM. Novel protective effect of O-1602 and abnormal cannabidiol, GPR55 agonists, on ER stress-induced apoptosis in pancreatic β-cells. Biomed Pharmacother. 2019;111:1176–86.
105. Kinney WA, et al. Discovery of KLS-13019, a Cannabidiol-derived neuroprotective agent, with improved potency, safety, and permeability. ACS Med Chem Lett. 2016;7:424–8.
106. Nelson KM, et al. The essential medicinal chemistry of Cannabidiol (CBD). J Med Chem. 2020;63:12137–55.
107. Rock EM, Sticht MA, Duncan M, Stott C, Parker LA. Evaluation of the potential of the phytocannabinoids, cannabidivarin (CBDV) and Δ9-tetrahydrocannabivarin (THCV), to produce CB1 receptor inverse agonism symptoms of nausea in rats. Br J Pharmacol. 2013;170:671–8.
108. Morales P, Reggio PH, Jagerovic N. An overview on medicinal chemistry of synthetic and natural derivatives of Cannabidiol. Front Pharmacol. 2017;8:422–39.
109. Bisogno T, et al. Molecular targets for cannabidiol and its synthetic analogues: effect on vanilloid VR1 receptors and on the cellular uptake and enzymatic hydrolysis of anandamide. Br J Pharmacol. 2001;134:845–52.
110. Fride E, Ponde D, Breuer A, Hanuš L. Peripheral, but not central effects of cannabidiol derivatives: mediation by CB1 and unidentified receptors. Neuropharmacology. 2005;48:1117–29.
111. Izzo AA, Muccioli GG, Ruggieri MR, Schicho R. Endocannabinoids and the digestive tract and bladder in health and disease. Handb Exp Pharmacol. 2015;231:423–47.

112. Kienzl M, Kargl J, Schicho R. The immune endocannabinoid system of the tumor microenvironment. Int J Mol Sci. 2020;21:8929–53.
113. McCoy KL. Interaction between cannabinoid system and toll-like receptors controls inflammation. Mediat Inflamm. 2016;2016:5831315–32.
114. Vemuri VK, Makriyannis A. Medicinal chemistry of cannabinoids. Clin Pharmacol Ther. 2015;97:553–8.
115. Svíženská I, Dubový P, Šulcová A. Cannabinoid receptors 1 and 2 (CB1 and CB2), their distribution, ligands and functional involvement in nervous system structures—a short review. Pharmacol Biochem Behav. 2008;90:501–11.
116. Laezza C, et al. The endocannabinoid system: a target for cancer treatment. Int J Mol Sci. 2020;21:747–67.
117. Dávila EM, et al. Interacting binding insights and conformational consequences of the differential activity of cannabidiol with two endocannabinoid-activated G-protein-coupled receptors. Front Pharmacol. 2022;13:945935–47.
118. Goldberger E, Tauber M, Ben-Chaim Y. Voltage dependence of the cannabinoid CB1 receptor. Front Pharmacol. 2022;13:1022275–84.
119. Izzo AA, Sharkey KA. Cannabinoids and the gut: new developments and emerging concepts. Pharmacol Ther. 2010;126:21–38.
120. Moldrich G, Wenger T. Localization of the CB1 cannabinoid receptor in the rat brain. An immunohistochemical study. Peptides. 2000;21:1735–42.
121. Onaivi ES. Neuropsychobiological evidence for the functional presence and expression of cannabinoid CB2 receptors in the brain. Neuropsychobiology. 2006;54:231–46.
122. Montero C, Campillo NE, Goya P, Páez JA. Homology models of the cannabinoid CB1 and CB2 receptors. A docking analysis study. Eur J Med Chem. 2005;40:75–83.
123. Luszczki JJ, et al. Effects of arachidonyl-2′-chloroethylamide (ACEA) on the protective action of various antiepileptic drugs in the 6-Hz corneal stimulation model in mice. PLoS One. 2017;12:e0183873.
124. Naderi N, Aziz Ahari F, Shafaghi B, Hosseini Najarkolaei A, Motamedi F. Evaluation of interactions between cannabinoid compounds and diazepam in electroshock-induced seizure model in mice. J Neural Transm. 2008;115:1501–11.
125. Miziak B, Walczak A, Szponar J, Pluta R, Czuczwar SJ. Drug-drug interactions between antiepileptics and cannabinoids. Expert Opin Drug Metab Toxicol. 2019;15:407–15.
126. Lafourcade CA, Sparks FT, Bordey A, Wyneken U, Mohammadi MH. Cannabinoid regulation of neurons in the dentate gyrus during epileptogenesis: role of CB1R-associated proteins and downstream pathways. Epilepsia. 2023;64:1432–43.
127. Ye M, et al. Coordinated regulation of CB1 cannabinoid receptors and anandamide metabolism stabilises network activity during homeostatic downscaling. eNeuro. 2022;9:0276–90.
128. Baker D, et al. Endocannabinoids control spasticity in a multiple sclerosis model. FASEB J. 2001;15:300–2.
129. Wallace MJ, Blair RE, Falenski KW, Martin BR, DeLorenzo RJ. The endogenous cannabinoid system regulates seizure frequency and duration in a model of temporal lobe epilepsy. J Pharmacol Exp Ther. 2003;307:129–37.
130. Guggenhuber S, Monory K, Lutz B, Klugmann M. AAV vector-mediated overexpression of CB1 cannabinoid receptor in pyramidal neurons of the hippocampus protects against seizure-induced Excitoxicity. PLoS One. 2010;5:e15707.
131. Lothman EW, Bertram EH III, Stringer JL. Functional anatomy of hippocampal seizures. Prog Neurobiol. 1991;37:1–82.
132. Wilson RI, Nicoll RA. Endocannabinoid signaling in the brain. Science. 2002;296:678–82.
133. De Petrocellis L, et al. Effects of cannabinoids and cannabinoid-enriched Cannabis extracts on TRP channels and endocannabinoid metabolic enzymes. Br J Pharmacol. 2011;163:1479–94.
134. Mackie K, Lai Y, Westenbroek R, Mitchell R. Cannabinoids activate an inwardly rectifying potassium conductance and inhibit Q-type calcium currents in AtT20 cells transfected with rat brain cannabinoid receptor. J Neurosci. 1995;15:6552–61.

135. Wang Z, et al. Cannabidiol regulates CB1-pSTAT3 signaling for neurite outgrowth, prolongs lifespan, and improves health span in Caenorhabditis elegans of Aβ pathology models. FASEB J. 2021;35:e21537.
136. Wang Z, et al. Cannabidiol induces autophagy and improves neuronal health associated with SIRT1 mediated longevity. GeroScience. 2022;44:1505–24.
137. Mackie K, Stella N. Cannabinoid receptors and endocannabinoids: evidence for new players. AAPS J. 2006;8:E298–306.
138. Ying S-W, et al. Brain-derived neurotrophic factor induces long-term potentiation in intact adult hippocampus: requirement for ERK activation coupled to CREB and upregulation of Arc synthesis. J Neurosci. 2002;22:1532–40.
139. He JC, Neves SR, Jordan JD, Iyengar R. Role of the Go/i signaling network in the regulation of neurite outgrowth. Can J Physiol Pharmacol. 2006;84:687–94.
140. Callén L, et al. Cannabinoid receptors CB1 and CB2 form functional heteromers in brain. J Biol Chem. 2012;287:20851–65.
141. Kimpinski K, Jelinski S, Mearow K. The anti-p75 antibody, MC192, and brain- derived neurotrophic factor inhibit nerve growth factor-dependent neurite growth from adult sensory neurons. Neurosci. 1999;93:253–63.
142. Jonsson T, et al. A mutation in APP protects against Alzheimer's disease and age-related cognitive decline. Nature. 2012;488:96.
143. Spires-Jones TL, Hyman BT. The intersection of amyloid beta and tau at synapses in Alzheimer's disease. Neuron. 2014;82:756–71.
144. Ittner LM, et al. Dendritic function of tau mediates amyloid-β toxicity in Alzheimer's disease mouse models. Cell. 2010;142:387–97.
145. Silva AR, Grosso C, Delerue-Matos C, Rocha JM. Comprehensive review on the interaction between natural compounds and brain receptors: benefits and toxicity. Eur J Med Chem. 2019;174:87–115.
146. Yarns BC, Holiday KA, Carlson DM, Cosgrove CK, Melrose RJ. Pathophysiology of Alzheimer's disease. Psychiatr Clin North Am. 2022;45:663–76.
147. Jucker M, Walker LC. Pathogenic protein seeding in Alzheimer disease and other neurodegenerative disorders. Ann Neurol. 2011;70:532–40.
148. Volloch V, Rits-Volloch S. The amyloid cascade hypothesis 2.0: on the possibility of once-in-a-lifetime-only treatment for prevention of Alzheimer's disease and for its potential cure at symptomatic stages. J Alzheimers Dis Rep. 2022;6:369–99.
149. Iyaswamy A, et al. Theranostic F-SLOH mitigates Alzheimer's disease pathology involving TFEB and ameliorates cognitive functions in Alzheimer's disease models. Redox Biol. 2022;51:102280.
150. Zhang W, et al. Impairment of the autophagy-lysosomal pathway in Alzheimer′s diseases: pathogenic mechanisms and therapeutic potential. Acta Pharm Sin B. 2022;12:1019–40.
151. Gupta SP, Patil VM. Recent studies on design and development of drugs against Alzheimer's disease (AD) based on inhibition of BACE-1 and other AD-causative agents. Curr Top Med Chem. 2020;20:1195–213.
152. Kim JY, et al. Distinct amyloid precursor protein processing machineries of the olfactory system. Biochem Biophys Res Commun. 2018;495:533–8.
153. Velazquez R, Shaw DM, Caccamo A, Oddo S. Pim1 inhibition as a novel therapeutic strategy for Alzheimer's disease. Mol Neurodegener. 2016;11:52.
154. Mal S, Malik U, Pal D, Mishra A. Insight γ-secretase: structure, function, and role in Alzheimer's disease. Curr Drug Targets. 2021;22:1376–403.
155. Engin AB, Engin A. Alzheimer's disease and protein kinases. Adv Exp Med Biol. 2021;1275:285–321.
156. Chong FP, Ng KY, Koh RY, Chye SM. Tau proteins and Tauopathies in Alzheimer's disease. Cell Mol Neurobiol. 2018;38:965–80.
157. Duan J, Marcellus KA, Qin X, Wang Y, Paudel HK. Cystatin C promotes tau protein phosphorylation and causes microtubule instability by inhibiting intracellular turnover of GSK3β in neurons. Mol Cell Neurosci. 2018;89:1–8.

158. Xia Y, et al. Tau Ser208 phosphorylation promotes aggregation and reveals neuropathologic diversity in Alzheimer's disease and other tauopathies. Acta Neuropathol Commun. 2020;8:88.
159. Garrido JL, Godoy JA, Alvarez A, Bronfman M, Inestrosa NC. Protein kinase C inhibits amyloid β peptide neurotoxicity by acting on members of the Wnt pathway. FASEB J. 2002;16:1982–4.
160. Yue H, Lu F, Shen C, Quan J-M. Structure-based design of benzo[e]isoindole-1,3-dione derivatives as selective GSK-3β inhibitors to activate Wnt/β-catenin pathway. Bioorg Chem. 2015;61:21–7.
161. Sun W, et al. Glycogen synthase kinase-3β is complexed with tau protein in brain microtubules. J Biol Chem. 2002;277:11933–40.
162. dos Santos MTA, Smidt MP. En1 and Wnt signaling in midbrain dopaminergic neuronal development. Neural Dev. 2011;6:23.
163. Esposito G, De Filippis D, Carnuccio R, Izzo AA, Iuvone T. The marijuana component cannabidiol inhibits β-amyloid-induced tau protein hyperphosphorylation through Wnt/β-catenin pathway rescue in PC12 cells. J Mol Med. 2006;84:253–8.
164. Vallee A, Lecarpentier Y, Guillevin R, Vallee J-N. Effects of cannabidiol interactions with Wnt/β-catenin pathway and PPARγ on oxidative stress and neuroinflammation in Alzheimer's disease. Acta Biochim Biophys Sin. 2017;49:853–66.
165. Zeng X, et al. Initiation of Wnt signaling: control of Wnt coreceptor Lrp6 phosphorylation/activation via frizzled, dishevelled and axin functions. Development. 2008;135:367–75.
166. Doble BW, Woodgett JR. GSK-3: tricks of the trade for a multi-tasking kinase. J Cell Sci. 2003;116:1175–86.
167. Moon RT, Bowerman B, Boutros M, Perrimon N. The promise and perils of Wnt signaling through β-catenin. Science. 2002;296:1644–6.
168. Mildew P, Mites R, Kit FDR, Blog M, Sequencing C. Neuroprotective effect of cannabidiol, a non-psychoactive component from Cannabis sativa, on b-amyloid-induced toxicity in PC12 cells. J Neurochem. 2004;89:134–41.
169. Uren RT, Iyer S, Kluck RM. Pore formation by dimeric Bak and Bax: an unusual pore? Philos Trans R Soc Lond Ser B Biol Sci. 2017;372:20160218.
170. Patel V, Abu-Hijleh F, Rigg N, Mishra R. Cannabidiol protects striatal neurons by attenuating endoplasmic reticulum stress. Cannabis Cannabinoid Res. 2023;8:299–308.
171. Alali S, et al. Cannabidiol inhibits tau aggregation in vitro. Cells. 2021;10:3521.
172. Hampson A, Grimaldi M, Axelrod J, Wink D. Cannabidiol and (−) Δ^9-tetrahydrocannabinol are neuroprotective antioxidants. Proc Natl Acad Sci. 1998;95:8268–73.
173. Wang Y, Tang H, Yang C, Zhao H, Jian C. Involvement of p75NTR in the effects of Aβ on L-type Ca2+ channel in cultured neuronal networks. Life Sci. 2020;243:117293.
174. Fernandez HH, et al. Effect of levodopa-carbidopa intestinal gel on resting tremors in patients with advanced Parkinson's disease. NPJ Parkinsons Dis. 2016;2:16015.
175. Kumaresan M, Khan S. Spectrum of non-motor symptoms in Parkinson's disease. Cureus. 2021;13:e13275.
176. Hornykiewicz O. Basic research on dopamine in Parkinson's disease and the discovery of the Nigrostriatal dopamine pathway: the view of an eyewitness. Neurodegener Dis. 2008;5:114–7.
177. Dickson DW, et al. Neuropathological assessment of Parkinson's disease: refining the diagnostic criteria. Lancet Neurol. 2009;8:1150–7.
178. Qamar MA, et al. Presynaptic dopaminergic terminal imaging and non-motor symptoms assessment of Parkinson's disease: evidence for dopaminergic basis? NPJ Parkinsons Dis. 2017;3:5.
179. Patricio F, Morales-Andrade AA, Patricio-Martinez A, Limon ID. Cannabidiol as a therapeutic target: evidence of its neuroprotective and neuromodulatory function in Parkinson's disease. Front Pharmacol. 2020;11:595635.
180. Ligresti A, et al. Antitumor activity of plant cannabinoids with emphasis on the effect of cannabidiol on human breast carcinoma. J Pharmacol Exp Ther. 2006;318:1375–87.

181. Howlett AC, Blume LC, Dalton GD. CB(1) cannabinoid receptors and their associated proteins. Curr Med Chem. 2010;17:1382–93.
182. Diaz-Alonso J, Guzman M, Galve-Roperh I. Endocannabinoids via CB1 receptors act as neurogenic niche cues during cortical development. Philos Trans R Soc B. 2012;367:3229–41.
183. Garcia MC, Cinquina V, Palomo-Garo C, Rabano A, Fernandez-Ruiz J. Identification of CB2 receptors in human nigral neurons that degenerate in Parkinson's disease. Neurosci Lett. 2015;587:1–4.
184. Mechoulam R, et al. Identification of an endogenous 2-monoglyceride, present in canine gut, that binds to cannabinoid receptors. Biochem Pharmacol. 1995;50:83–90.
185. Devane WA, et al. Isolation and structure of a brain constituent that binds to the cannabinoid receptor. Science. 1992;258:1946.
186. Sugiura T, et al. 2-Arachidonoylglycerol: a possible endogenous cannabinoid receptor ligand in brain. Biochem Biophys Res Commun. 1995;215:89–97.
187. Di Marzo V, Stella N, Zimmer A. Endocannabinoid signalling and the deteriorating brain. Nat Rev Neurosci. 2015;16:30–42.
188. Bisogno T, et al. Cloning of the first sn1-DAG lipases points to the spatial and temporal regulation of endocannabinoid signaling in the brain. J Cell Biol. 2003;163:463–8.
189. Fezza F, Mastrangelo N, Maccarrone M. Assay of NAPE-PLD activity. Methods Mol Biol. 2016;1412:123–30.
190. Maione S, et al. Elevation of endocannabinoid levels in the ventrolateral periaqueductal grey through inhibition of fatty acid amide hydrolase affects descending nociceptive pathways via both cannabinoid receptor type 1 and transient receptor potential vanilloid type-1 receptors. J Pharmacol Exp Ther. 2006;316:969–82.
191. Marinelli S, et al. N-arachidonoyl-dopamine tunes synaptic transmission onto dopaminergic neurons by activating both cannabinoid and vanilloid receptors. Neuropsychopharmacology. 2007;32:298–308.
192. Fernandez-Ruiz J, Hernandez M, Ramos JA. Cannabinoid-dopamine interaction in the pathophysiology and treatment of CNS disorders. CNS Neurosci Ther. 2010;16:e72.
193. Stella N. Cannabinoid and cannabinoid-like receptors in microglia, astrocytes, and astrocytomas. Glia. 2010;58:1017–30.
194. Lastres-Becker I, Molina-Holgado F, Ramos JA, Mechoulam R, Fernandez-Ruiz J. Cannabinoids provide neuroprotection against 6-hydroxydopamine toxicity in vivo and in vitro: relevance to Parkinson's disease. Neurobiol Dis. 2005;19:96–107.
195. Chagas MHN, et al. Effects of cannabidiol in the treatment of patients with Parkinson's disease: an exploratory double-blind trial. J Psychopharmacol. 2014;28:1088.
196. Chagas MHN, et al. Cannabidiol can improve complex sleep-related behaviours associated with rapid eye movement sleep behaviour disorder in Parkinson's disease patients: a case series. J Clin Pharm Ther. 2014;39:564–6.
197. Zuardi AW, et al. Cannabidiol for the treatment of psychosis in Parkinson's disease. J Psychopharmacol. 2009;23:979–83.
198. Putterman DB, Munhall AC, Kozell LB, Belknap JK, Johnson SW. Evaluation of levodopa dose and magnitude of dopamine depletion as risk factors for levodopa-induced dyskinesia in a rat model of Parkinson's disease. J Pharmacol Exp Ther. 2007;323:277–84.
199. Francardo V, et al. Impact of the lesion procedure on the profiles of motor impairment and molecular responsiveness to L-DOPA in the 6-hydroxydopamine mouse model of Parkinson's disease. Neurobiol Dis. 2011;42:327–40.
200. Espadas I, et al. Beneficial effects of the phytocannabinoid Δ9-THCV in L-DOPA-induced dyskinesia in Parkinson's disease. Neurobiol Dis. 2020;141:104892.
201. Grimm O, et al. Probing the endocannabinoid system in healthy volunteers: Cannabidiol alters fronto-striatal resting-state connectivity. Eur Neuropsychopharmacol. 2018;28:841–9.
202. Bhattacharyya S, et al. Modulation of mediotemporal and ventrostriatal function in humans by Δ9-tetrahydrocannabinol: A neural basis for the effects of Cannabis sativa on learning and psychosis. Arch Gen Psychiatry. 2009;66:442–51.

203. Winton-Brown TT, et al. Modulation of auditory and visual processing by delta-9-tetrahydrocannabinol and Cannabidiol: an fMRI study. Neuropsychopharmacology. 2011;36:1340–8.
204. Fusar-Poli P, et al. Distinct effects of Δ9-tetrahydrocannabinol and Cannabidiol on neural activation during emotional processing. Arch Gen Psychiatry. 2009;66:95–105.
205. Crippa JAS, et al. Neural basis of anxiolytic effects of cannabidiol (CBD) in generalised social anxiety disorder: a preliminary report. J Psychopharmacol. 2011;25:121–30.
206. Bhattacharyya S, et al. Effect of Cannabidiol on medial temporal, midbrain, and striatal dysfunction in people at clinical high risk of psychosis: A randomised clinical trial. JAMA Psychiatry. 2018;75:1107–17.
207. Wilson R, et al. Cannabidiol attenuates insular dysfunction during motivational salience processing in subjects at clinical high risk for psychosis. Transl Psychiatry. 2019;9:203.
208. Batalla A, Bos J, Postma A, Bossong MG. The impact of Cannabidiol on human brain function: A systematic review. Front Pharmacol. 2021;11:618184.
209. Bhattacharyya S, et al. Opposite effects of Δ-9-tetrahydrocannabinol and Cannabidiol on human brain function and psychopathology. Neuropsychopharmacology. 2010;35:764–74.

Index

P. A. Broderick (ed.), *Biomedical Imaging in Neurodegeneration*,
https://doi.org/10.1007/978-3-031-95878-6

MIX
Papier aus verantwortungsvollen Quellen
Paper from responsible sources
FSC® C105338

If you have any concerns about our products,
you can contact us on
ProductSafety@springernature.com

In case Publisher is established outside the EU,
the EU authorized representative is:
Springer Nature Customer Service Center GmbH
Europaplatz 3, 69115 Heidelberg, Germany

Printed by Libri Plureos GmbH
in Hamburg, Germany